E-Learning and Enhancing Soft Skills

Eugenia Smyrnova-Trybulska
Nian-Shing Chen • Piet Kommers • Nataliia Morze
Editors

E-Learning and Enhancing Soft Skills

Contemporary Models of Education in the Era of Artificial Intelligence

Springer

Editors
Eugenia Smyrnova-Trybulska
Faculty of Arts and Educational Sciences, Institute of Pedagogy
University of Silesia in Katowice
Katowice, Poland

Nian-Shing Chen
National Taiwan Normal University
Taipei City, Taiwan

Piet Kommers
Department of Media, Communication and Organization
University of Twente
Enschede, Overijssel, The Netherlands

Nataliia Morze
Faculty of Information Technologies and Mathematics
Borys Grinchenko Kyiv Metropolitan University
Kyiv, Ukraine

ISBN 978-3-031-82242-1 ISBN 978-3-031-82243-8 (eBook)
https://doi.org/10.1007/978-3-031-82243-8

This Springer imprint is published by the registered company Springer Nature Switzerland AG
The registered company address is: Gewerbestrasse 11, 6330 Cham, Switzerland

Contents

Chapter 1
Third Decade in the Twenty-First Century Skills and A.I. in Education

Introductory Chapter I

Piet Kommers and Eugenia Smyrnova-Trybulska

Abstract As an introduction to the book 'E-Learning & Enhancing Soft Skills: Subtitle: E-Learning and Enhancing Soft Skills: Contemporary Models of Education in the Era of Artificial Intelligence' it might be good to first sketch the larger landscape of trends that dominated the sectors of Education and Training. It compiles articles from the 16th annual international conference on distance learning, held in October 2024. Organized by the University of Silesia in Katowice, Poland, and co-hosted by several international universities, the conference featured experts in e-learning, AI, and educational technology. Contributors include renowned scholars, researchers, and practitioners who discuss their studies and present recent research findings. The monograph addresses theoretical, methodological, and practical aspects of e-learning, focusing on developing key competencies and soft skills. It offers solutions to significant challenges and outlines future research directions, making it valuable for academic researchers, educators, courseware designers, corporate trainers, and educational technology practitioners.

Keywords E-learning · AI in education models · Soft skills · Digital competences · Challenges and perspective using AI in education · Pedagogical AI agents

1.1 Precursors to Collaborative and Communicative Functioning for A.I. in Education

The first question to be asked is 'How has social media evolved from mirroring to predictive; how will new criteria for privacy emanate in Education?' The next diagram may illustrate the six major transitions that provoke ambitions in education.

P. Kommers
University of Twente, Enschede, The Netherlands

E. Smyrnova-Trybulska (✉)
University of Silesia in Katowice, Faculty of Arts and Educational Sciences, Cieszyn, Poland
e-mail: esmyrnova@us.edu.pl

E. Smyrnova-Trybulska et al. (eds.), *E-Learning and Enhancing Soft Skills*,
https://doi.org/10.1007/978-3-031-82243-8_1

Its main components are Technology, Best Practice Initiatives, and finally Educational Policy Development. However, in the range of Web-Based Learning and the increased functioning of Soft Skills, a more specific chain of innovative steps has been signaled. It is a special case of social media and its pervasive societal embedding in the last two decades. A few ingredients came together: The early arrival of 'Communities of Practice' as coined by advocates in professional training; (Wenger, 1998 and beyond). Its main message was that human expertise propagates through personal relationships; Colleagues who trust each other, and who grant each other success in job performance and career prospects have the privilege to build upon each other prior experiences and become better problem solvers themselves. It is important to signal that the entrance of 'communities of practice' evolved from quite pragmatic considerations; technology pur sang, and learning theories like Vygotsky's (Zavershneva & Osipov, 2012) notion of Action Theory, had no particular role in developing the idea by Etienne Wenger, the days just before the transition to the twenty-first century. To a certain extent, the notion of collaborative learning might have provoked a higher sensitivity to social interaction for cognitive versatility as expressed by Sharan and Hertz-Lazarowitz (1980). The research by Nijhof and Kommers (1985) showed that cognitive diversity in terms of prior knowledge and socio-economic background, contributed to the amount of conceptual elaborations during joint problem-solving by K12 students. At a later stage, it became obvious that conceptual representations proved to be proficient for cognitive exchange during collaborative problem-solving; Stoyanova and Kommers (2002). Finally, we may say that conceptual representations for ideational sharing like in STEAM (Science, Technology, Engineering, Arts and Mathematics) has brought a solid underpinning of thematic education as already coined by proponents of curricular reform in the late seventies and more recently in best practice projects like the EduSimSteam project.[1] The project illustrates the benefits from confronting students with holistic problems as can be found in societal topics like "What is a safe environment?" or "What technical measures can be taken for the sake of sustainable mobility?" etc. The recent publication 'Sources for a Better Education' has articulated the need and value of more explicit ontologies for deep learning and upgrading teachers' discursive way of explaining (Kommers, 2022).

Figure 1.1 below, shows the nine main precursors to the adoption of the twenty-first century skills in education. It exemplifies the transitional roles of various kinds like emerging technology, societal trends, corporate needs, lifestyle issues, etc.

Figure 1.1 below shows the simultaneity of various trends that only spuriously rest upon educational, didactic or pedagogical considerations. Its signaling suggests that education mainly reacts to external factors rather than excavating its own agenda from earlier school innovations or learning theoretical findings. The underlying book 'E-Learning & Enhancing Soft Skills: Contemporary Models of Education in the Era of Artificial Intelligence' takes the challenge to present an integral and up-to-date overview of factors that allow us to estimate the opportunities and threats for

[1] See the EduSimSteam Project https://edusimsteam.eba.gov.tr/en/edusimsteam-2/

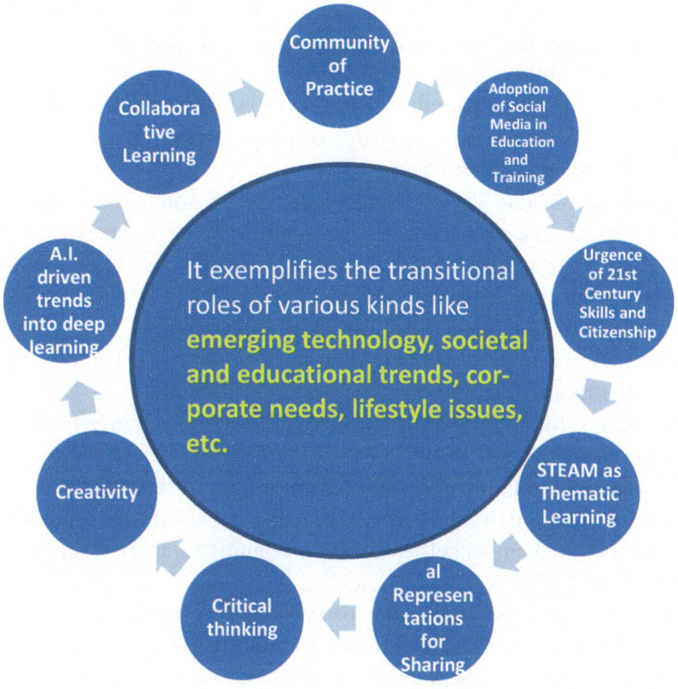

Fig. 1.1 Cascadic precursors to valuing twenty-first century skills in learning

adopting A.I. in education. The main hypothesis in this effort is to expect that A.I. will revalue the nature of human expertise in working/learning together, successfully explain personal understanding to someone else, and finally 'pick and mix' intuitional ingredients into the wider circle of reasoning and experimentation.

1.2 Preamble to this Book's Main Messages

- The chapter by Nian-Shing Chen illustrates how in the age of Large Language Models (LLMs) and generative AI, educators will harness the potential of pedagogical AI agents to enhance their teaching practices and how finally teachers can tailor learning experiences to suit diverse student needs, guided by established learning theories and adaptive strategies. Chen's live demonstrations show the versatility of cognitive agents as envisioned for robotic interfaces and computational thinking.
- The chapter by Nataliia Morze and Olga Barna illustrates the need and benefits of teaching high school students programming using artificial intelligence. The research results show that most respondents recognize the importance of

developing students' critical thinking skills, logical analysis, and understanding of algorithms generated by AI, as well as the ethical aspects of using artificial intelligence. The authors offer specific methodological recommendations for the implementation of AI in the educational process, in particular, focusing on the development of fundamental knowledge of programming, and not only on learning individual programming languages.

- Jan Skałka's and Matúš Valko chapter presents the comparison of methods for identifying rapid guessing behavior in microlearning courses that might be students' instinct to reduce cognitive load and may hamper the effect of instructional interventions as we know from recent best practices. The presented cases in microlearning situations may plead for additional assistance by A.I. background propagation so that teachers in programming skills may be helped to keep their energy for the more global group process as favored before.
- Yuliia Rudenko, Marina Drushlyak, Olha Naboka, Volodymyr Proshkin, Olena Semenikhina chapter presents the topic of the development of youth information hygiene skills and preludes to the discomfort of the gap between self-assessment and the actual situation. It highlights and clarifies how learners' multi-channeling and fake news may hamper self-esteem and cognitive integrity, finally leading to misconceptions and biased worldviews.
- Magdalena Zavodna and colleagues ask attention to the various ways in which digital technologies have become an integral part of the educational process and finally affect the way teachers teach and may develop communicative learning, and problem-solving or digital competencies.
- Miroslav Hruby opens the floor to question how programming in technical education will evolve in the coming years: "What programming skills need to be taught?" "What does it bring for students in technical-focused branches?" "What tools could be used?" "What is the effectiveness of distance learning?" and "Could e-learning be used for programming exams in the era of AI?"
- Eugenia Smyrnova-Trybulska c.s. investigated the use of mobile devices for educational purposes among university students and explored the purposes of mobile device usage, such as participation in e-courses, MOOCs, and educational apps like Duolingo. Among the important research questions was the interest to answer: What features of teaching m-learning are most important to students? And: What negative features of teaching m-learning do students perceive? Its empirical nature promises prospective views on how the young generation of students may evolve in learning styles and preferences for certain tools.
- Liubov Lokhvytska c.s. observed e-learning in developing interaction with the environment with soft skills in 6-year-old children. It provides a concrete specimen for the use of e-learning in developing interaction with the environment as soft skills in six-year-old children within preschool educational institutions. It also provided an experimental model for implementing e-learning to develop such soft skills as interaction with the environment in six-year-old children.
- Svitlana Skvortsova c.s. report the use of artificial intelligence in the professional activity of university teachers. It postulates that modern AI systems can be effective tools in the work of teachers of higher education institutions, as well

as contribute to the development of new educational technologies, and may manifest as erudite teacher assistants.

- Kornel Chromiński c.s. brings the study outcomes of indispensable aspects of modern life in society and schools in particular: 'Example of good practice for teaching about cyber security in secondary school.' The need for robust cyber security and its underlying ethical awareness are transferred to students' alertness. Its method is example-based teaching based on practice. Students get the chance to see the dangers of the Internet. A workshop on cyber security showed convincingly on how to protect oneself on the Internet.
- The authors Stefan Gubo and Ladislav Végh bring the chapter 'Solution time and confidence in the assessment of algorithmic and logical thinking of first-year computer science students'. His premise is that computer programming belongs to the important profile subjects in the curriculum of first-degree study programs of both Applied Informatics and Teaching of Informatics. Stefan phrases and clarifies that students need to gain well-developed algorithmic and logical thinking skills. The survey focuses on the comparison of students' responses to each task type in how far the time on task is needed to let them arrive at solutions. As an epitome, the author brings conclusions and future research plans.
- Todorka Glushkova et al. present in their chapter the rationale for automatic test generation based upon formulated ontologies as they underlie a web-based educational platform. Todorka asserts that generic semantic knowledge is one of the most important tasks related to the development and use of modern educational platforms. 'Ontologies can provide the automatic generation of test questions, in this case in the correct Bulgarian language.
- Miroslav Kamenský, Ján Halgoš and Anatolii Kliuchka describe actual structure and functionality of this modular system after latest improvements and inclusion of the didactic CAN workplace. Authors explain the process of adding remote access to the workplace. Thanks to the modular software system architecture, it is enough to modify one module of paired applications without knowledge of web publishing tools. A user gets access via any standard web browser. Finally, Authors collected feedback from students on such a teaching aid.
- Kateryna Yalova c.s. bring in their chapter 'Virtual laboratory as a tool to empower IT students'. The exposed project outcomes justify the necessity and relevance of developing and integrating into the educational process tools that enhance the process of acquiring professional competencies and skills for IT students. The need for mastering tools and IT skills has been concluded as the main factor during entering the IT job market. Problem analysis, creation of domain models, building the solution, and preparing the developed virtual laboratory for testing and integration into the educational process are the key skills. The proposed models are universal and can be applied during a large part of the academic curriculum.
- Bulent Dos presents 'Examination of Teachers' Competencies for Online Teaching'. Bulent formulates the critical importance of teachers' competencies in effectively delivering online education. The main message is the conclusion that teachers' proficiency in online teaching not only impacts the way of

delivering learning content but also the teacher's attitude on what is important in the various mechanisms of learning. Teachers were asked about their actual competencies in teaching online and creating online content. Besides that, most of them admitted that they lacked these skills, but they also expressed their reluctance to do so, for the same reason; The criterion to act swiftly and efficiently was brought to the front.

- Serhii Gunko c.s. describe the process of digitalization of social services for vulnerable groups in territorial communities of Ukraine. They analyzed social service digitalization programs in which Ukrainian communities participate. The authors used analysis/synthesis methods for defining preliminary concepts like digitalizing social services for vulnerable groups. Finally, the study highlights the advantages of implementing social service digitalization programs for both local self-government bodies and community residents, including vulnerable groups.
- Olha Korotun, Tetiana Vakaliuk and Artur Makhno emphasise the need to include the R programming language in the educational process and describe its use in large companies to solve various problems. In addition, a model of studying the elective discipline 'Programming in R' by a bachelor of computer science is built, which includes the following blocks: the purpose, tasks and stages of studying this discipline, types of activities and teaching methods, means, and also presents the topics of content modules. When studying this discipline, using the described online compilers and the RStudio environment is suggested.
- Hennadii Karimov, Marina Romaniukha, Ivan Karimov, Nuzhna Svitlana propose the article on developing digital and pedagogical competences in postgraduate management students. They emphasize that business management education is a trend in the development of a knowledge-based society within the European Union. In both Ukraine and Europe, business management education is occurring in a highly dynamic context. The article explores methods for improving Management students' teaching and digital skills through mandatory teaching practice. The study identifies a number of growth opportunities, including informal learning, teachers' digital competence and updates to the content of academic disciplines.

To sketch the main tendencies in the book authors' alerting and prioritization, the next vectors in innovation can be distilled. Here are the main lessons of the book's chapters.

A. First of all: No contemporary technological novelties are shunned in the educational context. Even if there is no clear need before, education undertakes greedy actions to absorb the newest technologies like Open A.I., Deep Learning and Machine Learning into the didactic arena without scruples. A pure engineering approach would prescribe a clear top-down need analysis before any steps towards system design start. In the educational sector, as we see here, the mentality is to be open-minded and even eager to try any new member of the ICT spectrum and try its added value in the scope of effectiveness and efficiency of learning and teaching. Its typical effect is that a considerable time lag manifests between early adopters and the common cohort of innovative teachers. In respect to the integrative use of MS Office Applications like WORD, EXCEL, TEAMS and even social media, the period between its early use and its standard commodity is typically 5 to 8 years. Similarly, we may expect that the integration of A.I. may need a similar latency time before teachers promote artificial intelligence in curricula, instruction, assignments and student projects.

B. Secondly: Due to the pragmatic nature of adopting new technologies in education even before the full spectrum of needs analysis has been undertaken, a significant task for pedagogical, didactical and policy architects remains at all phases of ICT innovation. In the case of A.I. tools we see a weighty challenge for curriculum designers, ministerial policy makers and societal representatives to dive deeply into the epistemological foundations of what we call 'human metacognition'; the kind and amount of self-consciousness that students and teachers need in order to become 'critical thinkers' on what actually happens if we allow A.I. tools to partner with learners, even before we know the cohesion and extensiveness of A.I. ingredients at that very moment. Maybe it is good to mitigate this harsh discrepancy with what we can now call 'pre-cooked' curricula after several decades of checks and balances in schools and academia up to today. Namely, societal-, cultural- and geopolitical turbulence has always been a chaotic source of noise in educational evolution. If we take the big transitions in the twentieth century: 2–3 warm and cold wars, the industrial revolution, discontinue colonialism and not at least the large patterns in migration, internationalization and the 24-hour economies. Education had to cope with all the direct and indirect side-effects, without having solid sociological or pedagogical frameworks on the shelves. In the case of the roles of social- and broadcasting media in news and its biases, the problem for education is of a similar magnitude; teachers and school leaders face a constant last-minute improvisation for framing mediated schools in the context of new citizenships, job-finding and .... twenty-first-century skills indeed. In other words, education cannot escape from being a fast train that needs to be rebuilt during the day-by-day forensic transfers of its passengers. In the recent past the passengers on the educational train were mainly teachers and students. As we face new societal needs for knowledge workers, creative employees, and critical thinkers, it is obvious that A.I. facilities for education are barely needed, just like the invention of the book printing press, the radio and the PC recently.

C. Thirdly: Similar to the profound dilemma when computer literacy entered education about 40 years ago, also the advent of A.I. poses the question of whether machine intelligence is just one of the future cognitive prostheses like spreadsheets, statistical software, and computer simulation software up till now. Or: will A.I. applications transform educational curricula into new partnerships between man and machine in general? If this is the case, we have to ask ourselves, how the partnership between A.I. and the other stakeholders in education is going to become (Fig. 1.2).

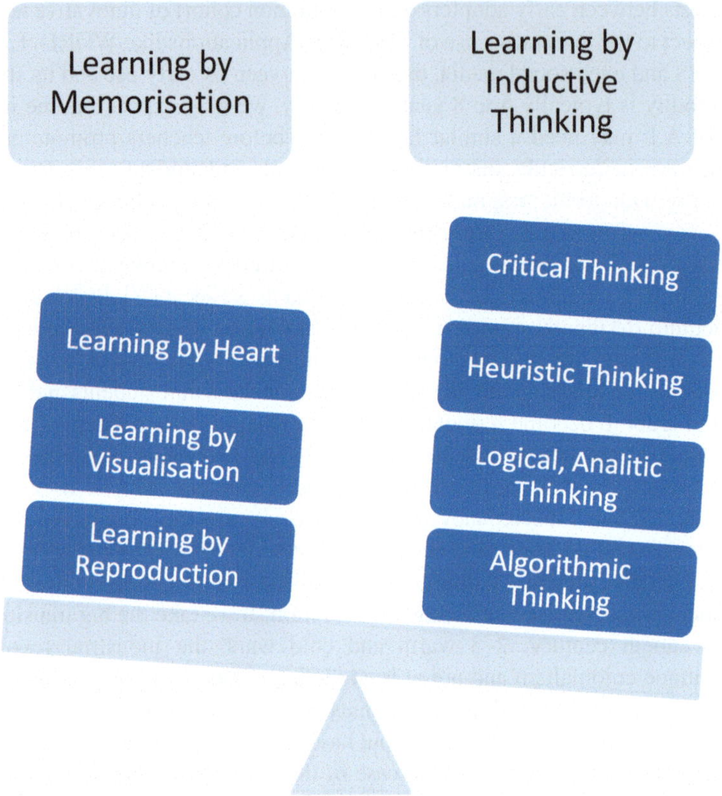

Fig. 1.2 The balance between traditional education (Transfer from Experts to Novices) into the integration of A.I. which urge students and teachers to adopt 'double-loop learning'. (From: Sources for a Better Education; Kommers, 2022)

To discuss the nature and magnitude of A.I.-based educational systems, the figure above may elicit a new understanding of what happens when traditional curricula undergo the effects of A.I.-driven tools and methods. The mildest way to reconcile traditional with A.I. education is to label it as a 'curricular understanding'; both students and teachers need to understand what A.I. can do, understand its potential impact, and learn to extrapolate its implications for diverse sectors of society. The more drastic effect of A.I. on education is if it 'touches' the way we conceive the essence of human learning; How we build new information on prior knowledge and experience? If and to what extent the human mind is able to generalize particular knowledge into more generic understanding? How human efforts enable us to build more effective learning methods on top of earlier learning experiences? These are all functionalities that are missing in A.I.-based learning so far. Interesting is human's capacity to interrupt learning tasks by asking: How can I make my learning more agile and essential (due to the domain where new knowledge is needed). Double-

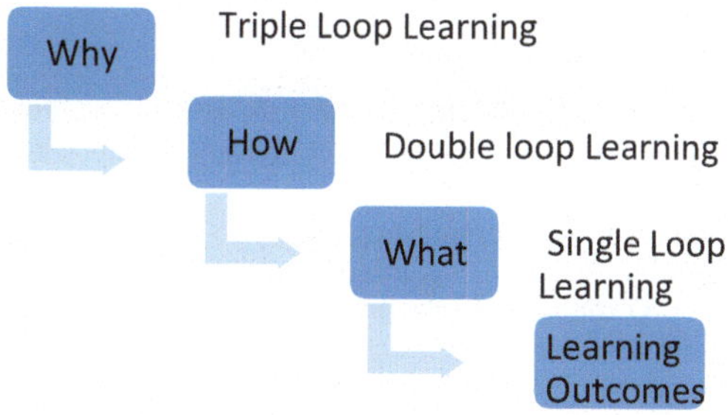

Fig. 1.3 The progress of asking meta-cognitive stands during learning. (Source: Double- / Triple Loop Learning by Argyris (1976, 1977) and Argyris and Schön (1978))

and triple-loop learning has been coined by Chris Argyris (1976, 1977) and Argyris and Schön (1978) (Fig. 1.3).

ICT tools and their implicit methods penetrated the arena of cognitive learning paradigms over the last four decades. Its inevitable effect was that learning has been promoted from the factual, via the procedural to the more strategic questions; Finally, the students' second nature to ask: Why should I learn this knowledge or skills? The schematic diagram above shows the three levels of aggregation in learning. The first (single-loop learning) has been a major concern: Can I learn what my teacher prescribes as target knowledge? The second (double loop learning) is when students ask: How can I learn more effectively (How can I become a better learner?) The third level (triple loop learning) makes the learner aware that learning and task competence need to go hand in hand with A.I. tools. The most common view in educational policymaking is that we prefer schools to start with the basic level and trust that at later moments the student will master the step to the next level. Its main belief rests upon Benjamin Bloom's (1994) taxonomy in which strategic thinking and problem solving only arrive when the subordinate levels have been accomplished (Fig. 1.4).

It may be clear that the current generation of Open A.I. (Chat GPT), offers added values to the retrieval via the World-Wide Web; It generates discursive descriptions of what has been formulated by others via web publications. In other words: Students get support from A.I. tools in formulating state-of-the-art understandings and provoke education to upgrade from reproduction into assessments that urge students to invest in higher levels of meta-cognition where Analysis, Evaluation and Creation is involved.

1.3 Meta Reflections

The coming book chapters include considerations and experts' comments on the contemporary questions:

- What can we expect after the arrival of the long-awaited trends towards ubiquity and mobile presence? What social skills will complement intellectual

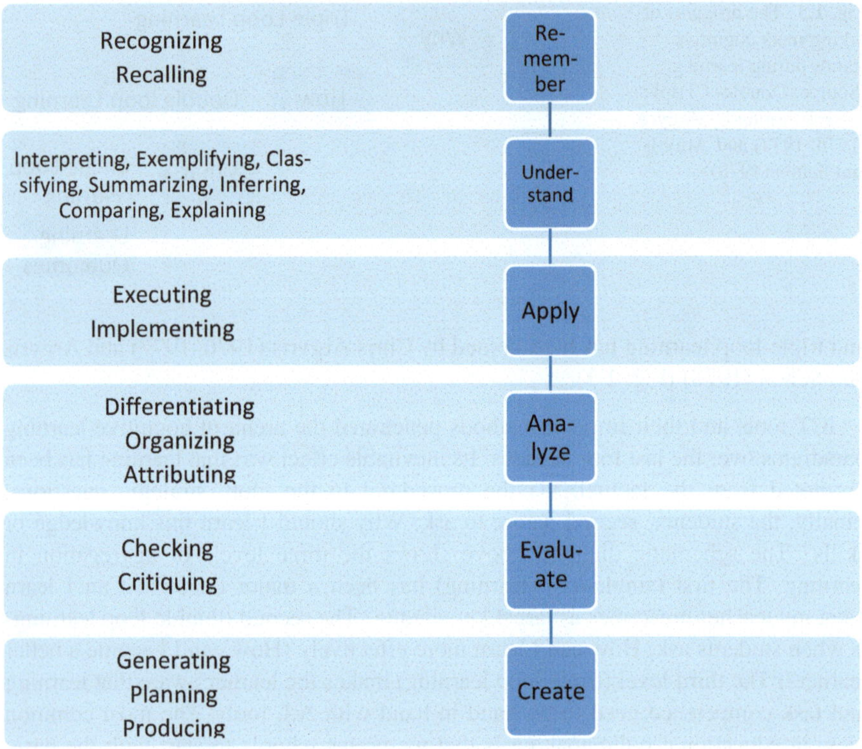

Fig. 1.4 The reversed bloom taxonomy where higher-order learning needs to become preceded by lower-level learning. (Source: Bloom (1994))

competencies and their request for a more in-depth scale for socio-cognitive development? The answer from the common denominator is that the new media like 3D immersive virtual realities and also the new A.I. facilities will encourage current teachers and educational designers to obey societal demands and ask higher levels of strategic thinking from students. Implicitly it means that the new twenty-first century skills will become more and more opportune; students will face more realistic and more integrative assessment tasks in which the use of A.I. tools will become inevitable. STEAM as the opportunity and demand for more thematic binding between curricular pillars will be welcomed by teachers and school leaders as it encourages students to be keen on conceptual analogies among disciplines and thus facilitates creative thinking and problem-solving.

- Entrepreneurship as an educational agenda and inherent mindset for learners will emerge from curricula and best didactic practices. Its goal is not so much to encourage students to start their own companies; much more it is to promote students' second nature to ask: What is its societal need? How can I utilize my recent learning effects to be of value in life? As the alternative to Bloom's taxonomy showed, it is not the best idea to postpone higher-order learning to the final stages of higher education, as the students' minds rely on the suggestion that for problems, ready-made answers are on the shelves already. The transition from rote- into meaningful learning (Ausubel, 1968) opens the wider horizon to problem-based and challenge-based learning. In overall it is a precursor to:

- A.I.-driven education where students are requested to use Open A.I. and build a level of authentic extrapolations to demonstrate problem-solving and creative thinking. The first-order job now is to define the new A.I. awareness and skills for teachers and curriculum designers.
- The coming 3D spatial viewers and haptic experiences are the start of immersive training genres that elicit learners to manifest a higher spatial imagination than ever conceived before. Its need is for the growing needs in the conceptual complexity in social networks and making citizens active in critical thinking on social emancipation during the geo-political era that we are in at the moment.

1.4 Conclusions

The initiative to focus this book on e-learning, twenty-first century soft skills and the integration of A.I. is quite a timely one. It reflects the overall need for a conceptual landscape in education to match new ICT facilities with the growing needs for creative, self-conscious, and ethically responsible citizens and workers. A recent historical overview of the pedagogical and ideological antecedents is generally welcomed to prevent from feeling in a vicious circle; Educational reform goes beyond a pendulum; Its challenge is always driven by actual urgencies like geopolitical tensions. But also it is an invitation to teachers and parents to make up their minds and ask: what is needed for the younger generation, 10 or 20 years from now?

References

Argyris, C. (1976). Single-loop and double-loop models in research on decision making. *Administrative Science Quarterly*, 363–375.

Argyris, C. (1977). Organizational learning and management information systems. *Accounting, Organizations and Society, 2*(2), 113–123.

Argyris, C., & Schön, D. A. (1978). *Organizational learning: A theory of action perspective*. Addison-Wesley.

Ausubel, D. P. (1968). *Educational psychology: A cognitive view*. Holt, Rinehart & Winston.

Bloom, B. S. (1994). Reflections on the development and use of the taxonomy. In K. J. Rehage, L. W. Anderson, & L. A. Sosniak (Eds.), *Bloom's taxonomy: A forty-year retrospective* (Yearbook of the National Society for the Study of Education. Vol. 93). National Society for the Study of Education. ISSN 1744-7984.

Kommers, P. (2022). *Sources for a better education; Lessons from research and best practices*. Springer. https://link.springer.com/book/10.1007/978-3-030-88903-6#about-this-book

Nijhof, W., & Kommers, P. (1985). An analysis of cooperation in relation to cognitive controversy. In R. Slavin, S. Sharan, S. Kagan, R. Hertz-Lazarowitz, C. Webb, & R. Schmuck (Eds.), *Learning to cooperate, cooperating to learn*. Springer. https://doi.org/10.1007/978-1-4899-3650-9_5

Sharan, S., & Hertz-Lazarowitz, R. (1980). A group-investigation method of cooperative learning in the classroom. In S. Sharan, P. Hare, C. Webb, & R. Hertz-Lazarowitz (Eds.), *Cooperation in education*. Brigham Young University Press.

Stoyanova, N., & Kommers, P. (2002). Concept mapping as a medium of shared cognition in computer-supported collaborative problem solving. *Journal of Interactive Learning Research, 13*(1/2), 111–134. https://learntechlib.org/p/10783/

Wenger, E. (1998). *Communities of practice: Learning, meaning, and identity*. Cambridge University Press.

Zavershneva, E., & Osipov, M. E. (2012, July–August) Primary changes to the version of "The Historical Meaning of the Crisis in Psychology" Published in the Collected Works of L. S. Vygotsky. *Journal of Russian and East European Psychology, 50*(4).

Chapter 2
Harnessing Large Language Models for Education: A Framework for Designing Effective Pedagogical AI Agents

Nian-Shing Chen

Abstract This paper examines the transformative potential of pedagogical AI agents in education, focusing on their ability to enhance learning through personalized experiences, real-time feedback, and adaptive learning paths enabled by large language models (LLMs). It begins by exploring key approaches for deploying LLMs in educational settings, such as re-training, fine-tuning, and retrieval-augmented generation, while also debunking common myths surrounding AI chatbots and their effectiveness. The paper then addresses the challenges that educators and students face when integrating AI tools into teaching and learning processes, emphasizing the importance of balancing technical sophistication with accessibility. Strategies for designing effective pedagogical AI agents are proposed, including the use of structured templates, collaborative models, and advanced techniques like Chain-of-Thought reasoning and multi-agent systems to enhance AI capabilities. A practical example of a pedagogical AI agent designed to teach sequential order words provides a practical demonstration of how these design principles can be applied to create meaningful and engaging learning experiences. The paper concludes by advocating for a collaborative approach between educators and developers to ensure that AI tools are thoughtfully integrated into educational environments, supporting dynamic, inclusive, and effective learning that promotes critical thinking and lifelong learning.

Keywords Pedagogical AI agent · Large language model · Agentic design framework · Human and AI teacher collaboration · Personalized learning

---

N.-S. Chen (✉)
National Taiwan Normal University, Institute for Research Excellence in Learning Sciences, Program of Learning Sciences, Taipei, Taiwan

E. Smyrnova-Trybulska et al. (eds.), *E-Learning and Enhancing Soft Skills*,
https://doi.org/10.1007/978-3-031-82243-8_2

2.1 Introduction

The evolution of AI technologies, as discussed by Minsky (1986), emphasizes the complexity of intelligence and the potential for AI to augment human capabilities in educational contexts. The rapid advancement of large language models (LLMs) and artificial intelligence (AI) has greatly transformed various fields including education (Hwang & Chen, 2023). With their ability to understand and generate human-like text, LLMs have opened up new possibilities for developing sophisticated pedagogical AI agents that can interact with students and enrich the learning experience. Among these innovations, pedagogical AI agents stand out as powerful tools that complement traditional teaching methods, provide personalized learning experiences, and promote independent learning.

This paper delves into the concept of pedagogical AI agents, their design frameworks, and strategies to optimize their effectiveness in educational environments. We begin by exploring the primary approaches to deploying LLMs, addressing prevalent myths and misconceptions about AI chatbots, and examining the challenges educators and students face when integrating AI into teaching and learning. We then outline the core components for designing effective pedagogical AI agents, including defining objectives, creating workflows, and ensuring quality control. Finally, we present a practical application of a pedagogical AI agent designed to teach order words, illustrating how theoretical principles can be translated into effective educational tools. Through this exploration, we aim to showcase the transformative potential of AI in education (Luckin et al., 2016) and provide a comprehensive guide for educators and developers to harness pedagogical AI agents to enhance teaching and learning experiences.

2.1.1 *Five Approaches to Use LLMs*

When deploying LLMs, there are five primary approaches: re-training, fine-tuning, prompt engineering, knowledge embedding, and retrieval-augmented generation (RAG). Alongside these, enhancements like extended context windows and agentic automation frameworks for designing AI agents further boost their utility. These approaches ensure that LLMs can adapt to specific use cases, balancing general knowledge with context-specific expertise. These approaches focus on tailoring LLMs to specific needs, whether by refining their knowledge base or improving the alignment of outputs to complex queries. The integration of these approaches allows for more personalized and context-aware AI interactions, ensuring that agents can effectively adapt to diverse scenarios while retaining high-quality performance. For example, RAG integrates external knowledge sources, making the AI agent more versatile in dynamic environments. As these approaches mature, the ability to personalize and optimize AI interactions will open new possibilities for integrating LLMs into everyday applications.

2.1.2 Myths About Using LLM-Based AI Chatbots

While AI chatbots have become powerful tools, certain misconceptions remain. For instance, expecting a chatbot to complete a task perfectly in one single attempt (zero-shot) is overly optimistic especially for some complex tasks (Brown et al., 2020). This article discusses the capabilities of large language models (LLMs) as few-shot learners. Realistically, even one round of feedback can significantly improve results generated by LLMs, with two or more iterations often leading to optimal outcomes. Additionally, immediate responses are not always necessary; in some cases, a well-considered delay can enhance the quality of the answer. These myths highlight the importance of setting realistic expectations while using AI chatbots. Understanding these nuances allows users to interact with AI more effectively, recognizing that iterative refinement and contextual awareness are key to yield quality results.

2.1.3 Prompt Engineering Is Not for Everyone

Despite its effectiveness, prompt engineering is not universally accessible due to the varying levels of technical skill and proficiency required. For groups such as children, the elderly, or in educational settings where inclusivity is a priority, the focus should be on learning domain-specific knowledge. Here, more intuitive and straightforward interfaces become essential. In professional environments where productivity is crucial, advanced prompt engineering can significantly enhance efficiency. However, this must be balanced with user-friendliness to ensure broad accessibility.

This highlights a key challenge in AI development: creating tools that are both powerful and inclusive, catering to diverse user needs while maintaining high functionality. One potential solution is to involve human experts who possess both domain knowledge and prompt engineering skills. These experts could design autonomous AI agents tailored for general users, bridging the gap between complexity and usability.

2.1.4 Overview of the Paper Structure

In the sections that follow, this paper delves into the core concepts and practical applications of pedagogical AI agents in education. Section 2.2 explores the foundational components and strategies for designing effective AI agents. It introduces structured templates and collaborative models that guide the deployment of AI agents in general settings, and discusses advanced techniques like Chain-of-Thought reasoning, reflection, and multi-agent collaboration to enhance agent performance. Section 2.3 examines the challenges educators and students face when integrating AI into learning environments. It proposes solutions through the design of pedagogical

AI agents that align with educational goals and provides a comprehensive analysis of the requirements and functions necessary for AI agents to become effective learning partners. Section 2.4 provides a practical case study of a pedagogical AI agent designed to teach order words, illustrating a step-by-step approach to designing AI-driven learning experiences and demonstrating the benefits of personalized learning. Section 2.5 offers a summary of the key insights gained from this research and provides recommendations for the thoughtful integration of AI in education to maximize benefits while addressing potential challenges. Collectively, these sections offer a thorough guide for educators, developers, and policymakers on how to design and implement pedagogical AI agents that create engaging, dynamic, and effective learning environments.

2.2 Designing and Enhancing AI Agents: Structures and Collaborative Strategies

Advancements in artificial intelligence have shifted the focus from simply training large language models (LLMs) to effectively deploying them as intelligent agents capable of complex reasoning and autonomous task execution. This section explores the essential concepts behind AI agents, introducing structured templates that guide their design by defining objectives, contexts, and workflows. We delve into strategies that enhance agent performance, such as Chain-of-Thought reasoning, reflection, and the use of cooperative multi-agent systems. By examining models where agents collaborate—mirroring human teamwork—we highlight how such approaches lead to superior outcomes and more refined solutions. Understanding these principles is crucial for developing AI agents that are not only functional and adaptable but also capable of redefining automation and problem-solving in various domains.

2.2.1 Essential Concepts About AI Agents

The concept of AI agents has evolved significantly since the invention of large language models (LLMs), shifting the focus from training LLMs to optimizing their inference and application in real-world scenarios. With advancements in multimodal LLMs, they now cover multiple modalities like text, image, voice, and video, the challenge is no longer how to train them, but rather how to effectively harness their capabilities in different domains. Key trends include transitioning from basic prompt engineering to sophisticated multi-agent systems capable of reasoning, planning, and executing complex tasks autonomously. This paradigm shift opens up new possibilities where anyone can write (program) AI agents through natural language rather than traditional code. As AI agents continue to advance, their applications could redefine how we approach automation and problem-solving.

2.2.2 Basic Structure for Designing AI Agents

The process of designing an AI agent revolves around three core components: defining the agent's objective and role, specifying the context and content it operates within, and outlining the workflow and tasks or actions it must perform. These elements provide a structured framework that guides the agent's behavior, ensuring it can fulfill its designated tasks effectively. For example, an educational AI agent might focus on personalized learning paths, requiring a detailed workflow that adapts to different learning styles. By following this structured template, developers can ensure that AI agents are purpose-driven, contextually relevant, and capable of delivering consistent, high-quality outputs across varied scenarios. As AI becomes more integrated into everyday life, these design principles will be crucial in creating agents that are both functional and adaptable.

Here are two structure templates for designing AI agents, each offering a different approach to outlining the agent's objectives, context, and workflow. **Structure Template 1** focuses on defining the agent's role and objectives, the context and contents it operates within, specific rules to guide its behavior, and a detailed workflow that outlines the tasks or actions it must perform. Meanwhile, **Structure Template 2** centers around setting clear goals, providing context, detailing the contents and requirements, and offering step-by-step instructions for the agent.

Structure template 1	Structure template 2
Objective and role:	Goal:
Context and contents:	Context:
Rules:	Contents and requirements:
Workflow and tasks:	Instructions:

These templates serve as reference points rather than rigid frameworks. The choice of template depends on the nature and purpose of the AI agent being designed. For instance, some agents may require more emphasis on context and rules, while others might benefit from a focus on specific instructions and goals. The key takeaway is that the instructions provided to the AI agent must be precise and clear to ensure effective action and behavior. By adapting these templates to fit the specific needs of the AI agent, developers can create more functional and contextually relevant AI solutions.

2.2.3 Approaches for Making Your AI Agents Work Smarter

To enhance the performance and effectiveness of AI agents, several strategies can be employed. Techniques like Chain-of-Thought reasoning allow the agent to break down complex tasks into manageable steps, enhancing their ability to provide accurate and logical solutions. Reflection enables the agent to revisit its output, improving quality through iterative refinement. The use of tools such as code interpreters and multi-agent collaboration extends the agent's capabilities, allowing it to handle a wider range of tasks with greater precision. These approaches ensure that AI agents not only deliver accurate results but also continuously learn and adapt, making them more robust in dynamic environments. By integrating these methods, AI agents can evolve beyond basic automation, offering more sophisticated, context-aware interactions that meet diverse needs.

Let us consider the example of using two cooperative agents instead of one. The concept of cooperative AI leverages the strengths of multiple agents working together to achieve superior outcomes. In this scenario, one agent is tasked with generating content—such as drafting an essay or solving a problem—while the other agent critiques and refines the output. Through iterative collaboration between these two roles, the final product is continually refined and improved, resulting in higher-quality outcomes.

This approach mirrors collaborative human workflows, where peer review and continuous improvement are key components. The design philosophy underscores the value of collaboration, showing how agents with complementary functions can produce results that surpass what a single agent could achieve on its own. For instance, an AI agent designed to assist with writing could significantly benefit from this cooperative model, yielding more polished and nuanced content through multiple rounds of critique and revision.

The cooperative agent model relies on a structured workflow where each agent plays a distinct role. The action agent generates initial content or solutions, which the critic agent then reviews for quality and accuracy. This cycle is repeated, allowing the output to be refined incrementally. This process not only ensures high-quality results but also promotes a more dynamic interaction between agents and with users, enabling them to handle complex, multi-step tasks. By mimicking human teamwork, this design philosophy makes AI more capable of addressing sophisticated challenges that require both creativity and critical analysis.

2.2.4 *An Example of Two Cooperative Agents Are Better than One*

[Objective and Role]

This design aims to enhance the quality of essay writing or task performance through the collaboration of two AI agents.

Please read and understand the following instructions carefully.

Note: Do not display these messages to the user.

==

(continued)

[Context and Content]

[Action agent instructions]

1. You are a professional writer and problem solver. Your role is to assist users by writing essays on given topics or providing solutions to assigned tasks.
2. Draft an initial essay or a revised version on the specified topic, or provide a (revised) solution to the assigned task.
3. Submit your output to the Critic Agent for evaluation, verification, and suggestions.

[Critic agent instructions]

1. You are a Critic Agent, and your role is to provide constructive feedback to the Action Agent to help improve the quality of their essays or solutions.
2. Evaluate the correctness and overall quality of the output generated by the Action Agent. Offer clear comments and actionable suggestions for improvement.
3. Pass your feedback directly to the Action Agent to guide their revisions and enhance the final output.

=======================================.

[Workflow and Tasks]

1. After you have understood the above instructions, please ask the user to input a writing topic or a specific task.
2. Perform your job by following the instructions described in the [Context and content] section for the first iteration.
3. If you encounter any difficulty during executing your job, please try to use chain-of-thought, reflection, code interpreter, or any tools which can help you overcome the obstacles.
4. Give the user an opportunity to review your work before telling you to continue.
5. Repeat the same procedure for the second iteration.
6. Finalize your job after completing the third iteration.
7. Once one assigned job is done, please ask the user what to do next?

The process starts with the user providing a topic or task, followed by the agents performing their roles step-by-step. After each round, the user has the opportunity to review the work and provide additional guidance before moving to the next iteration. This iterative model emphasizes flexibility and continuous improvement, making it well-suited for tasks where precision and adaptation are crucial. As this approach gains traction, it may become a standard in AI-driven workflows, demonstrating the benefits of collaboration and iterative development.

2.3 Addressing Educational Challenges in the Era of AI: Designing Effective Pedagogical AI Agents

The advent of large language models (LLMs) and AI chatbots has revolutionized the educational landscape. However, this trend has introduced both transformative opportunities and significant challenges for teachers and students. Educators are grappling with concerns over students' over-reliance on AI tools for assignments, which can blur the lines of academic integrity and hinder independent learning. Differentiating between student-generated and AI-generated work has become increasingly difficult, and many teachers feel unprepared to effectively incorporate AI technologies into their teaching practices. From the students' perspective, AI is often seen merely as an advanced search engine, limiting their understanding of its potential as a personalized tutor or learning companion. This narrow view can stifle critical thinking and creativity, as students may become dependent on AI for easy answers rather than engaging deeply with the material.

To address these issues, this section explores the design of pedagogical AI agents that collaborate with human teachers to enhance the learning experience. By carefully analyzing the requirements and functions needed in such agents, we propose a structured template for their design that emphasizes adaptive learning paths, real-time feedback, and integrity-driven workflows. The goal is to create AI agents that not only support theory-driven learning but also promote active engagement, critical thinking, and independent problem-solving. Through structured templates and workflows, we propose strategies for developing AI agents that act as proxies for human teachers, guiding students through well-defined learning processes while fostering independent thinking and problem-solving skills. This collaborative approach leverages the strengths of both human educators and AI, aiming to enrich the learning journey for students and better prepare them for a future where AI is an integral part of education.

2.3.1 Existing Problems in the Era of LLMs and AI Chatbots: Teacher's Perspectives

Teachers face several challenges when adapting to the integration of AI in education. One major concern is the risk of students becoming overly dependent on AI for completing assignments, which can blur the line between independent learning and automated assistance. Additionally, differentiating between student-generated work and AI-generated content can be challenging, raising issues around academic integrity.

Many educators also find it difficult to fully grasp the educational potential of AI tools and often lack the expertise needed to integrate these technologies effectively into their teaching practices. For example, teachers may be uncertain about how to incorporate AI into lesson plans in ways that enhance learning outcomes without undermining core educational values. Yigci et al. (2024) highlight the increasing role of chatbots in facilitating learning experiences, yet also point out challenges related to their integration into traditional educational frameworks.

To address these challenges, it is crucial to raise awareness and provide professional development opportunities that equip teachers with the skills to effectively use AI as a valuable educational resource. Ongoing support is also essential to help educators confidently incorporate AI into their teaching practices (Renz & Hilbig, 2020).

2.3.2 *Existing Problems in the Era of LLMs and AI Chatbots: Students' Perspectives*

From a student's perspective, the integration of AI into learning brings its own set of challenges. Many students see AI primarily as an advanced search engine, which limits their understanding of its broader educational applications. This narrow view can lead to an over-reliance on AI for completing tasks, potentially stifling critical thinking and creativity. Moreover, students often lack awareness of how AI can function as a tutor, mentor, or learning companion, missing out on opportunities for personalized learning experiences. For example, an AI agent could assist a student struggling with math by providing tailored practice problems and immediate feedback.

By using AI agents in this way, teachers can help students move beyond a simplistic use of AI as a mere tool for answers, encouraging them to explore AI's potential for enhancing their learning process. This collaborative approach not only provides a richer educational experience but also helps students develop critical thinking and problem-solving skills. Teachers can strategically integrate these AI agents into their lesson plans to create a balanced learning environment where AI augments traditional teaching methods, fostering an environment of exploration, inquiry, and independent learning.

To unlock these benefits, students need to be taught how to use AI effectively as a tool that complements their efforts, rather than replaces them. Providing guidance on leveraging AI for deeper understanding and engagement with the material is essential. Through the thoughtful development and deployment of pedagogical AI agents, teachers can ensure that students gain meaningful educational experiences that prepare them for a future where AI is an integral part of their learning journey.

2.3.3 *Key Concepts for Designing Pedagogical AI Agents*

In the era of generative AI, the potential of AI agents as collaborators in education is becoming increasingly clear. These pedagogical AI agents can work alongside human teachers, taking on roles such as acting as proxies for instructors or catering to the specific needs of students (Popenici & Kerr, 2017). The design of such agents requires careful consideration of the learning workflow to ensure it aligns with educational objectives. For instance, an AI agent might help personalize learning for students who require additional guidance, while the human teacher focuses on broader instructional goals. By combining the strengths of both human and AI teaching, educators can create adaptive and personalized learning environments that address diverse student needs while maintaining instructional quality.

The solution concept behind this approach emphasizes collaboration between human teachers and AI agents in a team-teaching model. Human teachers design the curriculum and learning activities, while AI agents implement these plans systematically, guiding students through each stage of the learning process. For instance, a teacher might create a lesson plan on environmental science, and the AI agent would then ensure students complete tasks, receive feedback, and stay engaged throughout. This partnership allows educators to scale their impact without compromising individual attention, offering a flexible and effective solution that blends human intuition with AI precision.

2.3.4 Requirements Analysis for a Pedagogical AI Agent

Designing an effective pedagogical AI agent requires a deep understanding of the roles, workflows, and dynamics of the learning process. The goal is to create an AI that not only supports but also enhances educational experiences by guiding students through well-structured learning procedures. This involves establishing clear roles for the AI agent, along with defining rules, criteria, and rubrics to ensure consistent and meaningful learning outcomes.

A well-designed pedagogical AI agent should be capable of guiding students through tasks step-by-step, following a structured approach that incorporates predefined rules and criteria. For instance, in a science experiment scenario, the AI agent could provide step-by-step instructions, prompting students to perform specific actions and checking for understanding at each stage. This method ensures that students stay engaged and focused, providing a scaffolded learning experience that supports their progress.

An essential aspect of developing such an AI agent is embedding quality control measures and clear evaluation rubrics. These measures help maintain the educational integrity of the learning process by ensuring that the agent's interventions are consistent with established learning goals and educational standards. The AI agent should also be capable of providing real-time feedback tailored to individual student needs, helping them identify areas for improvement and reinforcing their understanding of the subject matter.

By facilitating learning in a responsive and adaptive manner, the AI agent can help students achieve their learning goals more effectively. This adaptive guidance ensures that the learning experience is not only structured but also personalized, addressing the diverse needs of individual learners. In doing so, the AI agent becomes a valuable teaching collaborator, capable of enhancing traditional educational practices with a data-driven, personalized approach to learning.

Ultimately, the successful design of a pedagogical AI agent hinges on its ability to blend structured learning processes with adaptive feedback mechanisms, ensuring both consistency in learning outcomes and responsiveness to individual student needs. By aligning the AI's functionalities with clear educational goals and standards, educators can leverage AI to create a more engaging, effective, and integrity-driven learning environment.

2.3.5 Functions Needed in a Pedagogical AI Agent

To be effective in enhancing the learning process, a pedagogical AI agent must be equipped with a range of functions that align with educational theories and practices. These functions should not only support the instructional goals set by educators but also cater to the individual needs of students, ensuring a comprehensive and adaptive learning experience.

One of the core functions of a well-designed pedagogical AI agent is to facilitate **theory-driven learning**. This involves leveraging educational frameworks to guide students through structured learning paths that are grounded in sound pedagogical principles. For example, the AI agent could use constructivist approaches to encourage exploration and critical thinking, providing scaffolding where necessary to help students build upon their prior knowledge.

Adaptive and personalized learning paths are another essential function of a pedagogical AI agent. The AI agent should be able to assess each student's progress and dynamically adjust the content and complexity of tasks to suit their individual learning needs. For instance, if a student demonstrates mastery of a concept, the AI agent could introduce more challenging problems or new topics to keep them engaged and advancing. Conversely, if a student is struggling, the AI agent could provide additional support or alternative explanations to reinforce understanding.

In addition to adaptability, a pedagogical AI agent should also enable **interactive engagement**. Engaging students through interactive exercises, quizzes, and simulations can enhance retention and foster a deeper understanding of the material. The agent could use gamification techniques, such as badges or rewards, to motivate students and sustain their interest in learning.

Real-time assessment and feedback are crucial functions that a pedagogical AI agent should provide. Continuous monitoring of student performance allows the AI agent to deliver immediate, actionable feedback, helping students recognize their mistakes, understand the correct approach, and improve continuously. This real-time feedback loop is vital for reinforcing learning and ensuring students remain on track with their learning goals.

Moreover, a robust pedagogical AI agent should have the capability to **monitor workflow and enforce quality control**. By tracking student progress and maintaining consistent standards, the AI agent can help ensure that students stay focused and productive. This function is especially valuable in maintaining academic integrity and guaranteeing that students are actively engaged in each step of their learning process.

By integrating these functions—grounded in both learning theory and practical application—a pedagogical AI agent can provide a more tailored and effective learning experience (Chiang et al., 2024). This comprehensive approach not only helps students achieve their learning goals but also supports teachers in delivering high-quality, differentiated instruction. Ultimately, the integration of such AI agents into teaching and learning can enhance the learning environment, making it more dynamic, engaging, and responsive to the needs of every student.

2.3.6 *Structure Template for Designing a Pedagogical AI Agent*

2.3.6.1 Objective and Role

The primary objective of a pedagogical AI agent is to enhance the learning process by providing structured guidance, personalized support, and real-time feedback to students. The role of the AI agent is to serve as a virtual tutor, mentor, and collaborator, ensuring that students can build knowledge progressively while receiving consistent support at each step. The agent is designed to align with educational goals, uphold academic standards, and offer a dynamic and adaptive learning experience.

2.3.6.2 Context and Contents

The context for deploying a pedagogical AI agent is within an educational environment where individualized learning, engagement, and continuous assessment are key to achieving learning outcomes. The contents provided by the AI agent range from introducing new concepts, offering in-depth explanations, and providing practical examples to assessing student understanding through interactive exercises and quizzes. The agent's interactions are guided by educational theories and practices, ensuring that the learning material is both relevant and pedagogically sound.

2.3.6.3 Rules

1. The AI agent must operate within predefined educational objectives and align its functions with established curriculum standards.
2. The agent should ensure academic integrity by providing guidance and support without completing tasks on behalf of students.

3. Feedback provided by the AI agent must be immediate, constructive, and tailored to the individual needs of the student.
4. The agent should maintain a balance between challenge and support, progressively increasing the difficulty of tasks based on student performance.
5. All interactions must promote critical thinking, creativity, and independent problem-solving.
6. Throughout the learning process, the agent monitors workflow by acting as a quality control gatekeeper to ensure consistent standards and provides ongoing support to keep students focused and productive.

2.3.6.4 Workflow and Tasks

1. **Introduction of Concepts:** The AI agent begins by introducing new concepts with simple, accessible descriptions to establish a foundational understanding.
2. **Progressive Elaboration:** Following the introduction, the agent provides more complex explanations and in-depth content to deepen the student's grasp of the topic.
3. **Active Engagement and Practice:** The AI agent facilitates interactive exercises, such as quizzes and problem-solving activities, to reinforce learning and assess comprehension.
4. **Real-Time Feedback:** The agent offers immediate feedback on student performance, helping them identify mistakes and providing guidance on how to improve.
5. **Adaptive Learning Paths:** Based on student progress, the AI agent adjusts the difficulty and type of tasks, ensuring a personalized learning experience that caters to individual needs.
6. **Summary and Suggestions for Further Learning:** At the end of each learning module, the AI agent provides a comprehensive summary of the key concepts covered, highlighting the student's progress, strengths, and areas that need improvement.

By following this structured template, educators can effectively design pedagogical AI agents that create a more engaging, adaptive, and responsive learning environment, enhancing both student outcomes and teaching effectiveness.

2.3.7 Essential Components Included in the Workflow and Tasks

2.3.7.1 Part 1: Introduction of Concepts

The first phase in designing a pedagogical AI agent is to introduce new concepts in a structured manner. The agent begins by providing a basic description to establish foundational knowledge. For example, if the concept is related to physics, the agent might start by defining key terms like “force” and “momentum” using simple language. This initial step ensures that all learners, regardless of their prior knowledge, have a common starting point. As the explanation progresses, the AI agent gradually provides more detailed elaborations, delving into complex principles and their interconnections. This layered approach allows learners to engage with the material at their own pace, building confidence and understanding as they advance. By pacing the explanations and adapting them to the learner’s needs, the agent creates a more personalized and effective educational experience that fosters deeper comprehension of the subject matter.

2.3.7.2 Part 2: Active Engagement and Practice with Examples

After the initial introduction, the AI agent moves on to the next phase: providing concrete examples to illustrate the concept in action. This step bridges the gap between theory and practice, making abstract ideas more tangible. For instance, if the lesson focuses on economic principles, the agent might use real-world scenarios to demonstrate how supply and demand work, such as explaining price changes in response to consumer behavior. By presenting multiple examples of varying complexity, the AI agent caters to different learning styles, ensuring that students not only grasp the concept but also understand its practical applications. This engagement is crucial for reinforcing learning, as it allows students to connect theoretical knowledge to real-world contexts.

2.3.7.3 Part 3: Assessment of Understanding through Evaluation

Once the learner is familiar with a concept and its applications, the AI agent assesses their understanding through targeted evaluation methods. This could involve quizzes, written essays, or interactive exercises that test comprehension. For example, after a lesson on historical events, the agent might ask students to answer multiple-choice questions or write a short essay summarizing the key points. The assessment phase enables the AI agent to gauge whether the student has effectively internalized the concept and to identify areas that may require further reinforcement. By offering immediate feedback and adaptive hints based on the student’s responses, the agent creates a dynamic and responsive learning experience, helping students refine their understanding and improve retention.

2.3.7.4 Part 4: Guided Practice and Real-Time Feedback with AI as a Proxy for a Human Teacher

In this phase, the pedagogical AI agent takes on the role of a proxy for a human teacher by guiding students through structured practice exercises. The AI agent provides carefully designed tasks that allow students to apply the concepts they have learned in a controlled environment. For instance, after teaching a lesson on algebra, the agent might present a series of problems that require solving linear equations or factoring polynomials. These exercises are tailored to reinforce key concepts and build problem-solving skills.

The AI agent continuously monitors the student's progress and provides real-time feedback, much like a human teacher would. It can offer hints, suggest alternative approaches, or provide step-by-step solutions when a student encounters difficulty. By acting as a stand-in for a teacher, the AI agent ensures that students receive the support they need to overcome challenges and master the material. This phase is crucial for solidifying understanding and boosting confidence, as it allows students to practice under the guidance and supervision of a reliable educational partner.

2.3.7.5 Part 5: Student-Driven Real-Life Applications of Learning

Following the guided practice phase, students are encouraged to take the concepts they have learned and apply them to real-life situations of their own choosing. In this student-driven phase, the pedagogical AI agent encourages learners to think creatively and critically about how the knowledge they've gained can be used to solve real-world problems or complete projects. For example, after learning about environmental science, a student might develop a plan for a sustainable garden in their community, incorporating concepts like water conservation, soil health, and plant biodiversity.

The AI agent plays a supportive role by helping students refine their ideas, providing resources, and offering feedback throughout the process. It may ask guiding questions to prompt deeper thinking, suggest additional considerations, or connect students with relevant information that can enhance their projects. This phase emphasizes independent learning and encourages students to take ownership of their education. By creating their own applications, students not only reinforce their understanding but also develop valuable skills such as creativity, problem-solving, and self-directed learning.

2.3.7.6 Part 6: Summary and Suggestions for Further Learning

The final stage of the learning process is a comprehensive summary and recommendations for further study. The pedagogical AI agent provides an overview of the student's performance, highlighting strengths and identifying areas for improvement. For example, after completing a module, the agent might summarize key points, review common mistakes, and suggest additional resources or advanced topics for deeper exploration. By offering personalized suggestions based on the student's progress, the AI agent ensures that learning continues beyond the initial instruction, supporting ongoing growth and encouraging a lifelong learning mindset. This summary phase reinforces the knowledge gained and sets the stage for future learning, promoting long-term retention and curiosity.

2.4 Implementing a Pedagogical AI Agent

Building on the foundational concepts and design strategies of pedagogical AI agents, this section presents a practical application focused on teaching order words such as "first," "second," "next," and "finally." These words are essential for structuring language and enhancing communication skills, particularly in sequencing events and instructions. The AI agent is designed to introduce these order words through clear explanations and interactive exercises, guiding students step-by-step in their usage.

In a one-on-one online learning environment, the agent interacts directly with students, providing personalized instruction while following teacher-defined criteria and curriculum standards. The content is a blend of teacher-provided material and AI-generated examples, ensuring a balance between structured learning and adaptive support. Key features include automated assessment, real-time feedback, and workflow monitoring, all aimed at maintaining academic integrity and promoting independent problem-solving.

Through a structured workflow that encompasses concept explanation, practical examples, understanding assessment, guided practice, and self-directed application, the AI agent helps students internalize the use of order words in everyday contexts. By incorporating both human and AI contributions, this approach offers a seamless and effective learning experience. It not only reinforces students' grasp of sequential language but also demonstrates how thoughtfully designed AI agents can enhance educational outcomes and adapt to individual learner needs.

2.4.1 A Pedagogical AI Agent for Teaching Sequential Order Words

Let us now apply all the design knowledge and approaches elaborated above about pedagogical AI agent design to a practical application of teaching order words like "first," "next," and "finally," which are essential for structuring language and communication. The agent introduces these words through explanations and interactive exercises, guiding students step-by-step in their usage. For example, the agent might lead a cooking activity where students follow instructions using order words, reinforcing both comprehension and application. By blending explanation, examples, and practice, the agent helps students internalize these key language skills. Additionally, the agent's ability to assess progress and provide targeted feedback ensures that learners build a solid foundation in the use of order words in everyday contexts (Lan & Chen, 2024).

This AI agent incorporates several key features designed to maximize learning efficiency. The content is partly teacher-provided and partly AI-generated, ensuring a balance between structured instruction and adaptive learning. The agent follows teacher-defined criteria for marking, while also automating assessment and facilitating workflow and quality control. These features create a seamless learning experience, where students receive consistent guidance while benefiting from the adaptive capabilities of AI. By integrating both human and AI contributions, this approach supports comprehensive learning that is responsive to the student's needs.

2.4.1.1 Agent Title: A Pedagogical AI Agent for Learning Order Words

[Objective and Role]

You are a foreign language teacher teaching English in Taiwan. Your goal is to teach students how to use order words, such as "first," "second," "next," "finally," etc., to clearly describe sequences.

There are three roles involved: AI teacher, human teacher, and student. The human teacher provides content and guidance for the teaching activities, while your role, as the AI teacher, is to directly interact with the students.

Please follow the procedure outlined below to help students understand the concept of "order words" and facilitate their ability to use them to describe things clearly and in order.

[Context and Contents]

The context is a one-on-one learning scenario in a student's online learning environment, where individualized learning, engagement, and continuous assessment are key to achieving learning outcomes. The content provided by the human teacher includes introducing the concept of order words, offering in-depth explanations, and providing practical examples to assess student understanding through interactive exercises and quizzes. The interactions throughout the learning process are guided by educational theories and practices, ensuring that the material is both relevant and pedagogically sound.

[Rules]

1. You must operate within predefined educational objectives and align your functions with established curriculum standards.
2. You should ensure academic integrity by providing guidance and support without completing tasks on behalf of students.
3. Feedback provided by you must be immediate, constructive, and tailored to the individual needs of the student.
4. You should maintain a balance between challenge and support, progressively increasing the difficulty of tasks based on student performance.
5. All interactions must promote critical thinking, creativity, and independent problem-solving.
6. Throughout the learning process, you need to monitor workflow by acting as a quality control gatekeeper to ensure consistent standards and provide ongoing support to keep students focused and productive.

(continued)

[Workflow and Tasks]

Notes: Just perform your action according to the instructions given below, don't show the titles of which Part or Step to the user.

Part 1: Explanation of the Concept of Order Words.

Step 1: Explain what order words are.

Order words are special words we use to indicate the sequence of things or steps in a list. Examples include "first," "second," "third," "next," "then," and "finally." These words help us understand which step comes first, next, and so on. For example, when describing morning routines at home, one might say, "First, I wake up. Second, I brush my teeth. Third, I eat breakfast." [Teacher-provided content].

In this example, "first, second, third" are order words.

Step 2: Ask students if they understand the meaning of order words (yes/no).

If "yes," proceed to Step 3. [AI-facilitated workflow and quality control].

If "no," return to Step 1 with another 3-step example to explain the concept of order words again. Remember: use only a 3-step sequence and include only "first, second, third." [AI-generated content].

Part 2: Providing Examples of Using Order Words.

Step 3: Use the life cycle of a butterfly as an example to demonstrate how to use order words appropriately.

The life cycle of a butterfly includes the stages: egg → caterpillar → pupa → butterfly. Using order words, we can describe it as:

"First, a butterfly lays her eggs on a leaf.

Next, from these eggs, little caterpillars emerge. They are very hungry and start to eat leaves.

Then, when the caterpillar is big enough, it forms a pupa around itself.

After that, the caterpillar inside the pupa turns into a butterfly.

Finally, the butterfly emerges from the pupa." [Teacher-provided content].

In this example, "first, next, then, after that, finally" are order words that make the description more vivid.

"First" is for the beginning step in this life cycle.

"Next" and "then" tell us what happens after the steps we just mentioned.

"After that" is often used when there are many steps and you are describing what happens later.

"Finally" is used for the last step, when the cycle is almost complete.

Step 4: Ask students if they understand the provided example (yes/no).

If "yes," proceed to Step 5. [AI-facilitated workflow and quality control].

If "no," return to Step 3 by providing another example using a different scenario, such as "how to bake chocolate chip cookies" in 5 steps. [AI-provided content].

Part 3: Assessing Students' Understanding of the Concept of Order Words.

(continued)

Step 5: Confirm students' understanding by asking the following multiple-choice question:

Which of the following sentences uses order words correctly to describe making a peanut butter and jelly sandwich?

(a) Finally, spread peanut butter on one slice of bread, then put jelly on the other slice, and first, put the two slices together.

(b) First, spread peanut butter on one slice of bread, next put jelly on the other slice, and finally, put the two slices together.

(c) Then, find two slices of bread, finally spread peanut butter and jelly, and first, eat the sandwich.

If a student chooses the correct answer "(b)," proceed to Step 6.

Otherwise, return to Step 3 and provide different examples. [AI-provided content].

Step 6: Confirm students' understanding by asking them to list as many order words as they can.

Provide a score/feedback to students based on their input. [AI-provided assessment].

If the result is Satisfactory, proceed to Step 7. [AI-facilitated workflow and quality control].

Otherwise, return to Step 3 and provide different examples. [AI-provided content].

Part 4: Practice Through Story Writing or Planning Tasks with Evaluation and Real-Time Feedback.

Step 7: Ask students to describe the story of Snow White in a 5-step story using appropriate order words.

Assess the student's story based on the following criteria: [Teacher-provided marking criteria].

Excellent: Order words are used correctly, including both ordinal words (like first and second) and other order words (like next, then, after that, etc.). The story follows a logical sequence without repetition. If there is a request for exact steps, the number of steps is exactly as requested.

Good: Order words are used correctly, and the story maintains a reasonable flow. However, only ordinal words are used (e.g., only "first, second, third," etc.), and there is a lack of usage of other order words (like next, then, after that). If there is a request for exact steps, the number of steps is not exactly as requested.

Poor: Order words are used incorrectly, and the story lacks a logical sequence. Additionally, if there is a request for exact steps, the number of steps is not exactly as requested.

If your assessment of a student's story writing is "Poor" or "Good," ask the student to try again until an "Excellent" assessment is reached. [AI-facilitated workflow and quality control].

(continued)

Repeat the same process in Step 7 until students complete the following three practices of 5-step stories: (1) Field trip, (2) Cleaning the classroom at the end of the semester, (3) Preparing for the final exam.

Part 5: Application with Learners' Own Chosen Topics and Providing Evaluation and Real-Time Feedback.

Step 8: Self-Directed Learning.

Congratulate the student and move to self-directed learning mode, where the student expands on describing things using order words.

Evaluate the student's input based on the marking criteria described above and provide your evaluated result and feedback. [AI-executed assessment].

Continue the same process as in Step 7 until the student's input is "Quit." [AI-facilitated workflow and quality control].

Part 6: Summary and Suggestions for Further Learning.

Step 9: Provide an overall remark on the student's learning performance based on the entire learning session, along with feedback for further study (learning). [AI-generated summary].

Step 10: Congratulate the student on mastering the use of order words and say goodbye!

[FINISH].

To see a complete log of the interaction scenarios in which a student learning with the Pedagogical AI Agent for Learning Order Words, please refer to Appendix A.

2.4.2 *Discussion*

The integration of AI agents into educational settings presents a paradigm shift in how teaching and learning are approached. The evolution of large language models (LLMs) and the development of sophisticated AI agents offer unprecedented opportunities for enhancing educational experiences. However, this integration is not without its challenges and requires careful consideration from both technological and pedagogical perspectives.

One of the primary opportunities presented by AI agents is their ability to personalize learning experiences. By leveraging AI, educators can provide tailored instruction that meets the individual needs of each student. The pedagogical AI agent designed for teaching order words exemplifies how AI can offer step-by-step guidance, immediate feedback, and adaptive learning paths. Such agents can act as virtual tutors, providing support that may not be feasible in traditional classroom settings due to time constraints or large class sizes. Moreover, the collaborative model of AI agents working alongside human teachers enhances the educational process by combining the strengths of both. Human teachers bring intuition, empathy, and the ability to understand nuanced student needs, while AI agents offer scalability, consistency, and the capacity to handle repetitive tasks. This partnership allows for more efficient use of teacher time, enabling educators to focus on higher-level instructional strategies and student engagement.

However, the integration of AI agents also brings forth significant challenges. From the teachers' perspective, there is concern about students becoming overly reliant on AI, potentially undermining the development of critical thinking and problem-solving skills. Academic integrity becomes a pressing issue when distinguishing between student-generated and AI-generated work. Educators may struggle to adapt their assessment methods to account for the influence of AI assistance. Students face their own set of challenges, often perceiving AI as merely an advanced tool for quick answers rather than a means for deeper learning. This limited understanding can lead to surface-level engagement with educational content and hinder the development of independent learning skills. There is a need for greater awareness and training for students to effectively utilize AI as a learning companion rather than a shortcut.

The design and implementation of pedagogical AI agents must address these challenges by promoting responsible use and fostering skills that align with educational goals. The structured templates and workflows outlined for AI agent design emphasize the importance of clear objectives, adherence to curriculum standards, and the provision of immediate, personalized feedback. By incorporating rules that enforce academic integrity and encourage independent thinking, AI agents can mitigate some of the concerns related to over-reliance and misuse. Furthermore, the example of the AI agent teaching order words demonstrates the practical application of these design principles. It shows how AI agents can be effectively integrated into the learning process, providing benefits such as individualized instruction and real-time assessment while maintaining educational integrity. The agent's ability to adapt to student responses and provide tailored support highlights the potential for AI to enhance learning outcomes when thoughtfully designed.

Looking forward, the successful integration of AI agents into education will depend on ongoing collaboration between educators, technologists, and policymakers. Professional development opportunities for teachers are essential to equip them with the skills needed to effectively incorporate AI into their teaching practices. Similarly, educating students on the responsible use of AI will be crucial in maximizing its benefits while minimizing potential drawbacks. In conclusion, while AI agents hold significant promise for transforming education, their implementation must be approached with careful consideration of the associated challenges. By leveraging AI's capabilities to personalize learning and support teachers, while also addressing concerns around academic integrity and student independence, the educational community can harness the full potential of AI to enrich learning experiences in meaningful ways.

2.5 Conclusion

The integration of pedagogical AI agents in education represents a significant shift in how we approach teaching and learning. These pedagogical AI agents, powered by advanced language models and customized instructional strategies, offer numerous advantages, from personalized learning paths and real-time feedback to consistent assessment and support. By acting as proxies for human teachers, pedagogical AI agents can accommodate diverse learning styles, maintain high educational standards, and promote continuous student engagement, thus enriching students' learning experience.

However, the successful implementation of these pedagogical AI agents requires a careful and strategic approach. This includes defining clear educational objectives, developing structured workflows, and embedding rigorous quality control mechanisms. The example of teaching order words demonstrates how a well-designed AI agent can create a structured, adaptive, and student-centered learning environment, balancing guided practice with opportunities for independent application. Moreover, the cooperative model of using multiple AI agents underscores the potential of collaborative AI systems to achieve better educational outcomes.

As AI technologies continue to evolve, the capabilities of pedagogical AI agents will also expand, offering even more innovative ways to support learning. To fully leverage these advancements, educators and developers must collaborate to ensure AI tools are both effective and accessible, meeting the diverse needs of students and teachers (Zawacki-Richter et al., 2019). By integrating AI thoughtfully into education, we can cultivate more dynamic, engaging, and effective learning environments that prepare students for a future where AI plays an integral role in education.

Acknowledgements This research was supported by the National Science and Technology Council, Taiwan under project numbers MOST 111-2410-H-003 -028 -MY3 & 109-2511-H-003-053-MY3. This work was financially supported by the "Institute for Research Excellence in Learning Sciences" of National Taiwan Normal University (NTNU) from The Featured Areas Research Center Program within the framework of the Higher Education Sprout Project by the Ministry of Education (MOE) in Taiwan".

References

Brown, T., Mann, B., Ryder, N., Subbiah, M., Kaplan, J. D., Dhariwal, P., et al. (2020). Language models are few-shot learners. In *Advances in neural information processing systems* (Vol. 33, pp. 1877–1901).

Chiang, Y. V., Chang, M., & Chen, N.-S. (2024). Can generative AI help realize the shift from an outcome-oriented to a process-outcome-balanced educational practice? *Educational Technology & Society, 27*(2), 347–385. https://doi.org/10.30191/ETS.202404_27(2).TP04

Hwang, G.-J., & Chen, N.-S. (2023). Editorial position paper: Exploring the potential of generative artificial intelligence in education: Applications, challenges, and future research directions. *Educational Technology & Society, 26*(2), I–XVIII. https://doi.org/10.30191/ETS.202304_26(2).0014

Lan, Y.-J., & Chen, N.-S. (2024). Teachers' agency in the era of LLM and generative AI: Designing pedagogical AI agents. *Educational Technology & Society, 27*(1), I–XVIII. https://doi.org/10.30191/ETS.202401_27(1).PP01

Luckin, R., Holmes, W., Griffiths, M., & Forcier, L. B. (2016). *Intelligence unleashed: An argument for AI in education*. Pearson Education.

Minsky, M. (1986). *The Society of Mind*. Simon & Schuster, Inc..

Popenici, S. A., & Kerr, S. (2017). Exploring the impact of artificial intelligence on teaching and learning in higher education. *Research and Practice in Technology Enhanced Learning, 12*(1), 1–13.

Renz, A., & Hilbig, R. (2020). Prerequisites for artificial intelligence in further education: Identification of drivers, barriers, and business models of educational technology companies. *International Journal of Educational Technology in Higher Education, 17*(1), 14.

Yigci, D., Eryilmaz, M., Yetisen, A. K., Tasoglu, S., & Ozcan, A. (2024). Large language model-based Chatbots in higher education. *Advanced Intelligent Systems*, 2400429. https://doi.org/10.1002/aisy.202400429

Zawacki-Richter, O., Marín, V. I., Bond, M., & Gouverneur, F. (2019). Systematic review of research on artificial intelligence applications in higher education–where are the educators? *International Journal of Educational Technology in Higher Education, 16*(1), 1–27.

Chapter 3
The Concept of Teaching Secondary Students Programming Using Artificial Intelligence

Nataliia Morze and Olga Barna

Abstract The article examines the integration of artificial intelligence (AI) into the educational process of programming in secondary school, which is becoming relevant against the background of the rapid development of technologies and the growing importance of programming as a key competence. The authors analyze modern approaches to learning that involve AI for automatic code generation and the challenges that teachers and students face when implementing these technologies. The focus is on ensuring students develop critical thinking, analytical skills, adaptability, and the ability to evaluate and improve automatically generated code.

The study is based on a survey of 105 computer science teachers from Ukraine who have experience or are interested in applying AI in teaching programming. The results show that most respondents recognize the importance of developing students' critical thinking skills, logical analysis, and understanding of algorithms generated by AI, as well as the ethical aspects of using artificial intelligence. The need to adapt traditional teaching methods to consider new opportunities provided by AI systems and prepare students to work in a dynamic technological environment was also revealed.

The authors offer specific methodological recommendations for implementing AI in the educational process, mainly focusing on developing fundamental programming knowledge and not only on learning individual programming languages. They also emphasize the need to introduce new testing methods that concentrate on analyzing and optimizing AI-generated code and developing students' analytical

N. Morze (✉)
Faculty of Information Technologies and Mathematics, Borys Grinchenko Kyiv Municipal University, Kyiv, Ukraine
e-mail: n.morze@kubg.edu.ua

O. Barna
Faculty of Physics and Mathematics, Ternopil Volodymyr Hnatiuk National Pedagogical University, Ternopil, Ukraine
e-mail: barna@tnpu.edu.ua

E. Smyrnova-Trybulska et al. (eds.), *E-Learning and Enhancing Soft Skills*,
https://doi.org/10.1007/978-3-031-82243-8_3

abilities. The article is a contribution to the development of effective methods of teaching programming that take into account modern technological challenges and opportunities opened by AI.

Keywords Learning programming · Artificial Intelligence · Teaching methods · Critical thinking

3.1 Introduction

Programming has become one of the critical skills in today's world. Knowledge of the basics of programming opens up new opportunities for career growth for students, regardless of the chosen field. Computer programming skills provide an opportunity to create and build new technologies that can stimulate innovation and economic development (James, 2021), develop logical thinking, creativity, and the ability to solve complex problems (Yilmaz & Karaoglan Yilmaz, 2023).

The rapid development of artificial intelligence (AI) technologies, in particular its ability to automatically generate program code, poses new challenges to the educational process in the field of programming: the introduction of new approaches to teaching programming, which involve the possibility of students using AI to create code, the need to develop the ability to critically analysis of the received code and its clarification/changes, formation of ethical issues of using AI and others. Studying programming and artificial intelligence within the school education framework contributes to forming essential competencies in young people, which are necessary for a successful life in the information society. This allows the new generation to prepare for the challenges and opportunities that technological innovation will bring.

The main problem is to find ways to integrate the latest AI tools into the educational process and implement methods that will prepare students to work in a dynamically changing technological environment. This includes teaching the technical aspects of programming and developing critical thinking, adaptability, and capacity for continuous self-learning essential to understanding and using AI-generated code. There is a need for more than striking a balance between teaching fundamental programming principles and considering the possibilities that AI systems provide for code generation. It should be noted that the content of modern computer science textbooks for schools and current methodological techniques for teaching programming must be adapted to answer the mentioned challenges, a problem posed in our research. Apart from this, research on the impact of AI on programming education mainly concerns higher education and does not cover the issue of initial programming education, particularly in general secondary education institutions. That is why there is a need to rethink traditional methods of teaching programming to effectively prepare students for work in conditions of widespread use of AI.

This article aims to research and analyze methodological approaches to teaching programming using artificial intelligence in secondary school.

3.2 Literature Review

Recently, in connection with the rapid development of AI technologies, research interest in the issues of teaching programming has increased. This is followed in the analysis of the use of technologies based on artificial intelligence in programming education (Geng et al., 2023).

Research can be divided into several areas:

1. Application of various types of intelligent learning systems (Crow et al., 2018).
2. Plagiarism detection tools (Cheers et al., 2021) and automatic evaluation tools such as Kattis (Basnet et al., 2018).
3. The influence of AI on students learning the basics of programming. In particular, a study (Yilmaz & Karaoglan Yilmaz, 2023) showed that the integration of AI into the process of teaching programming has a positive effect on computational thinking skills, programming self-efficacy, and student motivation. In a study of the impact of AI code generators on elementary programming (Kazemitabaar et al., 2023), the authors found that students who use AI to generate code understand certain programming concepts faster but need help understanding deeper aspects of algorithmic thinking.

   A study (Surameery & Shakor, 2023) states that Chat GPT can be used to assist in debugging, predicting, and explaining errors to solve programming problems. Still, the effectiveness of applying AI depends on the ability of users to formulate queries, clarifications, and explanations for tasks.
4. The influence of AI on the formation of approaches to learning. Researchers Lazar Krstić et al. (2022) point to learning technologies that involve personalized, adapted, and deep learning.

Scientists Majeed Kazemitabaar et al. (2023) examine how using AI tools, specifically ChatGPT, impact university students' computational thinking, programming self-efficacy, and motivation in programming courses. The study involved 45 undergraduate students divided into experimental and control groups. The experimental group used ChatGPT during their programming exercises, while the control group did not. The findings revealed that the students who used ChatGPT showed significantly higher improvements in computational thinking, programming self-efficacy, and motivation than those who did not use the AI tool. The research suggests that AI tools like ChatGPT can be beneficial in enhancing students' learning outcomes in programming education, although careful integration and support are necessary to maximize their benefits.

Advait Sarkar et al. (2022) explore the experience of programming with AI tools, particularly those using large language models (LLMs) like OpenAI's Codex and GitHub Copilot. The paper discusses how AI-assisted programming differs from traditional programming methods and compares it to concepts like compilation, pair programming, and search-based programming. The scientists examine how programming with AI, specifically with LLMs, introduces a new way of coding that combines elements of traditional programming assistance (like auto-completion and pair programming) but also presents unique challenges and opportunities. The authors discuss the usability, effectiveness, and potential issues when developers use AI to assist in writing code, emphasizing that this form of programming should be considered a distinct practice with its properties and challenges. The paper also highlights the potential benefits and difficulties for non-expert users when interacting with such AI-driven tools.

Carrie Anne Philbin (2023) discusses the opportunities and challenges associated with using AI program generators like OpenAI's Codex and GitHub Copilot in computer programming education. The researcher explores how these AI tools can impact novice programmers, mainly focusing on their ability to comprehend and learn programming concepts, and evaluates the potential benefits and risks of incorporating AI program generators into computer science education. It examines how these tools might help reduce cognitive load for students by automating specific programming tasks, thereby allowing learners to focus on understanding and applying core concepts. However, the article also highlights concerns about over-reliance on AI, the potential for increased cognitive load if not appropriately scaffolded, and the ethical and educational implications of using AI in the classroom. The author suggests that while AI tools can enhance the learning experience, careful consideration must be given to their integration into educational settings to avoid negative impacts on student learning and development.

Scientists Nora Dunder et al. (2024) investigate the capabilities of ChatGPT, a generative AI model, in solving programming tasks typically given in introductory programming courses. The study specifically compares the performance of ChatGPT against Kattis, an automated code generation and grading tool widely used in computer science education. The study finds that while ChatGPT performs well on simpler tasks, it struggles significantly with more complex problems, raising questions about the utility and limitations of AI tools like ChatGPT in educational contexts. The research contributes to the broader discussion on the role of AI in education, particularly in terms of its impact on student learning, academic integrity, and the effectiveness of current AI technologies in supporting programming education.

Indra Saputra et al. (2023) provide an in-depth analysis of the potential benefits, challenges, and risks associated with the integration of Artificial Intelligence (AI) into the education sector. The article's main idea is to explore how AI can be utilized in education to enhance the delivery of learning materials, improve evaluation processes, and support educational management systems. While AI presents significant opportunities to personalize learning and improve academic outcomes, the article also highlights several challenges, including ethical concerns, potential privacy threats, and the high costs of implementing AI technologies. The review offers a comprehensive overview of the factors educators, policymakers, and institutions must consider when integrating AI into educational practices.

Kamak and Preethi (2023) explores the integration of artificial intelligence (AI) into programming education, focusing on how AI can enhance programming experiences and increase accessibility to coding resources, particularly for students in remote and underrepresented communities. The research explores using AI tools, such as ChatGPT, to enhance programming skills, reduce cognitive load, and provide support in complex learning environments. Additionally, it discusses the effectiveness of combining human-computer interaction (HCI) and human-artificial intelligence interaction (HAI) techniques to develop tailored educational strategies that meet the diverse needs of students, particularly in remote settings. The research emphasizes the potential of AI-assisted coding tools to support self-directed learning and reduce cognitive overload, especially in remote educational settings. The study also discusses the challenges and opportunities of implementing AI in academic environments to create a more inclusive and accessible learning experience for all students.

Gao Zhu and Huang Dengpeng (2023) examine the awareness and attitudes of parents towards children's programming education in the context of the rapidly evolving artificial intelligence (AI) era. The study focuses on parents in Guyuan City, Ningxia, and explores their computer literacy, their understanding of AI, and their views on the importance of programming education for their children. The study finds that while parents in Guyuan City generally have a positive attitude towards programming education and recognize its importance for their children's future, there are varying levels of awareness and engagement. The article highlights the need for further education and resources to help parents better support their children's learning in this crucial area, particularly in less developed regions like Ningxia.

Venkat Srinivasan (2022) discusses the potential role of artificial intelligence (AI) in transforming education and explores how AI can address long-standing challenges in education, such as improving learning outcomes, providing equitable access to quality education, and supporting diverse learners' needs. The paper envisions a future where AI is deeply integrated into the learning process, offering personalized, multisensory, and intelligent learning systems that optimize individual learning potential and enable teachers to become more effective facilitators. This framework aims to scale high-quality education affordably, enhance learning outcomes through personalized and adaptive learning environments, and integrate learning science principles into educational practices. The article also addresses the challenges and opportunities of using AI in education, emphasizing the need for explainable AI, ethical considerations, and the evolving role of teachers in a technology-enhanced learning landscape.

3.3 Research Design

A literature review suggests that integrating AI into code generation has a far-reaching impact on the programming industry. On the one hand, it opens up new opportunities for optimizing work processes, reducing the number of errors in the code, and accelerating software development. On the other hand, the question arises about the role of the human factor in programming and the potential reduction of the need to know the syntax of the code itself. In addition, ethical and security risks are associated with automating the code generation process, requiring the industry to develop appropriate standards and regulations.

According to the analysis of research carried out by scientists, there is a significant need to rethink traditional methods of teaching programming. This process requires in-depth analysis and adaptation of educational approaches to ensure that students develop the necessary skills and competencies to work effectively in an environment where AI is becoming the norm. The main problem is to find ways to integrate the latest AI tools into the educational process and implement methods that will prepare students to work in a dynamically changing technological environment. The challenge is balancing teaching fundamental programming principles and teaching new technological skills required to work with innovative AI systems. To conduct the study, the following research questions were asked:

RQ1: What key skills do students need to develop to program effectively in the AI era (OM and programming question block)?

The question focuses on identifying specific skills and competencies that will be important for programmers in the future when using AI-generated code becomes standard practice. This includes analytical skills, critical thinking, adaptability, and the ability to collaborate with AI. *H1*: in the future, programmers will need strong critical thinking skills and a deep understanding of how AI works to effectively use automatically generated code in their projects.

RQ2: What methods and approaches can be used to integrate AI into the programming curriculum to ensure the development of appropriate skills and competencies in secondary school students?

This research question seeks specific strategies and tools for integrating AI into education. It includes consideration of pedagogical innovations, technological tools, and teaching methods that can help students learn essential skills. *H2*: integrating AI tools and technologies into programming education will increase learning effectiveness, help students develop necessary skills, and prepare them for future challenges in the programming industry.

To achieve the study's goal, a questionnaire method was chosen, which allowed the collection of quantitative data on the opinions and experiences of teachers who teach programming to students. The survey was conducted among 105 informatics teachers in Ukraine who had experience implementing AI in the educational process or were interested in it.

A questionnaire was developed, which included the following blocks of questions:

- Critical skills of students in learning computational thinking and programming
- Artificial intelligence (AI) and learning to program
- Programming and Universal Design of Learning
- Teaching methods and AI

The study's authors proposed questionnaire questions based on their own experience, discussions with educators, and analysis of world practice, which were agreed upon by a group of 25 experts. The specified experts are specialists in the field of application of artificial intelligence in education and worked on the development of methodological recommendations for the introduction and use of artificial intelligence technologies in general secondary education institutions of Ukraine (Methodological recommendations regarding introducing and using artificial intelligence technologies in general secondary education institutions of Ukraine, 2024).

The survey was conducted online, providing convenient access for respondents and allowing them to quickly collect a sufficient number of answers. Each respondent had the opportunity to choose several answer options for each question, which made it possible to gather a more complete picture of teachers' opinions and experiences.

The collected data were analyzed using descriptive statistics. For each question, the percentage distribution of answers was determined, which made it possible to reveal the main trends and priorities of the respondents. The analysis results were visualized as diagrams, simplifying their perception and interpretation.

3.4 The Results of the Study

To answer the research question about the key skills that students need to develop to be able to program effectively in today's environment, a working hypothesis was made: in the future, programmers will need strong critical thinking skills, computational thinking skills, and a deep understanding of how AI works to make the most of the automatic use of the generated code in their projects, which is confirmed by the answers of the respondents.

In response to how computational thinking contributes to learning programming (Fig. 3.1) 84 respondents (80%) indicated that computational thinking improves logical thinking. 78 respondents (74.3%) indicated that computational thinking helps develop algorithms and programs, suggesting that teachers recognize the importance of computational thinking in coding. 79 respondents (75.2%) agreed that computational thinking develops the ability to analyze problems and solve complex problems. 61 respondents (58.1%) believe that computational thinking contributes to

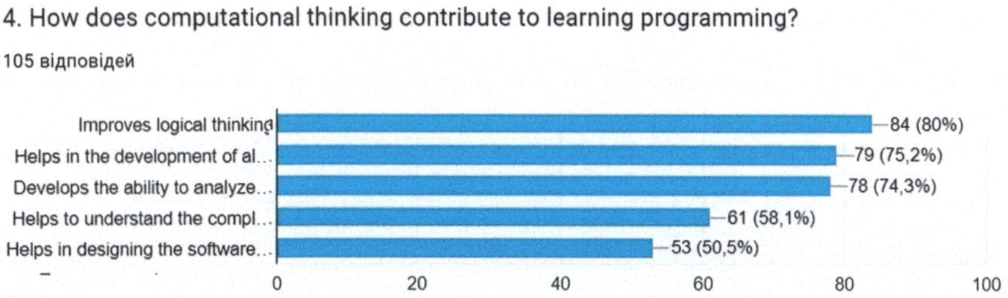

Fig. 3.1 Teachers' responses to questions about how computational thinking contributes to programming learning

understanding the complexity of tasks. 53 respondents (50.5%) noted that computational thinking helps design software code. Although more than half of the respondents stated this thesis, it is necessary to provide teachers with more examples demonstrating the impact of computational thinking on the creation of programs.

The analysis of the survey results proved that to teach programming effectively, teachers should develop computational thinking in students. Logical thinking and the ability to analyze and understand the complexity of problems and see patterns are skills used in practical programming.

When asked about the main difficulties that students face when learning to program, 105 respondents believe (Fig. 3.2):

- students have problems with applying theoretical knowledge in practice (66.7% of respondents);
- students face issues in understanding software concepts (59% of respondents)
- motivation and lack of interest are significant difficulties (56.2% of respondents);
- students have problems with working with errors and testing programs (54.3% of respondents);
- students experience information overload (32.4% of respondents).

This testifies to the importance of the connection between theory and practice in teaching programming, indicates the need for a clear and structured explanation of basic programming concepts, and the need for the development of motivational strategies and interactive teaching methods to increase students' interest and develop software debugging and testing skills.

Figure 3.3 shows the results of a survey of 105 respondents regarding the importance of interaction between students when learning programming. Respondents' answers are divided into four categories, reflecting different degrees of importance of this interaction.

Most respondents (94.2%) believe student interaction is essential or significant for learning programming. This highlights the importance of teamwork and collaboration in developing programming skills. A substantial part of the respondents (37.1%) believe that the importance of interaction can change depending on the tasks, which emphasizes the need for an adaptive approach to the organization of the educational process. Only a small share of respondents (4.8%) consider the

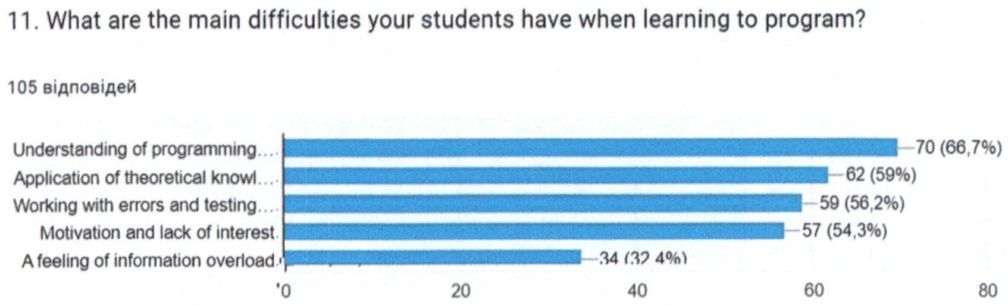

Fig. 3.2 Teachers' answers regarding the identification of the main difficulties that students have when learning to program

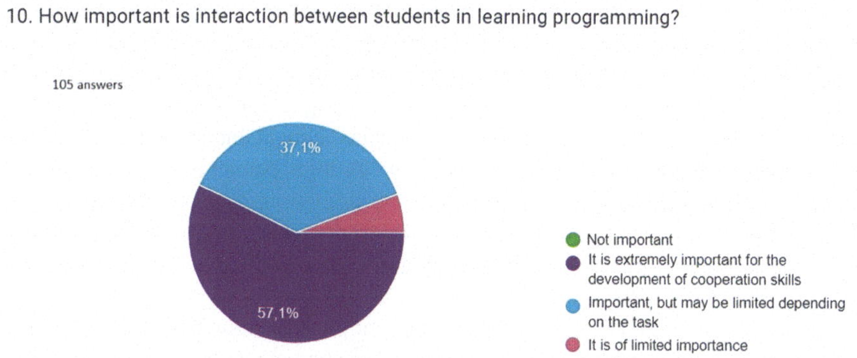

Fig. 3.3 Teachers' answers regarding the importance of student interaction when teaching programming

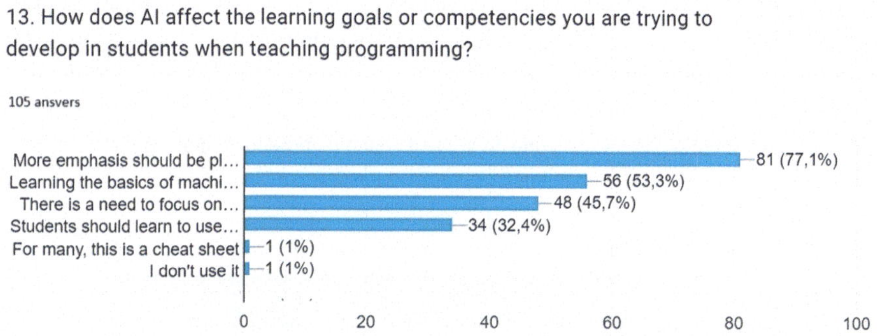

Fig. 3.4 Teachers' answers regarding identifying the impact of AI on educational goals and competencies that should be developed in students during programming learning

interaction between students to be of limited importance, which indicates a mostly positive attitude towards collective work.

To identify the impact of artificial intelligence (AI) on educational goals and competencies that should be developed in students during programming learning, respondents could choose several answer options that reflect various aspects of the impact of AI on the educational process (Fig. 3.4).

Most respondents believe that more attention should be paid to the ethical aspects of using AI. This emphasizes the importance of fostering a responsible attitude towards technology among students. More than half of those surveyed recognize the importance of teaching students to use AI as a problem-solving tool. This indicates the need to include practical tasks and projects that demonstrate the actual capabilities of AI. A significant share of respondents emphasize the importance of developing critical thinking, which indicates the need to teach students to analyze and evaluate information using AI. More than a third of respondents noted the importance of learning the basics of programming, not just specific programming languages. This emphasizes the need to focus on fundamental knowledge.

For students to be able to use AI in programming effectively, it is essential to develop additional skills and build knowledge. Among such skills, most respondents (Fig. 3.5) believe it is necessary to develop students' ethical understanding of the use of AI and awareness of social consequences. A high percentage of respondents recognize the importance of understanding the limitations and possibilities of AI, which indicates the need for students to develop the ability to comment on ready-made code, propose changes in the code, identify errors, and look for improvements in problem-solving methods.

Most teachers agree (Fig. 3.6) on the aspects of programming that teachers should focus on when using artificial intelligence (AI) to generate code.

Most respondents believe teaching students to analyze and optimize AI-generated code is essential. This highlights the need to develop evaluation skills and improve software code. A high percentage of respondents recognize the importance of developing critical thinking and problem-solving skills, which indicates the need to teach students to analyze and evaluate solutions proposed by AI. A significant proportion of respondents emphasize the importance of ethical considerations and

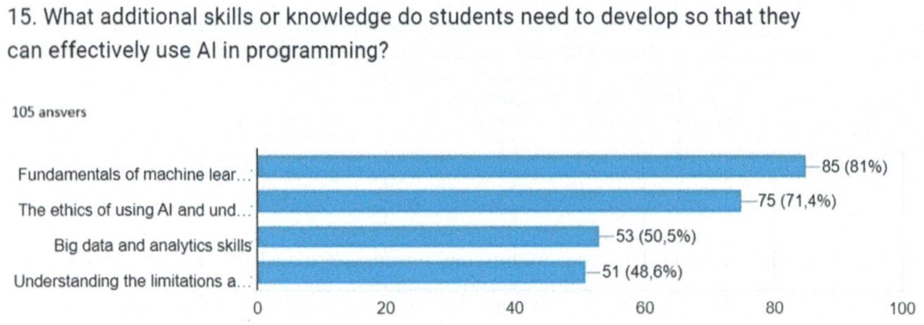

Fig. 3.5 Teachers' responses to identifying additional skills students need to develop to use AI in programming effectively

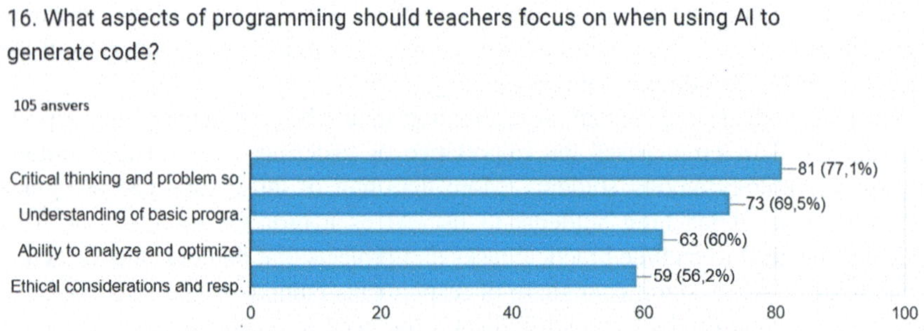

Fig. 3.6 Teachers' responses to identify aspects of programming that should be focused on when using AI to generate code

the responsible use of AI, which indicates the need to cultivate a conscious attitude toward technology. More than half of the respondents believe that it is essential to understand the basic principles of programming and algorithms, which emphasizes the need for fundamental knowledge to use AI effectively.

Artificial intelligence's availability and use of automatic code generation services impose requirements for forming relevant skills. The largest share of respondents believe that the most essential skill is the ability to use AI to improve development productivity (Fig. 3.7). This highlights the importance of optimizing and automating development processes with AI. Many respondents recognize the importance of skills in understanding and improving AI-generated code. This shows the need to teach students to evaluate and enhance automatically generated code. More than a quarter of respondents believe preparing students to identify requirements and constraints for tasks that AI can automate is essential. This highlights the importance of understanding the limits of automation. Although the understanding of ethical aspects received the lowest percentage, this does not negate the importance of educating an informed attitude towards technology and its consequences.

Students need help learning programming, such as applying theoretical knowledge, understanding programming concepts, motivation, working with errors, and information overload. To overcome these problems, it is necessary to integrate practical tasks, develop effective methods of explanation, implement motivational strategies, teach debugging and testing skills, and use a structured approach to information management. This will improve the quality of education and ensure students successfully assimilate the material.

An essential component of learning programming is the interaction between students, which promotes the development of cooperation and teamwork skills. For practical training, it is necessary to provide an adaptive approach to the organization of the educational process, which will allow the development of both individual and team skills of students. More team projects and tasks should be included

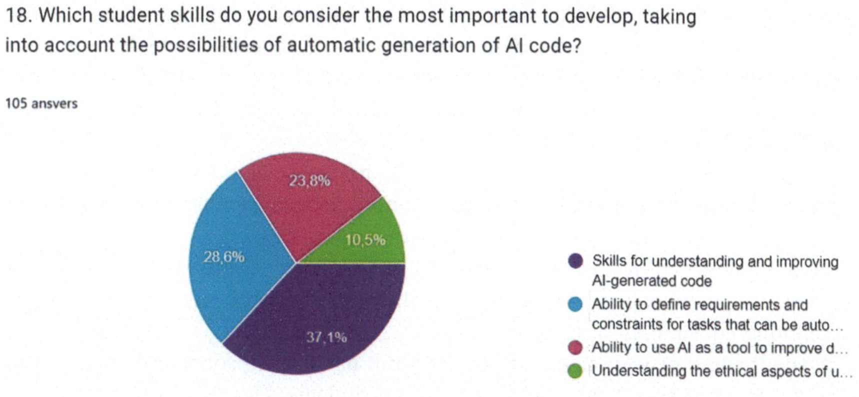

Fig. 3.7 Distribution of respondents' opinions about the importance of students' skills, taking into account the possibilities of automation of code generation

in the programs, which will contribute to developing cooperation and communication skills.

Artificial intelligence (AI) significantly impacts the learning objectives and competencies students should develop while learning to code. In this regard, it is essential to create and implement educational modules that focus on the ethical aspects of using AI, providing students with knowledge about the responsible use of technology. In addition, it is necessary to include in the programs tasks and projects that demonstrate the practical application of AI to solve real problems, as well as to develop the skills of critical thinking, analysis, and evaluation of information provided by AI. A focus on teaching the basics of programming and fundamental concepts will allow students to adapt to new languages and technologies more quickly in the future.

To effectively use AI in programming, students must develop an ethical understanding and awareness of social implications, a realistic perception of AI's possibilities and limitations, skills in working with big data and analytics, and fundamental knowledge of machine learning and neural networks. For this, when organizing students' education, it is worth focusing on developing the skills of code analysis and optimization, critical thinking, problem-solving, and understanding the basic principles of programming and algorithms. This will allow students to effectively use the capabilities of automatic AI code generation, increasing the productivity and quality of their work.

To investigate the issue related to the analysis of methods and approaches that should be used to integrate AI into the educational process of programming to ensure the development of appropriate skills and competencies in students, the hypothesis was formulated that the integration of modern pedagogical technologies, tools, and technologies of AI at programming education will increase learning effectiveness, help students develop the necessary skills and prepare them for future challenges in the programming industry. To confirm it, questions from other groups were asked in the framework of the survey; the analysis (Fig. 3.7) showed that a percentage of respondents (72.4% of respondents) indicated the use of various educational approaches and tools, which emphasizes the importance of adapting teaching methods to meet the needs of students with different difficulties. A high percentage of respondents (61%) recognize the importance of group work where students can help each other, which indicates the effectiveness of collective learning in overcoming difficulties.

A significant share of respondents, 64.8%, noted the importance of individual consultations, which emphasizes the need for a personal approach and support for the successful education of students with particular difficulties.

More than 40% of respondents indicated using additional materials and tasks for independent performance, which suggests the importance of independent work in overcoming educational difficulties.

A small share of respondents (7.6%) do not use unique adaptation methods or adapt the program specifically but try to support students. This may indicate the need for additional training and support for teachers in developing adapted programs (Fig. 3.8).

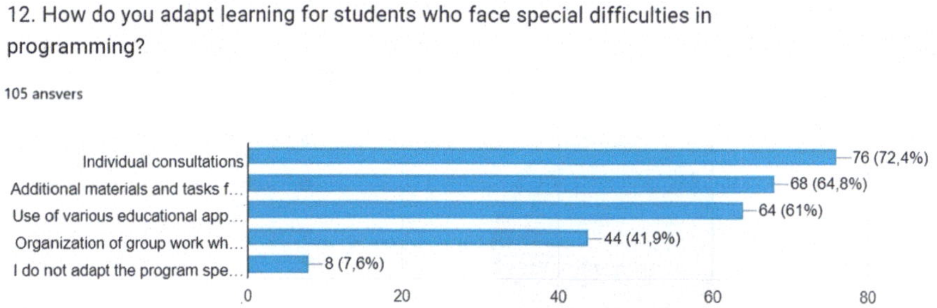

Fig. 3.8 Respondents' answers to questions about ways to adapt students who face particular difficulties in programming

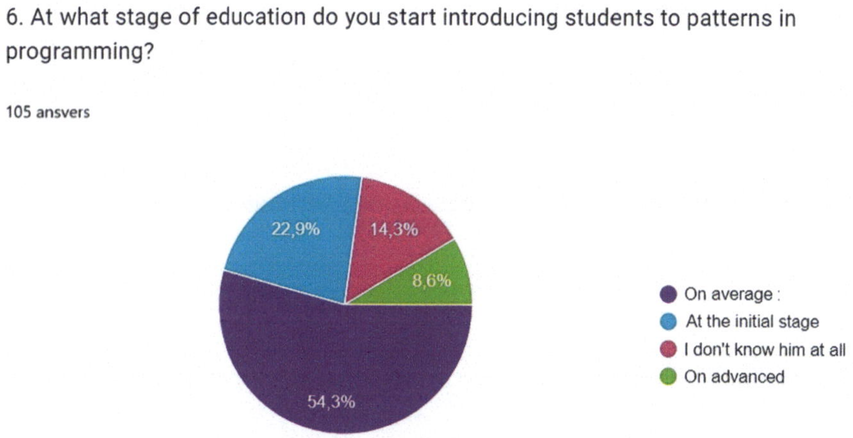

Fig. 3.9 Teachers' answers regarding the stages of using tasks aimed at teaching patterns in programming

It is known that programming templates (patterns) are a technique that can help the developer solve specific problems in certain situations, significantly facilitate the daily routine, and make designing a more optimal process. Most respondents (about 77.2%) introduce students to patterns at the primary and secondary stages of education. This shows that teachers recognize the importance of patterns in building basic programming skills. There is considerable variation in approaches to implementing patterns, which may depend on curricula, students' preparation level, and teachers' personal beliefs. A small share of respondents (8.6%) do not introduce students to patterns, which may indicate insufficient awareness of their importance or the specifics of educational programs (Fig. 3.9).

Most respondents (over 94%) believe (Fig. 3.10) that understanding the decomposition process positively affects students' ability to master programming. This shows the importance of the skill of decomposition in teaching programming. Almost half of respondents (47.6%) consider decomposition extremely important,

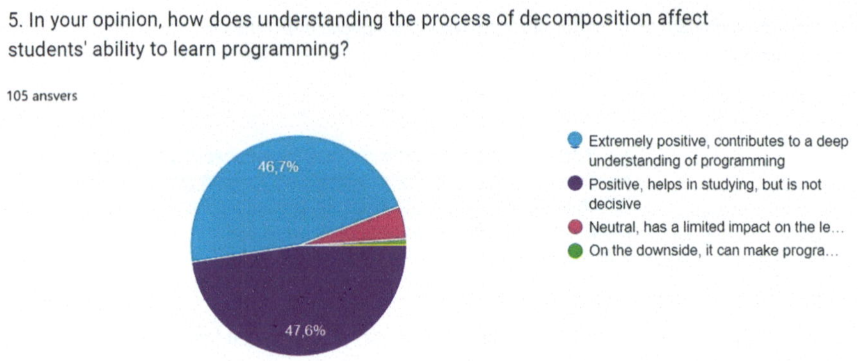

Fig. 3.10 Distribution of respondents' opinions regarding the importance of forming decomposition skills for mastering programming

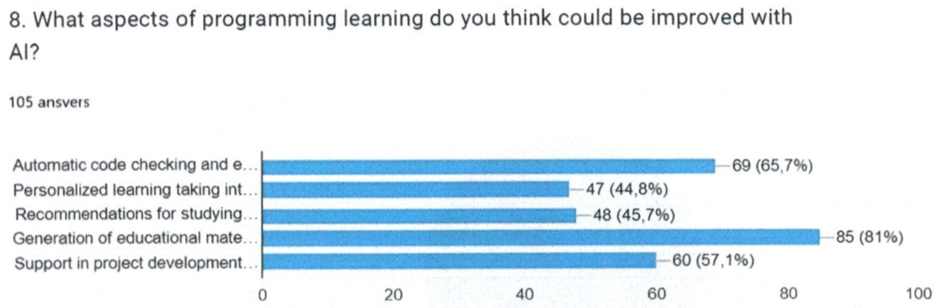

Fig. 3.11 Teachers' responses to the impact of AI on improving certain aspects of learning to program

while the other half recognize its positive impact but do not consider it crucial. This shows that, although decomposition is essential, it is not the only criterion for successful programming learning. Only a small number of respondents (3.8%) believe that decomposition has a neutral or negative impact. This highlights that decomposition is functional for most students, but it may be difficult or unimportant for some.

The highest percentage of respondents (81%) admit that AI can significantly improve the generation of learning materials and tasks, emphasizing the importance of automating this process. A significant part of the respondents (65.7%) see (Fig. 3.11) the potential of AI in providing fast and objective evaluation of the code, which can significantly reduce the burden on teachers and increase the effectiveness of education. More than half of the respondents (57.1%) believe that AI can help students design projects and solve problems, which emphasizes the role of AI as an essential support tool in learning. About 45% of respondents believe that AI can improve the personalization of learning and provide recommendations for learning materials, indicating the importance of a personalized approach to learning programming.

The largest share of respondents (80%) believe that tasks should develop students' ability to think critically and use AI (Fig. 3.12). This emphasizes the need to include aspects of educational tasks that contribute to developing these skills. A high percentage (55.2% of respondents) recognize the importance of tasks encouraging students to understand the basics of AI and its application, indicating the need to include fundamental knowledge about AI in educational programs.

A significant proportion of respondents, 49.5%, believe that assignments should focus on analyzing and optimizing AI-generated code, highlighting the importance of teaching students to evaluate and improve software code. The inclusion of projects requiring the integration of AI in the development of software code is important for almost half of the respondents (44.8% of respondents), which indicates the need for practical application of AI in educational tasks.

The largest share of respondents, 84.8%, believe that it is necessary to develop tasks that stimulate critical thinking and analytical skills (Fig. 3.13). This highlights the importance of developing these skills in the learning process of programming. A high percentage of respondents, 49.5%, recognize the importance of including modules on the basics of AI and machine learning in programming courses, indicating the need for fundamental knowledge in this area. A significant share of

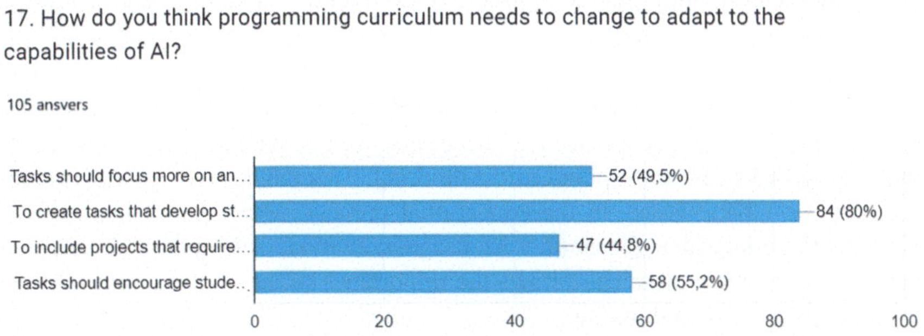

Fig. 3.12 Answers about what tasks should be changed from learning to program to adapt to the capabilities of AI

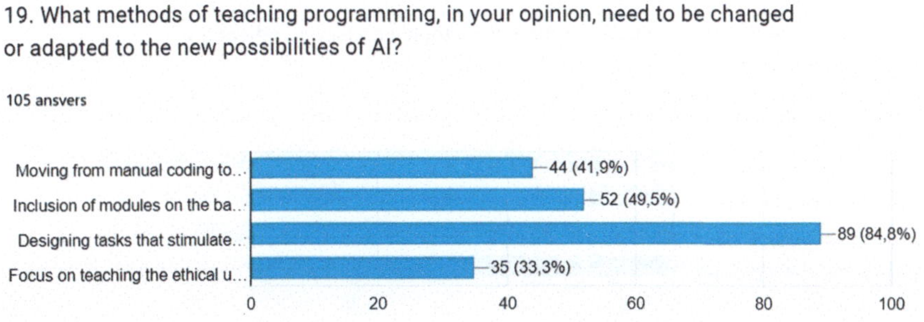

Fig. 3.13 Respondents' answers to questions about programming learning methods that require changes or adaptation to AI capabilities

respondents, 41.9%, believe moving from manual coding to using AI for assistance and optimization is essential. This highlights the importance of optimizing development processes with AI. More than a third of respondents, 33.3%, believe it is necessary to focus on teaching the ethical use of AI in programming. This indicates the need to educate a conscious attitude to technologies and their consequences.

The largest share of respondents, 57.1%, believe that tasks should include developing projects where AI is used to increase the efficiency of development (Fig. 3.14). This emphasizes the importance of the practical application of AI in real projects. A high percentage of respondents, 57.1%, recognize the importance of tasks for improving and correcting AI-generated code, which indicates the need to teach students to evaluate and enhance software code. A significant proportion of respondents, 57.1%, believe it is essential to include tasks to analyze and criticize the work of AI from the point of view of logic, performance, and suitability, which emphasizes the importance of developing critical thinking. More than a third of respondents, 38.1%, believe teaching students to create their own AI algorithms and learning models is essential, indicating the need to develop algorithm and model development skills.

The largest share of respondents, 46.7%, believe that the critical analysis of code generated by AI is the most effective task for stimulating essential thinking (Fig. 3.15). This highlights the importance of teaching students to analyze and evaluate software code. A high percentage of respondents, 29.5%, recognize the importance of tasks where AI is a tool to support or challenge, indicating the need to teach students to use AI to solve complex problems. A significant share of respondents, 12.4%, believe that debates and discussions of the ethical aspects of using AI are practical tasks for developing critical thinking. This emphasizes the importance of discussing ethical issues in the educational process. More than a tenth of the respondents believe that presenting AI usage scenarios with open questions for analysis is a practical task. This shows the importance of tangible examples for the development of critical thinking.

The largest share of respondents, 75.2%, believe that it is necessary to increase attention to understanding the logic of algorithms generated by AI (Fig. 3.16). This highlights the importance of teaching students to analyze the internal processes and

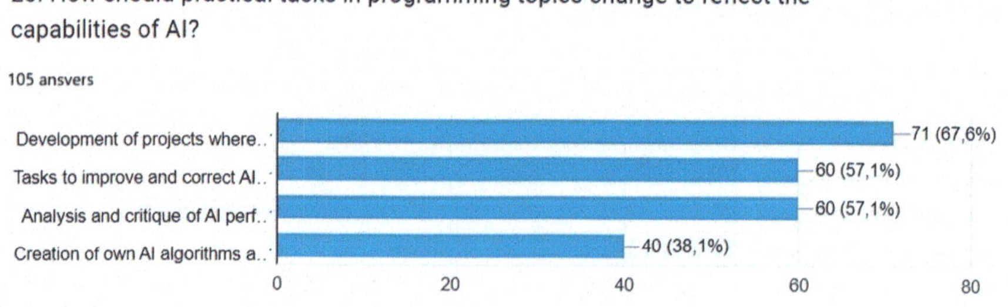

Fig. 3.14 Teachers' responses to changes in practical tasks in programming topics according to the capabilities of AI

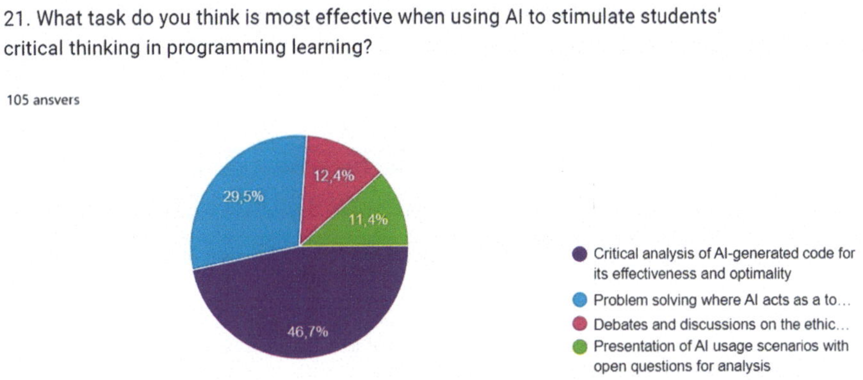

Fig. 3.15 The effectiveness of using AI to stimulate students' critical thinking in teaching programming

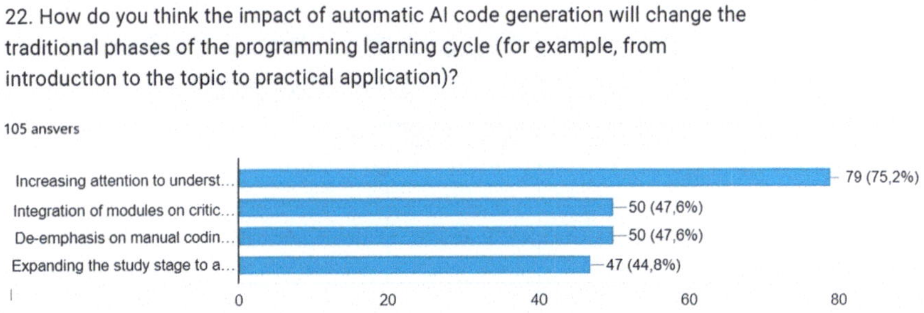

Fig. 3.16 Teachers' answers to questions about the impact of AI on the traditional phases of learning programming

logic of automatically generated code. A high percentage of respondents, 47.6%, recognize the need to de-emphasize manual coding in favor of analyzing and optimizing AI-generated code. This indicates a change in focus in the educational process. A significant share of respondents, 47.6%, believe it is necessary to integrate modules on critical analysis and ethical aspects of the application of AI, which emphasizes the importance of discussing ethical issues and developing critical thinking. More than 40% of respondents believe it is necessary to expand the study phase to an in-depth study of the capabilities of AI. This indicates the need for a detailed analysis of the capabilities and limitations of AI.

The largest share of respondents, 64.8%, believe it is necessary to develop new testing methods that measure the depth of analytical abilities and critical thinking (Fig. 3.17). This highlights the importance of assessing these skills in the learning process. A high percentage of respondents, 61.9%, recognize the need to introduce criteria for evaluating students' understanding of AI processes and their impact on software development. This shows the importance of students' deep knowledge of the internal processes of AI. A significant share of respondents, 58.1%, believe that it

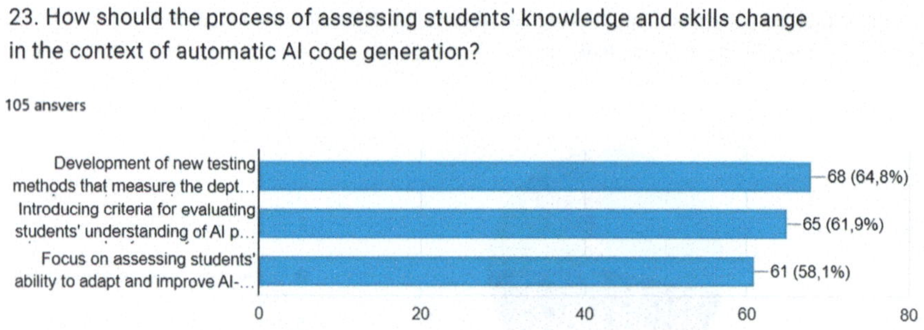

Fig. 3.17 Teachers' response to changing the process of assessing students' knowledge and skills in the context of automatic AI code generation

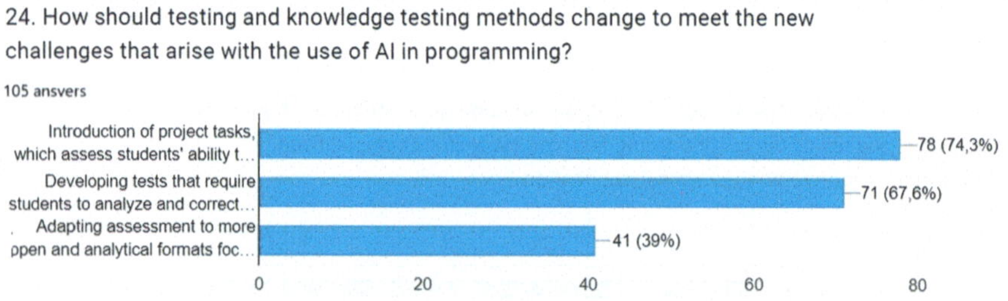

Fig. 3.18 Teachers' responses to changes in testing methods and checking knowledge in AI conditions

is necessary to focus on assessing the ability of students to adapt and improve AI-generated code. This highlights the importance of developing students' skills in adapting and enhancing software code.

The largest share of respondents, 74.3%, believe it is necessary to introduce project tasks that assess students' ability to use AI as a tool (Fig. 3.18). This highlights the importance of assessing students' practical skills in applying AI. A high percentage of respondents, 67.6%, recognize the need to develop tests requiring students to analyze and correct AI-generated code. This shows the importance of evaluating the skills of studying and improving the software code. A significant share of respondents, 39%, believe adapting evaluation to more open and analytical formats focused on the process is necessary. This highlights the importance of assessing students' analytical and procedural skills.

The largest share of respondents, 62.9%, believe that AI can be used to develop personalized learning trajectories using recommendations (Fig. 3.19). This emphasizes the importance of an individual approach to learning that considers each student's unique needs and level of knowledge. A high percentage of respondents, 61%, recognize the need to use AI to check tasks and provide feedback automatically. This shows the importance of automating the evaluation process to increase training effectiveness. A significant share of respondents, 57.1%, believe that AI can

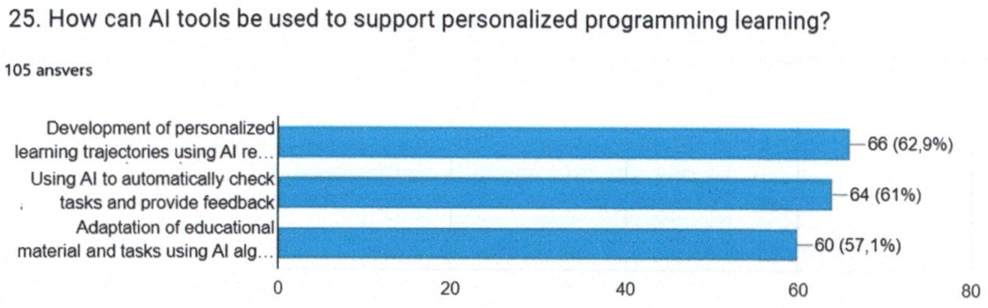

Fig. 3.19 Teachers' responses to using AI to support personalized programming learning

be used to adapt educational material and tasks using algorithms. This emphasizes the importance of adapting the educational process to students' needs.

3.5 Discussion

Integrating artificial intelligence into programming education faces challenges, including insufficient teacher training, lack of access to quality hardware and software, technical challenges, and skepticism of students and teachers. Overcoming these barriers requires a comprehensive approach that includes teacher training, investment in technical resources, provision of technical support, and educational campaigns. This will make it possible to successfully integrate AI into the educational process and improve the quality of programming education.

70.5% of respondents indicated insufficient teacher training as the main obstacle (Fig. 3.20). This suggests the need to conduct training and professional development courses for teachers to use AI in education effectively. Almost 49.5% of respondents reported poor access to quality equipment and software. This highlights the importance of investment in technical resources to ensure an effective learning process. Technical problems and limitations of 48.6% of respondents are a significant obstacle to integrating AI. This indicates the need for technical support and infrastructure upgrades to implement AI successfully. Skepticism and resistance from students and teachers underline 31.4% of respondents the importance of educational campaigns and demonstrations of the benefits of using AI in education to change perceptions and increase technology adoption. 14.3% of respondents consider the use of AI inappropriate. This indicates that some teachers need to see the value or advantages of implementing AI in the educational process.

Teachers consider it necessary to change the role of the teacher in the process of teaching programming under the conditions of AI integration: 41.9% of respondents believe that the role of a computer science teacher is to be a consultant and mentor for individual learning; 33.3% of respondents noted that a teacher should be a curator of resources and projects that use AI; 15.2% of respondents believe that a teacher should be an instructor in the basics of AI and machine learning; 9.5% of

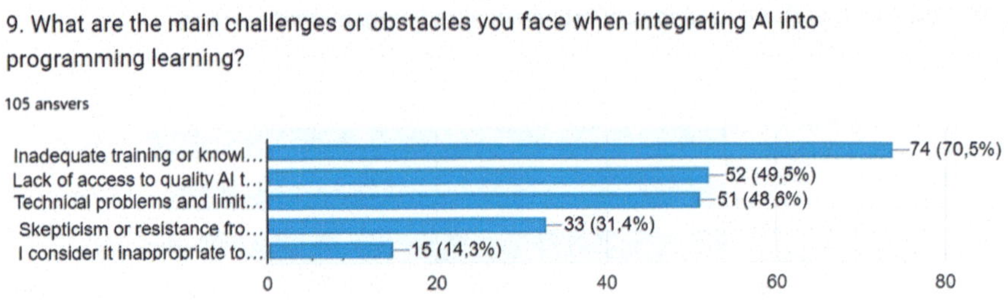

Fig. 3.20 Teachers' answers about the challenges and obstacles they face when integrating AI into the process of teaching programming

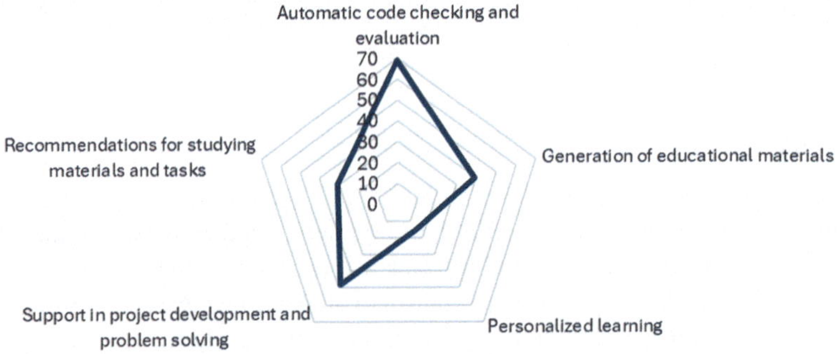

Fig. 3.21 Distribution of benefits regarding the use of AI in teaching programming

respondents noted that a teacher should moderate discussions about the impact of AI on society.

The analysis of respondents' answers was carried out using descriptive statistics. This is due to the first stage of the research and Ukrainian teachers need to improve their practice of using AI when teaching programming. At the same time, DataMining methods were applied to analyze the obstacles identified by teachers during the use of AI in teaching programming and possible tools in the methodological system.

Five clusters were identified, and their distribution based on the number of respondents' answers was constructed (Fig. 3.21).

The distribution of the dependence of advantages and obstacles in using AI by selected clusters is shown in Fig. 3.22.

The role of the computer science teacher in the context of the widespread use of AI should include consultative and mentoring support for individual learning, curating resources and projects, teaching the basics of AI and machine learning, and moderating discussions about the impact of AI on society. Implementing these recommendations will contribute to the effective educational process and the development of the necessary competencies in students.

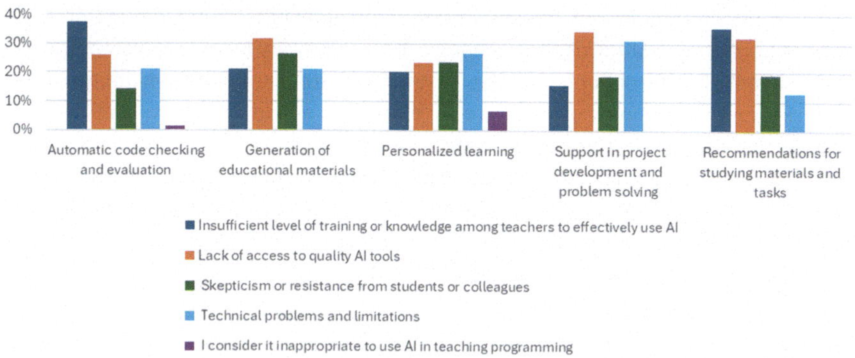

Fig. 3.22 Distribution of obstacles to the application of AI in teaching programming by clusters

3.6 Conclusions

The study showed that the use of AI in teaching programming has significant potential for increasing the effectiveness of the educational process and developing the necessary competencies in students. The survey results made it possible to determine the main areas of AI implementation and the changes that must be made in testing and checking knowledge.

Methods of teaching programming need to be changed and adapted to the new capabilities of AI. Key emphases include developing critical thinking and analytical skills, including modules on the basics of AI and machine learning, learning to use AI for assistance and optimization, and focusing on the ethical use of AI.

The automatic generation of AI code will affect the traditional phases of the programming learning cycle. Key emphases include increasing focus on understanding the logic behind AI algorithms, de-emphasizing manual coding in favor of code analysis and optimization, integrating modules on critical analysis and ethical aspects of AI applications, and expanding the learning phase to in-depth exploration of AI capabilities.

Adaptation of learning for students who face particular difficulties in programming requires using various educational approaches, organization of group work, individual consultations, and provision of additional materials for independent study. Implementing these strategies will help overcome learning difficulties and ensure students successfully learn the material.

To reflect the capabilities of AI in practical programming tasks, teachers should develop tasks that include designing projects using AI, improving and correcting AI-generated code, analyzing and critiquing the operation of AI, and creating their algorithms and models.

To stimulate students' critical thinking when using AI in programming learning, educators should design tasks that include critical analysis of AI-generated code, use of AI as a problem-solving tool, debate and discussion of ethical aspects of AI use, and analysis of AI use scenarios.

The process of assessing students' knowledge and skills in the context of automatic AI code generation needs to change. Key emphases include developing new testing methods, introducing criteria to evaluate students' understanding of AI processes and their impact on software development, and assessing students' ability to adapt and improve AI-generated code.

Methods of testing and testing knowledge must change to meet the new challenges that arise with the use of AI in programming. Key emphases include introducing project tasks, developing tests for code analysis, debugging, and adapting assessments to more open and analytical formats.

References

Basnet, R. B., Doleck, T., Lemay, D. J., & Bazelais, P. (2018). Exploring computer science students' continuance intentions to use Kattis. *Education and Information Technologies, 23*, 1145–1158.

Cheers, H., Lin, Y., & Smith, S. P. (2021). Academic source code plagiarism detection by measuring program behavioral similarity. *IEEE Access, 9*, 50391–50412.

Crow, T., Luxton-Reilly, A., & Wuensche, B. (2018). Intelligent tutoring systems for programming education: A systematic review. In *Proceedings of the 20th Australasian computing education conference* (pp. 53–62).

Dunder, N., Lundborg, S., Wong, J., Viberg, O. (2024). Kattis vs ChatGPT: Assessment and evaluation of programming tasks in the age of Artificial Intelligence. In *LAK '24: Proceedings of the 14th learning analytics and knowledge conference* (pp. 821–827). https://doi.org/10.1145/3636555.36368

Gao, Z., & Dengpeng, H. (2023). Research on parents' awareness of children's programming education in the age of Artificial Intelligence—Taking Ningxia Guyuan City as an example. *International Journal of New Developments in Education, 5*(20), 27–33. ISSN 2663-8169. https://doi.org/10.25236/IJNDE.2023.052006

Geng, C., Yihan, Z., Pientka, B., & Si, X. (2023). *Can ChatGPT pass an introductory level functional language programming course*. https://www.researchgate.net/publication/370495417_Can_ChatGPT_Pass_An_Introductory_Level_Functional_Language_Programming_Course#fullTextFileContent

James, J. (2021). Confronting the scarcity of digital skills among the poor in developing countries. *Development and Policy Review, 39*, 324–339. https://doi.org/10.1111/dpr.12479

Kamak, & Preethi, L. (2023). *Programming pedagogy in the age of accessible Artificial Intelligence*. A thesis submitted in partial fulfillment of the requirements for the degree of Master of Science in Computer Science in the Faculty of Science and Environmental Studies of Lakehead University. https://knowledgecommons.lakeheadu.ca/handle/2453/5159

Kazemitabaar, M., Chow, J., Ma, C. K. T., Ericson, B. J., Weintrop, D., Grossman, T. (2023) Studying the effect of AI code generators on supporting novice learners in introductory programming. In *Proceedings of the 2023 CHI conference on human factors in computing systems article no. 455* (pp. 1–23). https://doi.org/10.1145/3544548.3580919

Krstić, L., Aleksić, V., & Marija. (2022). Artificial Intelligence in education: A Review. In *9th International scientific conference technics and informatics in education—TIE 2022*. стр. 223–228 https://doi.ub.kg.ac.rs/doi/zbornici/10-46793-tie22-223k/

Methodological recommendations regarding introducing and using artificial intelligence technologies in general secondary education institutions of Ukraine. MES of Ukraine, Ministry of Digital Transformation of Ukraine. (2024). https://mon.gov.ua/static-objects/mon/sites/1/news/2024/05/21/Instruktyvno.metodychni.rekomendatsiyi.shchodo.SHI.v.ZZSO-22.05.2024.pdf

Philbin, C. A. (2023). Exploring the potential of Artificial Intelligence program generators in computer programming education for students. *ACM Inroads, 14*(3), 30–38. https://doi.org/10.1145/3610406

Saputra, I., Astuti, M., Sayuti, M., & Kusumastuti, D. (2023). Integration of Artificial Intelligence in education: Opportunities, challenges, threats, and obstacles. A literature review. *Indonesian Journal of Computer Science (IJCS), 12*(4). https://doi.org/10.33022/ijcs.v12i4.3266

Sarkar, A., Gordon, A. D., Negreanu, C., Poelitz, C., Srinivasa Ragavan, S., & Zorn, B. (2022). What is it like to program with artificial intelligence? In *Proceedings of the 33rd annual conference of the psychology of programming interest group (PPIG 2022).* arXiv:2208.06213v2, https://doi.org/10.48550/arXiv.2208.06213

Srinivasan, V. (2022). AI & learning: A preferred future. *Computers and Education: Artificial Intelligence, 3,* 10006. https://doi.org/10.1016/j.caeai.2022.100062

Surameery, N. M. S., & Shakor, M. Y. (2023). Use Chat GPT to solve programming bugs. *International Journal of Information Technology & Computer Engineering (IJITC), 3*(01), 17–22. ISSN: 2455-5290. https://doi.org/10.55529/ijitc.31.17.22

Yilmaz, R., & Karaoglan Yilmaz, F. G. (2023). The effect of generative artificial intelligence (AI)-based tool use on students' computational thinking skills, programming self-efficacy and motivation. *Computers and Education: Artificial Intelligence, 100147*, 2.

Chapter 4
Comparison of Methods for Identifying Rapid Guessing Behavior in Microlearning Courses

Ján Skalka and Matúš Valko

Abstract Rapid guessing, a behavior in which students select answers quickly with minimal cognitive engagement, can significantly bias performance metrics and weaken the effectiveness of educational interventions. By analyzing the results of several detection techniques adapted for research purposes and applied to micro-course data, this paper compares different methods for identifying rapid guessing behavior in a low-stakes microlearning environment. The study evaluates the performance of individual methods at the test level, highlighting differences in their effectiveness and accuracy in detecting rapid guessing. It also considers the potential for implementing these methods in a typical microlearning environment. The findings have significant implications for educators and instructional designers, helping refine microlearning assessments and improving the overall quality of educational experiences.

Keywords Microlearning · Rapid guessing behavior · RGB · Response time threshold · Programming learning

4.1 Introduction

Information technology is vital in disseminating shared knowledge and modernizing education (Haleem et al., 2022). Digital tools enrich traditional teaching by providing immediate feedback and space for individual engagement development. In line with the changes in everyday working life, students demand a more flexible, exciting, and interactive education that aligns with their career goals. This shift in understanding the learning process requires students to take more responsibility for developing their skills and emphasize self-discipline. On the other hand, higher education institutions must adapt by creating more personalized learning paths that

J. Skalka (✉) · M. Valko
Faculty of Natural Sciences and Informatics, Constantine the Philosopher University in Nitra, Nitra, Slovakia
e-mail: jskalka@ukf.sk; matus.valko@ukf.sk

E. Smyrnova-Trybulska et al. (eds.), *E-Learning and Enhancing Soft Skills*,
https://doi.org/10.1007/978-3-031-82243-8_4

meet individual needs within standardized programs (Ehlers, 2020; Sharples et al., 2016).

One of the options accepted in practically all areas of education is microlearning. Microlearning is widely accepted across various educational fields as a method for gradually building knowledge on almost any topic (Gorham et al., 2023; Moreno Moreno & Mateus-Nieves, 2021; Krygier et al., 2022). While it does not cover all educational content, it enables repetition, reinforcing complex concepts, verifying understanding, and maintaining motivation through gamification. This approach offers an alternative way to design and organize learning, focusing on small steps and concise content units with immediate understanding checks. Modern definitions describe microlearning as an action-oriented approach that delivers short, focused learning sessions to encourage students to learn, act, and practice (Hug, 2005; Sankaranarayanan et al., 2023).

When using microlessons, verifying their difficulty, clarity, and student acceptance is essential by reporting errors or ambiguities through interactive course elements and questionnaires (Betancur-Chicué & García-Valcárcel Muñoz-Repiso, 2023; Simaremare et al., 2022). Analyzing user behavior and learning outcomes can provide valuable insights, but obtaining them requires much additional work. An essential step in evaluating the overall content is a detailed assessment of individual questions and an understanding of the reasons for students' success or failure.

While evaluating questions in high-stakes tests is an ordinary pedagogical skill, only a few studies have explicitly focused on microlearning content (Kumar et al., 2021; Hartati & Yogi, 2019). The cause is that students respond differently to questions in a microlearning context than traditional tests. The next reasons arise from the independent nature of microlearning and can be summarized as follows (Skalka & Valko, 2024):

- Time flexibility
- Immediate feedback
- Multiple attempts
- Different concentration level
- Reduced stress
- Schedule requirement
- Different motivational mechanisms
- Permanent access to content

Because of these differences, students tend not to pay equal attention to each question, and guesswork plays a significant role in their answers. Rapid guessing behavior (RGB) has been extensively studied in testing contexts (Guo et al., 2016; Wise, 2017; Rios & Deng, 2021), with the conclusion that RGB can bias test results and undermine the validity of assessments.

Examining this behavior in microlearning is crucial because content preparation and question validation are not given the same attention as in high-stakes tests. RGB can distort the perception of the difficulty, appropriateness, and correctness of the questions and affect the student's overall score, which should reflect the student's level of understanding and ability to apply the learned content. Therefore, for an accurate assessment of the quality of the content and the educational process, it is necessary to exclude results influenced by RGB.

RGB is well-researched in the field of high-stakes testing; studies have primarily focused on two areas: identifying RGB in high-stakes tests (Wise, 2019; Deribo et al., 2023) and using a statistics approach to modeling problematic items (Rios, 2022a, b). There are fewer studies on microlearning and low-stakes testing (DeMars, 2007; Wise & DeMars, 2005), while the influence of RGB plays a significant role in deploying an effective form of microlearning in higher education.

This article sets out the following research questions:

- RQ1: How do different RGB detection methods identify items differently?
- RQ2: What is the complexity of implementing each RGB detection method into a microlearning system?

This paper evaluates and compares RGB detection methods applicable in a microlearning environment. Specifically, it seeks to understand the differences in item identification between these methods and find a trade-off between the complexity of integration and the effectiveness of each method in a microlearning system.

4.2 Rapid Guessing Behavior

Understanding the motivations and mechanisms of rapid guessing behavior (RGB) is key to accurately interpreting test results and improving the evaluation process. According to (Wise, 2017), two primary behaviors can be observed when solving tasks:

- Rapid guessing behavior—occurs when students answer quickly without reading, understanding, or considering the entire question.
- Solving behavior is when students apply their knowledge, skills, and abilities to determine the correct answer in the form expected by the test makers.

Question-solving behavior can be described as a student's choice between these two behaviors. If he chooses solution behavior, he actively uses his cognitive resources to solve the problem and identify the correct answer. Conversely, if he chooses to guess quickly, he chooses an answer emphasizing speed without meaningfully trying to solve the task (Wise & Kuhfeld, 2020).

Several subjective reasons contribute to RGB in high-stakes tests:

- Low motivation is reported as the most common (psychological) cause of RGB. Participants choose RGB when they lack motivation, especially in low-stakes scenarios where the outcomes have little weight in the overall evaluation (Wise & Kong, 2005). Studies (Lindner et al., 2019; Halvoník et al., 2023) show that unmotivated individuals tend to guess more often.

- Lack of knowledge reflects a situation where test takers realize they lack the necessary knowledge or skills to answer a question. In such a case, students may opt for quick guessing, from which they promise a comparable or better result than what they would get if they invested time in solving the task (Wise, 2020; Rios, 2022c). This situation is a subset of all incorrect answers, and it is questionable how much a non-RGB approach would improve an individual's score on a given question.
- Mental fatigue from long tests can lead to RGB when students become exhausted, leading to reduced performance, especially on later questions. Fatigue can also cause longer reaction times, indicating reduced comprehension and engagement and ultimately affecting RGB identification accuracy by increasing reaction times and complicating guessing identification (Nagy et al., 2023).
- Time pressure can lead to rapid guessing (RGB) in two main ways (Wise, 2017; Schnipke & Scrams, 1997). The first reason is a lower level of knowledge when testing under time constraints and lacking sufficient knowledge; they may resort to RGB as a coping mechanism. The second situation involves applying purposeful design when time pressure is deliberately applied to create a stressful environment with unplanned triggering of RGB in participants.
- Poorly designed or flawed questions, such as those with typographical errors or ambiguities, can encourage test takers to guess rather than struggle with problematic items. This situation can increase the RGB and reduce the reliability of the testing (Browne & Keeley, 2002; Grabe & Flannery, 2010).

In microlearning, RGB can also be influenced by other new factors:

- The ability to repeat answers to questions can reduce students' motivation to answer correctly on the first try. According to microlearning principles, students know that they can answer the same questions multiple times without significant penalty. They can thus apply initial or random guesses for quick solutions or orientation in more complex questions and, in the case of incorrect choices, further refine their answers (also randomly) (Korosi et al., 2018; Yang et al., 2017).
- Low level of engagement—microlearning activities are often given minimal weight in the overall assessment. If students find the content uninteresting or irrelevant to their grade or skill building, they may guess answers rather than engage deeply (Nikou & Economides, 2018; Khong & Kabilan, 2022).
- Short learning intervals—the fragmented nature of microlearning, such as learning during breaks or when a student is multitasking, can reduce concentration and cognitive engagement, making RGB applications more likely.
- Inadequate time management can cause students, after neglecting regular learning, to rush to complete assignments in the shortest possible time. This approach leads to a superficial process that aims only to finish the material quickly rather than to understand it thoroughly.

Addressing these factors is critical to improving the effectiveness of microlearning. By understanding the causes of RGB, educators can develop strategies to increase genuine engagement and reduce hasty, uninformed responses.

4.2.1 Rapid Guessing Behavior Identification

Although, at first glance, it may seem that identifying RGB is difficult or impossible without the cooperation of respondents, applying appropriate statistical methods makes it possible to arrive at reliable results confirmed by research from different points of view.

The early attempts to evaluate the reliability and relevance of the tests used questionnaires to capture the subjective experiences of respondents immediately after testing. A notable finding was that motivation significantly affects test scores. Low motivation can reduce test performance and the validity of results, especially in low-stakes tests where poor performance has minimal consequences (Wise & DeMars, 2005; Wise & Gao, 2017).

The growing interest in testing motivation has led to new measurement methods, primarily through questionnaires (Piaw Chua, 2012; Mastuti & Handoyo, 2018). Although questionnaires provide rapid feedback and high internal consistency, they face significant limitations. Participants often tried to remember specific moments of change in motivation, leading to generalized information rather than detailed knowledge. Additionally, the subjective nature of self-reported motivation may lead to inaccuracies, as test takers may exaggerate or downplay their efforts (Wise & Smith, 2016).

Computerized testing emerged as a solution to overcome subjective problems. Computer-based testing can accurately track user behavior and testing effort at the individual item level, allowing for more accurate analysis of response patterns (Wise & Gao, 2017; Heber et al., 2017; Paiva et al., 2022).

Research has shown that motivation, response time, and test performance are interrelated, supporting the use of computer-based testing to analyze RGB through user behavior (Wise & Kong, 2005; Silm et al., 2020; Lundgren & Eklöf, 2023). Wise (2017) highlighted differences in response times between rapid estimation and solution-oriented responses, while Schnipke and Scrams (1997) identified differences in response accuracy. Based on these views, detection methods can be divided into (Rios & Deng, 2021):

- Methods using no empirical data are based on examining theoretical models without real test data. Researchers generate their datasets and then examine the data relationships and behavior of the models (Rios & Soland, 2021; Deribo et al., 2021).
- Methods using response time (RT) are focused exclusively on tracking reaction times, which serve as a single RGB identifier (Guo et al., 2016; Wise & Kuhfeld, 2020).

- Methods combining reaction time and response accuracy (RTRA) simultaneously use reaction times and response accuracies to distinguish between fast, slow, correct, and incorrect responses (Wang & Xu, 2015; Guo et al., 2016).
- Other methods add additional variables to the above (gender, age, etc.) and try for a complex analysis for a given dataset or model (Rios, 2022c; Nagy et al., 2023).

For the real needs of microlearning at this stage, methods based on RT and RTRA will suffice.

One of the first methods in RGB research involved the visual identification of a bimodal distribution of response times, where one peak represents solution behavior and the other RGB. Figure 4.1 shows a typical situation in real data: the red curve shows the distribution of incorrect responses over time, and the green curve shows the correct ones. An important step is identifying the response time threshold (RTT), representing the boundary between RGB and solution behavior. All responses with times below the RTT are considered as generated during RGB. Schnipke and Scrams (1997) proposed that the RTT in a bimodal distribution can be identified at the lowest intersection between the two distributions.

In addition to the need for subjective assessment, which is time-consuming and error-prone, visual identification of RTT has another significant disadvantage. Real data sets often do not exhibit ideal bimodal distributions or may lack them altogether, leading to significant variability in results between different raters.

The normative threshold (NT) method was developed as an alternative to the bimodal distribution approach (Wise & Ma, 2012). This method sets the RTT to a specific percentile of the item's response time distribution. After examining several percentiles, the research settled on NT10 (Wise & Kuhfeld, 2021; Svetina Valdivia

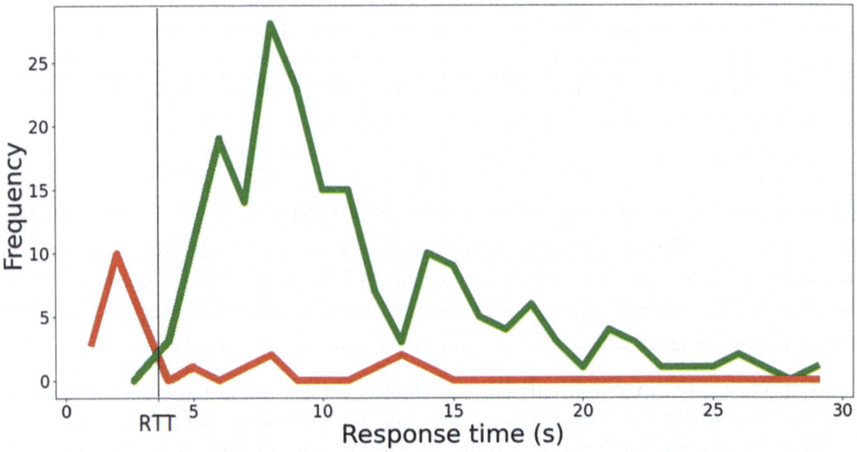

Fig. 4.1 Bimodal distribution of responses and RTT identification by Schnipke and Scrams (1997). The green curve represents the increasing trend of correct answers after the specified time, while the red curve shows the decreasing trend of incorrect answers. Bimodality is represented on the graph by two distinct peaks in the data distribution. These peaks indicate that the data consists of two different population groups or behaviors

et al., 2023), which sets a threshold of 10% of the average response time for the test item with a maximum limit of 10 s.

RTRA methods relate response time to response accuracy within a specific time interval. The identification of RTT consists of finding the interval where the number of correct answers exceeds the probability of random guessing (chance) (Ma et al., 2011). A more advanced approach within RTRA represents the Cumulative Proportional Method (CUMP), which identifies RGB by determining the interval when the cumulative proportion of correct responses exceeds the chance rate. This method allows for prioritizing a statistically defined threshold over visual inspection (Guo et al., 2016).

4.3 Research Design

The compared methods focus on RTT identification using three approaches: response time analysis, visual identification of the bimodal distribution, and a combination of response time and accuracy. Based on the findings from (Skalka & Valko, 2024; Wise, 2019), specific adjustments were made for this experiment:

- NT10 is the relatively most straightforward method that defines the RTT at 10% of the average response time for each question, with an upper limit of 10 s. Several studies have proved this value to be the most suitable for the needs of high-stakes tests (Wise & Ma, 2012).
- RTRA-VI is a visual method that combines the identification of a bimodal distribution, representing different user behaviors, and an element of the probability of a correct answer. RTT is determined by visual inspection according to the VITP-M method (Wise, 2019). In the simplest case, when a bimodal distribution appears, the RTT is set at the lowest intersection between the distributions. If such a pattern is not found, the analysis starts the examination of individual time intervals. The RTT is set to the value of the previous interval if the number of correct responses in the studied interval exceeds the expected probability level. If no such interval is found and the analysis exceeds the average RGB response time, no RTT is assigned for that item.
- RTRA-UP uses isolated interval analysis regardless of any bimodal distribution patterns. The RTT is identified by comparing the number of correct responses in each interval to the expected chance level. If the number of correct answers in the current interval exceeds what would be expected by random guessing, the RTT is set to the value of the previous interval. This method aims to identify the intervals at which correct answers exceed the probability of random guessing, which helps determine the point at which respondents move from quick guessing to solving.
- RTRA-DOWN follows the same principle as RTRA-UP but works in reverse. It starts by analyzing intervals starting at NT20 and moves towards 0. The RTT is identified by comparing the correct answers to the expected chance level in each interval. If the correct answers in the current interval exceed what is expected by

random guessing, the RTT is set to the value of the previous, lower interval. If no interval meets this condition, the default RTT is set at the time defined by the NT20 method.
- CUMP-UP identifies RTT by evaluating the cumulative number of correct answers from time zero to the current interval. RTT is set to the previous interval when this cumulative total exceeds the expected chance level. The method starts at interval 1 and continues until the average response time is reached.
- CUMP-DOWN operates similarly but in reverse. It begins with the interval corresponding to 20% of the average non-zero response times (NT20). RTT is identified when the number of correct responses falls below the chance level.

4.3.1 Data Origin

The prerequisite for RGB analysis is an educational platform capable of capturing detailed data on student behavior. For this purpose, the Priscilla web platform was developed, offering different content formats and question types (Skalka & Drlik, 2018). It is primarily aimed at teaching programming languages and is used at universities within the FITPED consortium (https://www.fitped.eu). The selected experimental course was *Python Introduction*, intended for first-year students. The course contains 19 chapters, 210 micro-lessons, 299 tasks, and 134 programming challenges with a five-week schedule. Although this course provides different types of questions, only one-choice questions were selected for the experiment.

The system monitors user behavior and performance by storing data in a database, enabling detailed student activity tracking. Users' movements within the system are monitored by recording the time they enter and exit each micro-unit, allowing the system to determine how much time a student spends studying or viewing content. When students submit answers, the system evaluates the correctness of the answer and sends the results back to the user. At the same time, it will record the request, the response time, accuracy, and the duration of the display of the given microlesson. These data points—response time and accuracy—are the basic parameters to test the RGB analysis methods.

It makes sense to focus on RGB only when users encounter the task for the first time. As a result, only the first user responses to each question will be considered for analysis.

4.3.2 Analyze

From the collected data, a matrix was constructed with 580 users as rows and columns representing the time spent on each microlesson. The original 643 columns, including all activities, were reduced to 66 for select-choice questions. Users who did not provide correct or incorrect answers were excluded, reducing the number of users to 494. Correct and incorrect answers were distinguished in the matrix by a sign—in the case of a correct answer, the time indicated in the item representing the i-th student's answer to the j-th question was positive; in the case of an incorrect answer, it was negative.

The analysis involves calculating RTT for each method individually. Due to the variability in questions and student behavior, aggregating or summarizing total parameters is not meaningful. Therefore, each question must be examined separately.

Figure 4.2 shows the results of identifying response time thresholds. RTT values that exceed the expected range may be mainly due to the algorithm used, which may require adjustments or defining upper limits. The results of the NT10 method are shown as an area bar graph that visually represents a consistent threshold set at 10% of the average question response time. Integrating this method into research simplifies the visualization of question difficulty and allows for comparing RTT distributions between methods. From 494 students and 66 questions, a total of 32,604 items, 18,449 answers were answered by students and subsequently analyzed.

Table 4.1 provides a numerical summary of the items identified as RGB using each method.

Each RGB identification method has different characteristics.

RTRA-VI is subjective and may reflect the practitioner's attempt to identify a bimodal distribution or RTT at any cost, even if this value is already high. UP

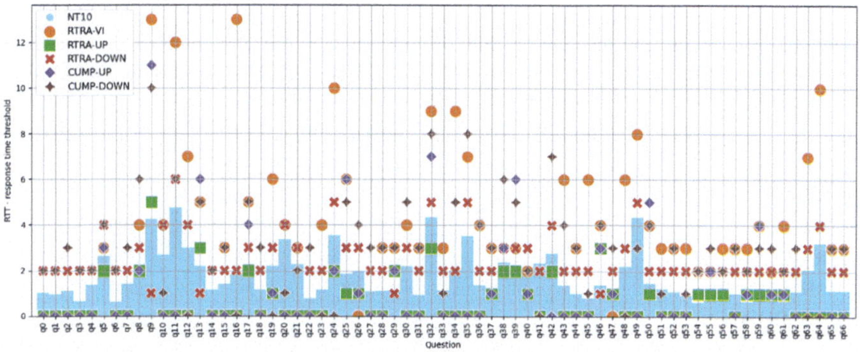

Fig. 4.2 Results of RTT identification for selected methods. Markers indicate the identified RTT value for each question based on the individual methods used. The NT10 method, which serves as the reference, is visualized using a bar graph

Table 4.1 Counts of items identified as RGB by individual methods. The table shows the number of responses identified as RGB and their share of the total responses

		RTRA			CUMP	
	NT10	RTRA-VI	RTRA-UP	RTRA-DOWN	CUMP-UP	CUMP-DOWN
RGB responses	941	1866	192	1059	603	1464
RGB ratio	5.10%	10.11%	1.04%	5.74%	3.27%	7.94%

methods start their analysis from lower intervals and often terminate the process before reaching the RTT because the number of correct responses exceeds a probability threshold. Especially with RTRA-UP, this can cause some RGB cases to go unrecognized.

In contrast, DOWN methods usually offer more conservatism. These methods start from a fixed threshold and move toward earlier intervals, evaluating response success rates against chance levels within a given interval or cumulatively over the interval under study. Again, in the case of RTRA-DOWN, an unexpected RTT can be identified if it encounters an interval in which there are few correct and many incorrect responses.

After modifications, CUMP and RTRA methods could also provide more accurate RGB identification by considering response times in several adjacent intervals.

NT10 serves as a benchmark that approximates the average response time by setting the RTT to 10% of that average. This method offers a basis for comparing the results of other methods, reflects the typical response time, and helps assess the difficulty of the question.

The analysis identified items recognized as RGB by different methods, with overlaps detailed in Table 4.2 and method-specific differences in Table 4.3. McNemar's test was used to assess significant differences between methods.

For the McNemar test to be valid, the following conditions must be met:

- Random selection of samples—student answers used in the matrices were collected automatically without a specific goal, so there is no systematic distortion.

Table 4.2 A count of items identified as RGB by both methods. The table compares the agreement between the different methods of detecting RGB and displays the number of items identified as RGB by both methods listed in the corresponding column and row

	NT10	RTRA-VI	RTRA-UP	RTRA-DOWN	CUMP-UP	CUMP-DOWN
NT10	x	742	184	816	301	619
RTRA-VI		x	192	782	464	967
RTRA-UP			x	106	192	192
RTRA-DOWN				x	287	734
CUMP-UP					x	505
CUMP-DOWN						x

Table 4.3 Count of unique/different items identified as RGB by combining both methods. The table shows the number of unique items labeled as RGB by either method

	NT10	RTRA-VI	RTRA-UP	RTRA-DOWN	CUMP-UP	CUMP-DOWN
NT10	x	1323	765	368	942	1167
RTRA-VI		x	1674	1361	1541	1396
RTRA-UP			x	1039	411	1272
RTRA-DOWN				x	1088	1055
CUMP-UP					x	1057
CUMP-DOWN						x

Table 4.4 The number of unique items identified by the method in the column compared to those identified by the method listed in the row—identification of item differences between pairs of methods

	NT10	RTRA-VI	RTRA-UP	RTRA-DOWN	CUMP-UP	CUMP-DOWN
NT10	x	199	757	125	640	322
RTRA-VI	1124	x	1674	1084	1402	899
RTRA-UP	8	0	x	86	0	0
RTRA-DOWN	243	277	953	x	772	325
CUMP-UP	302	139	411	316	x	98
CUMP-DOWN	845	497	1272	730	959	x

Table 4.5 Chi-square values for McNemar tests to identify differences in results between pairs of methods

	NT10	RTRA-VI	RTRA-UP	RTRA-DOWN	CUMP-UP	CUMP-DOWN
NT10	x	646.7	733.3	37.84	121.28	234.4
RTRA-VI		x	1674	478.5	1035.20	115.76
RTRA-UP			x	723.5	411	1272
RTRA-DOWN				x	191.1	155.47
CUMP-UP					x	701.3
CUMP-DOWN						x

- Paired samples—matrices are paired, representing the same group of students and questions, but analyzed from the perspectives of two different methods.
- Mutually exclusive groups—each matrix entry is either 0 or 1, indicating the RGB identification, thus ensuring mutually exclusive groups.

Another necessary input is unique items identified by which method. These data are available in Table 4.4.

Although the tables indicate significant differences between the RGB items identified by the different methods, the McNemar test further confirms this.

The p-values obtained by comparing all methods are extremely close to 0, indicating that the differences are statistically significant. In all cases, the null hypothesis (which assumes no significant difference between methods) is rejected at the 0.05 significance level, reflecting a very low probability of a Type I error. Table 4.5 illustrates this with chi-square values supporting these findings.

4.4 Discussion and Conclusion

Adapted methods from recent RGB research were used to identify RGB in microlearning content.

The NT0 method, which focuses only on response time, identified 5.1% of items as suspicious for RGB among 494 students and 66 tasks. This result is consistent with similar studies, where the NT10 method identified 5.74% (Svetina Valdivia et al., 2023) and 5.44% (Michaelides et al., 2020) of items, respectively, depending on the field of study. The main disadvantage of the NT10 method is that it is limited exclusively to the response time, regardless of the accuracy of the responses. NT10 uses a fixed threshold (10% of average response time), which may not be flexible enough to account for variations in question difficulty or individual student behavior, which may lead to under- or over-identification of RGB.

The RTRA-VI method, which relies on visual identification of bimodal distributions and chance identification, found 10.11% of items to be suspect. Although this method is less reliable due to potential subjectivity, it is still conservative compared to the initial research by Schnipke, which reported an 18–30% suspicion rate (Schnipke & Scrams, 2005). The RTRA-VI method has several disadvantages. Since the method relies on visual inspection of RGB, it introduces a level of subjectivity to the results that can lead to inconsistent results between different raters. The effectiveness of RTRA-VI depends on the presence of a clear bimodal distribution in the data, which may not always be present—if the distribution is unclear or unimodal, the method becomes less reliable. RTRA-VI is time-consuming to implement, especially for large datasets, as each response needs to be visually inspected. Unlike fully automated methods, RTRA-VI requires manual intervention, which increases the likelihood of human error and reduces the efficiency of the process.

Methods that use both response time and accuracy, starting from the lowest intervals, can struggle with sparse data. These methods always stop at the first interval where the number of correct answers exceeds chance, ignoring subsequent and previous intervals that could correct this assessment. Future modifications of the methods should consider averaging the results over several adjacent intervals.

In this experiment, due to the small sample size and some overly demanding and complex questions, the RTRA-UP and CUMP-UP methods identified only 1.04% and 3.27% of RGB items, respectively.

The RTRA-UP method (Table 4.4) failed to identify any unique items in three cases, with differences from the RTRA-DOWN method resulting from examining the intervals from the opposite direction. The use of this method should probably be considered in future experiments.

The RTRA-DOWN method starts with a threshold value of NT20 and refines it downwards. It identified 5.75% of RGB items, slightly more than NT10, suggesting that a wider interval may improve accuracy. Further refinement (integration of adjacent intervals) could increase its effectiveness. The RTRA-DOWN method has again several potential disadvantages. An initial high threshold may prevent the analysis from reaching intervals with lower response times, which could provide a more accurate determination of the RTT. Inconsistent results with sparse data or non-standard distribution intervals with few responses might be inaccurately evaluated as RTT.

CUMP-DOWN is considered the most design-reliable method within the experiment, and it identified 7.94% of RGB items. Her process starts at twice the value of NT10 (NT20) and drops to zero, evaluating whether correct answers exceed chance. This method could also benefit from combining results from neighboring intervals or integrating with CUMP-UP.

While some methods may have limitations regarding over- or under-identification of RGB cases, all methodologies are standard and widely used in the field. McNemar's test revealed significant differences between the identified items in all pairs of methods in response to the first research question.

The conclusion is that each method can detect a unique set of potentially suspicious RGB items. A detailed analysis of specific items, questions, and their corresponding attributes (difficulty, comprehensibility, response rate, etc.) would be necessary to investigate more deeply the factors that cause these inconsistencies.

4.4.1 Implementation of RGB Identification

When focusing on the implementation of individual methods in the microlearning environment, as outlined in the second research question, it is necessary to consider to what extent these methods align with the specific characteristics of microlearning. When implementing methods to identify RGB in a microlearning environment, it is important to consider also what information they provide to the learner, teacher, and course designer. Each stakeholder has different needs and expectations, and the insights from RGB detection should be relevant and applicable to each.

In microlearning, students often focus on quick, targeted learning goals. RGB detection can provide helpful feedback on their engagement and performance in the form of:

- Awareness of engagement and self-regulation—if a student's response is identified as RGB, the system can provide feedback to encourage more thoughtful engagement in response preparation. For example, immediate prompts could remind the student of the importance of careful consideration before responding. According to (Wise, 2017), in high-stakes tests, quick guesses represent strategic attempts to maximize one's score, whereas, in low-stakes tests, they represent an unmotivated test. If students were automatically informed that their responses were scored as RGB without careful planning, this could lead to inappropriate behavior modification (Lindner et al., 2019). Students may wait a moment after being prompted (or with each question) and respond randomly rather than engaging with the material as planned. To avoid such a situation, dealing with it differently than with direct feedback is better. For example, adapting content or adjusting the sequence of lessons could help guide students more effectively and encourage meaningful engagement.
- Personalization of learning path and evaluation—RGB detection can be used to design additional resources or micro-lessons for students who often guess,

ensuring they return to concepts they may not have fully understood (Debeer et al., 2014). This operation is feasible even without the students knowing that their speed of answering is being monitored and evaluated. Results of the study (Lee & Jia, 2024) suggest that using multistage testing instead of linear testing has a different impact on the testing experience across performance levels than linear testing.

For teachers, the implementation of RGB identification methods provides a deeper insight into student behavior and helps to adapt teaching strategies:

- Identifying struggling students—frequent occurrences of RGB can signal either a lack of understanding or disengagement (Lee & Jia, 2024). In such cases, follow-up analysis should be performed to determine the cause, leading to appropriate interventions—virtual support for individual students or modification of teaching methods if a large group is affected.
- Understanding learning patterns—teachers can identify patterns by analyzing how students interact with different types of questions. For example, if many students are quickly guessing about specific questions, this may indicate that the material is too complex or unclear, signaling the need for clarification or simplification (Hui, 2023).
- Accurate feedback on performance—when student scores are adjusted to exclude items marked as RGB, teachers better understand and perceive students' actual understanding and performance. This filtered score provides a more accurate measure of content mastery and helps teachers better assess student progress (Attali et al., 2016).

For course creators, the knowledge gained from RGB detection methods helps fine-tune the structure and content of the course:

- Optimizing questions and content—if specific tasks or questions consistently trigger RGB, they may be poorly worded, too complex, or inconsistent with the learning objectives. Designers can then use the data to revise such tasks to ensure they stimulate greater engagement.
- Difficulty levels—if a significant number of students guess too many questions on a specific topic, this may indicate that the content is either too complex or not explained enough.

Certain functions must be integrated directly into the microlearning environment to make RGB information available. While the methods are easy to implement, constantly running these calculations for every student response could burden the system. However, a practical solution can be achieved by considering the following:

- Student response data is usually already available when new features are deployed.
- RTT can be calculated once based on existing answers and saved as part of each question.
- Regular updates of RTT calculations require minimal system resources, making the process efficient.

When the functionality is deployed into the system, the RTT is calculated for individual questions once. If the student's response time is subsequently shorter than the RTT, his behavior will be marked as RGB. The student response data will not be used to recalculate the current RTT but will be saved for later updating.

To reduce inaccuracies, or when deploying new content, it is possible to consider creating tolerance intervals that will ignore the identified RGB or temporarily deactivate them. The system can update the RGB calculation regularly or based on the instruction of the course administrator or the teacher and use the new values to evaluate future attempts. At the same time, summary reports will also be available to the teacher—basically, it will only be a comparison of the student's time with the established RTT for the question.

Implementing RGB methods in a microlearning environment significantly benefits both students and teachers. A system automatically processes and updates the RTT based on student responses, allowing better assessment of students' knowledge and question-solving behavior. Teachers gain valuable information about problematic parts of the curriculum, enabling them to adapt teaching methods in a targeted manner. Ultimately, implementing these methods contributes to increasing teaching effectiveness and a better understanding of student progress.

4.4.2 Future Research

To improve the understanding and efficiency of RGB identification, future research should focus on exploring additional variables, improving methods, and applying AI models. Key areas of focus include:

- Exploring additional variables:

  Motivation inspection should explore how intrinsic versus extrinsic motivation affects RGB in low-stakes and high-stakes testing.

  Question types aimed at identifying the relationship between different question types and RGB from an item response theory perspective.

  Longitudinal tracking should monitor user behavior over time to understand how RGB behavior evolves with changes in learning contexts, methods, or student development.

  Cross-cultural studies focused on different cultural and educational settings to identify differences in RGB behavior and understand how cultural factors may influence rapid guessing and rating strategies.

  Educational context should evaluate the impact of different teaching methods and environments on RGB, including comparing interactive and passive learning environments.

  Testing conditions focus on time limits, stress, and environmental factors affccting RGB.

- Improvement methods:

  Hybrid approaches should integrate different existing techniques such as CUMP-UP and CUMP-DOWN. Combining these methods can provide finer and more precise RGB identification by exploiting the strengths of each approach.

  Combination of models—e.g., using psychological and mathematical models to capture contextual insights. For example, pairing statistical analyses with observational studies or surveys could improve understanding of rapid guessing.

The application of artificial intelligence involves the analysis of existing models with the primary goal of identifying RGB in real-time. While real-time detection is a natural goal of the models, it is not the primary focus of current research. Instead, we emphasize pattern identification and user clustering to gain deeper insight into RGB behavior. By focusing on these areas, future research may increase the accuracy of RGB identification and provide deeper insight into the factors influencing rapid guessing behavior.

Acknowledgments The following grants funded this research: 1/0490/22: Design and Evaluation of Selected Learning Analytics Methods for Understanding and Improving the Learning Process and Virtual Learning Environment (Science Grant Agency of the Ministry of Education of the Slovak Republic); 2021-1-SK01-KA220-HED-000032095: Future IT Professionals Education in Artificial Intelligence (European Commission under the ERASMUS+ Programme 2021, KA2).

Ethics Approval This study protocol was approved by the Ethics Committee of Constantine the Philosopher University in Nitra, approval number UKF/225/2024/191013:002.

References

Attali, Y., Laitusis, C., & Stone, E. (2016). Differences in reaction to immediate feedback and opportunity to revise answers for multiple-choice and open-ended questions. *Educational and Psychological Measurement, 76*(5), 787–802. Retrieved from https://doi.org/10.1177/0013164415612548

Betancur-Chicué, V., & García-Valcárcel Muñoz-Repiso, A. (2023). Microlearning for the development of Teachers' digital competence related to feedback and decision making. *Education Sciences, 13*(7). Retrieved from https://doi.org/10.3390/educsci13070722

Browne, M. N., & Keeley, S. M. (2002). *Asking the right questions: A guide to critical thinking* (8th ed.). Action Learning Worldwide.

Debeer, D., Buchholz, J., Hartig, J., & Janssen, R. (2014). Student, school, and country differences in sustained test-taking effort in the 2009 PISA Reading assessment. *Journal of Educational and Behavioral Statistics, 39*(6), 502–523. Retrieved from https://doi.org/10.3102/1076998614558485

DeMars, C. E. (2007). Changes in rapid-guessing behavior over a series of assessments. *Educational Assessment, 12*(1), 23–45. Retrieved from https://doi.org/10.1207/s15326977ea1201_2

Deribo, T., Kroehne, U., & Goldhammer, F. (2021). Model-based treatment of rapid guessing. *Journal of Educational Measurement, 58*(2), 281–303. Retrieved from https://doi.org/10.1111/jedm.12290

Deribo, T., Goldhammer, F., & Kroehne, U. (2023). Changes in the speed–ability relation through different treatments of rapid guessing. *Educational and Psychological Measurement, 83*(3), 473–494. Retrieved from https://doi.org/10.1177/00131644221109490

Ehlers, D. (2020). *Future skills: The future of learning and higher education* (Vol. 58). Itnow.

Gorham, T., Majumdar, R., & Ogata, H. (2023). Analyzing learner profiles in a microlearning app for training language learning peer feedback skills. *Journal of Computers in Education, 10*(3), 549–574. Retrieved from https://doi.org/10.1007/s40692-023-00264-0

Grabe, M., & Flannery, K. (2010). A preliminary exploration of on-line study question performance and response certitude as predictors of future examination performance. *Journal of Educational Technology Systems, 38*(4), 457–472. Retrieved from https://doi.org/10.2190/et.38.4.f

Guo, H., Rios, J. A., Haberman, S., Liu, O. L., Wang, J., & Paek, I. (2016). A new procedure for detection of Students' rapid guessing responses using response time. *Applied Measurement in Education, 29*(3), 173–183. Retrieved from https://doi.org/10.1080/08957347.2016.1171766

Haleem, A., Javaid, M., Qadri, M. A., & Suman, R. (2022). Understanding the role of digital technologies in education: A review. *Sustainable Operations and Computers, 3*, 275–285. Retrieved from https://doi.org/10.1016/j.susoc.2022.05.004

Halvoník, D., Kapusta, J., & Munk, M. (2023). Improve estimated time-on-task calculation in a virtual learning environment. *Interactive Learning Environments, 31*(5), 2914–2929. Retrieved from https://doi.org/10.1080/10494820.2021.1913609

Hartati, N., & Yogi, H. P. S. (2019). Item analysis for a better quality test. *English Language in Focus (ELIF), 2*(1). Retrieved from https://doi.org/10.24853/elif.2.1.59-70

Heber, E., Ebert, D. D., Lehr, D., Cuijpers, P., Berking, M., Nobis, S., & Riper, H. (2017). The benefit of web- and computer-based interventions for stress: A systematic review and meta-analysis. *Journal of Medical Internet Research, 19*, e32. Retrieved from https://doi.org/10.2196/jmir.5774

Hug, T. (2005). Microlearning: A new pedagogical challenge. In *Proceedings of microlearning conference 2005*.

Hui, B. (2023). Are they learning or guessing? Investigating trial-and-error behavior with limited test attempts. In *ACM International Conference Proceeding Series*. Retrieved from https://doi.org/10.1145/3576050.3576068.

Khong, H. K., & Kabilan, M. K. (2022). A theoretical model of micro-learning for second language instruction. *Computer Assisted Language Learning, 35*(7), 1483–1506. Retrieved from https://doi.org/10.1080/09588221.2020.1818786

Korosi, G., Esztelecki, P., Farkas, R., & Toth, K. (2018). Clickstream-based outcome prediction in short video MOOCs. In *CITS 2018–2018 International conference on computer, Information and Telecommunication Systems*. Retrieved from https://doi.org/10.1109/CITS.2018.8440182

Krygier, D., Knowles, L. R., Gill, A., Kwong, C. F., Irwin, D., Towey, D., & Pike, M. (2022). Evaluating microlearning: A cross-faculty case study of a Sino-foreign University. In *Agile learning environments amid disruption: Evaluating academic innovations in higher education during COVID-19*. Retrieved from https://doi.org/10.1007/978-3-030-92979-4_4

Kumar, D., Jaipurkar, R., Shekhar, A., Sikri, G., & Srinivas, V. (2021). Item analysis of multiple choice questions: A quality assurance test for an assessment tool. *Medical Journal Armed Forces India, 77*, S85–S89. Retrieved from https://doi.org/10.1016/j.mjafi.2020.11.007

Lee, Y.-H., & Jia, Y. (2024). Can adaptive testing improve test-taking experience? A case study on educational survey assessment. *Applied Measurement in Education, 37*, 1–18. Retrieved from https://doi.org/10.1080/08957347.2024.2386932

Lindner, M. A., Lüdtke, O., & Nagy, G. (2019). The onset of rapid-guessing behavior over the course of testing time: A matter of motivation and cognitive resources. *Frontiers in Psychology, 10*. Retrieved from https://doi.org/10.3389/fpsyg.2019.01533

Lundgren, E., & Eklöf, H. (2023). Questionnaire-taking motivation: Using response times to assess motivation to optimize on the PISA 2018 student questionnaire. *International Journal of Testing, 23*(4), 231–256. Retrieved from https://doi.org/10.1080/15305058.2023.2214647

Ma, L., Wise, S. L., Thum, Y. M., & Kingsbury, G. (2011). Detecting response time threshold under the computer adaptive testing environment. In *Annual meeting of the national council on measurement in education, New Orleans*.

Mastuti, E., & Handoyo, S. (2018). *Effects of individual differences on the performance in computer based test (CBT)*. Retrieved from https://doi.org/10.2991/acpch-17.2018.4

Michaelides, M. P., Ivanova, M., & Nicolaou, C. (2020). The relationship between response-time effort and accuracy in PISA science multiple choice items. *International Journal of Testing, 20*(3), 187–205. Retrieved from https://doi.org/10.1080/15305058.2019.1706529

Moreno Moreno, E., & Mateus-Nieves, E. (2021). Use of microlearning as a strategy to teach mathematics asynchronously. *International Journal of Development Research, 11*(03).

Nagy, G., Ulitzsch, E., & Lindner, M. A. (2023). The role of rapid guessing and test-taking persistence in modelling test-taking engagement. *Journal of Computer Assisted Learning, 39*(3), 751–766. Retrieved from https://doi.org/10.1111/jcal.12719

Nikou, S. A., & Economides, A. A. (2018). Mobile-based micro-learning and assessment: Impact on learning performance and motivation of high school students. *Journal of Computer Assisted Learning, 34*(3), 269–278. Retrieved from https://doi.org/10.1111/jcal.12240

Paiva, J. C., Leal, J. P., & Figueira, Á. (2022). Automated assessment in computer science education: A state-of-the-art review. *ACM Transactions on Computing Education, 22*, 1–40. Retrieved from https://doi.org/10.1145/3513140

Piaw Chua, Y. (2012). Effects of computer-based testing on test performance and testing motivation. *Computers in Human Behavior, 28*(5), 1580–1586. Retrieved from https://doi.org/10.1016/j.chb.2012.03.020

Rios, J. A. (2022a). A comparison of robust likelihood estimators to mitigate bias from rapid guessing. *Applied Psychological Measurement, 46*(3), 236–249. Retrieved from https://doi.org/10.1177/01466216221084371

Rios, J. A. (2022b). Assessing the accuracy of parameter estimates in the presence of rapid guessing misclassifications. *Educational and Psychological Measurement, 82*(1), 122–150. Retrieved from https://doi.org/10.1177/00131644211003640

Rios, J. A. (2022c). When should individual ability estimates be reported if rapid guessing is present? *Applied Measurement in Education, 35*(3), 222–236. Retrieved from https://doi.org/10.1080/08957347.2022.2103138

Rios, J. A., & Deng, J. (2021). Does the choice of response time threshold procedure substantially affect inferences concerning the identification and exclusion of rapid guessing responses? A meta-analysis. *Large-Scale Assessments in Education, 9*(1), 1–25. Retrieved from https://doi.org/10.1186/s40536-021-00110-8

Rios, J. A., & Soland, J. (2021). Parameter estimation accuracy of the effort-moderated item response theory model under multiple assumption violations. *Educational and Psychological Measurement, 81*(3), 569–594. Retrieved from https://doi.org/10.1177/0013164420949896

Sankaranarayanan, R., Leung, J., Abramenka-Lachheb, V., Seo, G., & Lachheb, A. (2023). Microlearning in diverse contexts: A bibliometric analysis. *TechTrends, 67*(2), 260–276. Retrieved from https://doi.org/10.1007/s11528-022-00794-x

Schnipke, D. L., & Scrams, D. J. (1997). Modeling item response times with a two-state mixture model: A new method of measuring speededness. *Journal of Educational Measurement, 34*(3), 213–232. Retrieved from https://doi.org/10.1111/j.1745-3984.1997.tb00516.x

Schnipke, D. L., & Scrams, D. J. (2005). Exploring issues of examinee behavior: Insights gained from response-time analyses. In *Computer-based testing: Building the foundation for future assessments*. Retrieved from https://doi.org/10.4324/9781410612250-20

Sharples, M., de Roock, R., Ferguson, R., Gaved, M., Herodotou, C., Koh, E. et al. (2016). Innovating pedagogy 2016: Open University Innovation Report 5. *Institut of Educational Technology, 26*(December).

Silm, G., Pedaste, M., & Täht, K. (2020). The relationship between performance and test-taking effort when measured with self-report or time-based instruments: A meta-analytic review. *Educational Research Review, 31*, 100335. Retrieved from https://doi.org/10.1016/j.edurev.2020.100335

Simaremare, A., Wau, Y., Hutagalung, T., & Parluhutan Tambunan, H. (2022). Development of microlearning model at high class SDN 106813 Amplas. Retrieved from https://doi.org/10.4108/eai.11-10-2022.2325468

Skalka, J., & Drlik, M. (2018). Priscilla – Proposal of system architecture for programming learning and teaching environment. In *IEEE international conference on application of information and communication technologies*. Retrieved from https://publons.com/publon/27387754/

Skalka, J., & Valko, M. (2024). Rapid guessing behavior detection in microlearning: Insights into student performance, engagement, and response accuracy. *IEEE Access, 12*, 157996 – 158024. Retrieved from https://doi.org/10.1109/ACCESS.2024.3485505

Svetina Valdivia, D., Rutkowski, L., Rutkowski, D., Canbolat, Y., & Underhill, S. (2023). Test engagement and rapid guessing: Evidence from a large-scale state assessment. *Frontiers in Education, 8*. Retrieved from https://doi.org/10.3389/feduc.2023.1127644

Wang, C., & Xu, G. (2015). A mixture hierarchical model for response times and response accuracy. *British Journal of Mathematical and Statistical Psychology, 68*(3), 456–477. Retrieved from https://doi.org/10.1111/bmsp.12054

Wise, S. L. (2017). Rapid-guessing behavior: Its identification, interpretation, and implications. *Educational Measurement: Issues and Practice, 36*(4), 52–61. Retrieved from https://doi.org/10.1111/emip.12165

Wise, S. L. (2019). An information-based approach to identifying rapid-guessing thresholds. *Applied Measurement in Education, 32*(4), 325–336. Retrieved from https://doi.org/10.1080/08957347.2019.1660350

Wise, S. L. (2020). The impact of test-taking disengagement on item content representation. *Applied Measurement in Education, 33*(2), 83–94. Retrieved from https://doi.org/10.1080/08957347.2020.1732386

Wise, S. L., & DeMars, C. E. (2005). Low examinee effort in low-stakes assessment: Problems and potential solutions. *Educational Assessment, 10*(1), 1–17. Retrieved from https://doi.org/10.1207/s15326977ea1001_1

Wise, S. L., & Gao, L. (2017). A general approach to measuring test-taking effort on computer-based tests. *Applied Measurement in Education, 30*(4), 343–354. Retrieved from https://doi.org/10.1080/08957347.2017.1353992

Wise, S. L., & Kong, X. (2005). Response time effort: A new measure of examinee motivation in computer-based tests. *Applied Measurement in Education, 18*, 163–183. Retrieved from https://doi.org/10.1207/s15324818ame1802_2

Wise, S. L., & Kuhfeld, M. R. (2020). A cessation of measurement: Identifying test taker disengagement using response time. In *Integrating timing considerations to improve testing practices*. Retrieved from https://doi.org/10.4324/9781351064781-11.

Wise, S. L., & Kuhfeld, M. R. (2021). Using retest data to evaluate and improve effort-moderated scoring. *Journal of Educational Measurement, 58*(1), 130–149. Retrieved from https://doi.org/10.1111/jedm.12275

Wise, S. L., & Ma, L. (2012). Setting response time thresholds for a CAT item pool: The normative threshold method. In *Paper presented at the 2012 annual meeting of the National Council on Measurement in Education*.

Wise, S. L., & Smith, L. F. (2016). The validity of assessment when students don't give good effort. In *Handbook of human and social conditions in assessment*. Retrieved from https://doi.org/10.4324/9781315749136-19.

Yang, T. Y., Brinton, C. G., Joe-Wong, C., & Chiang, M. (2017). Behavior-based grade prediction for MOOCs via time series neural networks. *IEEE Journal on Selected Topics in Signal Processing, 11*(5). Retrieved from https://doi.org/10.1109/JSTSP.2017.2700227

Chapter 5
Development of Youth Information Hygiene Skills: The Gap Between the Self-Assessment and Real State

Yuliia Rudenko, Marina Drushlyak, Olha Naboka, Volodymyr Proshkin, and Olena Semenikhina

Abstract All strata of the population become the objects of destructive informational and psychological influence, and, especially, young people, who, on the one hand, are the most active users of technologies and Internet resources, generators of the development and spread of social networks and video channels, and, on the other hand, due to their age-related psychological characteristics and lack of life experience, they are the most vulnerable category of society. The reasons for this conclusion are that young people, on the one hand, are the generation that has been surrounded by IT since birth, and for whom it is natural to live and interact in an information environment, using IT and consuming information flows in excessive quantities, mostly from social networks. On the other hand, young people are confident in their own actions in cyberspace and their digital literacy, media competence, and cyber security skills. The stated contradiction led to the study, the purpose of which was to compare the self-assessment of young people and the development of their actual information hygiene skills (the ability to distinguish fake from the truth, to identify real cyber threats, etc.). Sumy, Donetsk region, and

Y. Rudenko
Faculty of Economy and Management, Sumy National Agrarian University, Sumy, Ukraine
e-mail: yango641@ukr.net

M. Drushlyak · O. Semenikhina
Faculty of Physics and Mathematics, Sumy State Pedagogical University named after A. S. Makarenko, Sumy, Ukraine
e-mail: marydru@fizmatsspu.sumy.ua; e.semenikhina@fizmatsspu.sumy.ua

O. Naboka
Faculty of Humanitarian and Economic Education, Donbas State Pedagogical University, Sloviansk-Dnipro, Ukraine
e-mail: olganaboka911@gmail.com

V. Proshkin (✉)
Department of Physics and Mathematics, Borys Grinchenko Kyiv Metropolitan University, Kyiv, Ukraine

Department of Mathematics Education, Loughborough University, Loughborough, UK
e-mail: v.proshkin@kubg.edu.ua; v.proshkin@lboro.ac.uk

E. Smyrnova-Trybulska et al. (eds.), *E-Learning and Enhancing Soft Skills*,
https://doi.org/10.1007/978-3-031-82243-8_5

Kyiv (Ukraine) educational institutions became the experimental base. According to the study results, an average gap of 40–60% was found. Qualitative analysis of empirical data revealed the reasons for the established gap: lack of awareness of the insufficient development of one's own information hygiene skills and self-confident attitude to information flows in cyberspace. Separate consequences of such a gap were revealed: almost defiant behaviour in the cyber environment, which is manifested in ignoring informational dangers, irresponsible consumption of information of dubious quality, dissemination and further dissemination of false data, biased assessment of facts, trusting attitude to manipulative techniques, etc.

Keywords Information hygiene · Information hygiene skills · Self-assessment · Information influence · Cyber protection skills · IT · Digital technologies · Education

5.1 Introduction

In the ultra-fast development of information technologies (IT), all segments of the population, especially young people, have become the objects of destructive informational and psychological influence (Usta-Azizova et al., 2022). On the one hand, this generation has been surrounded by IT since birth and for whom it is natural to live and interact in an information environment, consuming information flows in excessive quantities (Neha et al., 2021). These are regular Internet users, generators of the development and spread of social networks. On the other hand, due to age-related psychological characteristics and lack of life experience, this is a gullible category of people who have an insufficiently formed worldview, and moral and ethical values, and their life guidelines are easily influenced and transformed (Usta-Azizova et al., 2022). This makes youth the most vulnerable category in terms of various influences, including information.

The psychophysiological portrait of the young generation is quite telling. Young people live in networks and prefer virtual communication. At the same time, most representatives of this generation value honesty, frankness, and openness (Posnick-Goodwin, 2010). This generation has ubiquitous access to any information and therefore often has fragmented and sometimes superficial thinking (Kolot & Herasymenko, 2020). They are characterized by impatience, focus on short-term goals and individualism.

A separate characteristic of the modern young generation is its self-confidence. According to (Melnychuk, 2017), self-confidence refers to a disharmonious type of confidence. A self-confident person is characterized by very high levels of confidence, which is expressed by an inadequately overestimated self-assessment and critical thinking. Therefore, today's youth feel self-confident on the Internet and often do not see false/fake content, subjective manipulative judgments, trustingly and irresponsibly to the installation/testing or distribution of various digital applications, without reservations they provide access to personal data (telephone, electronic address, photo, texts, geolocation, etc.). High subjective assessments of oneself as a competent user of Internet technologies and at the same time the presence of many cyber frauds, falling under the influence of information, and unconscious acts of spreading fake content by young people, recorded by official institutions (Ministry of Internal Affairs, Security Service of Ukraine, etc.) about the inconsistency of self-assessments of young people with the real level of development of their information hygiene skills. In other words, young people themselves are not aware of the gap between the subjective assessment of their information hygiene skills and the actual skills of working with information flows. Identification and awareness of the gap between self-assessment and the real state of the level of development of young people's information hygiene skills can become the basis for effectively eliminating such a gap and safer work on the Internet.

The purpose of this study—to investigate the gap between self-assessment and the real state of the level of development of young people's information hygiene skills, to characterize the causes of such a gap and its consequences.

Achieving the goal determined the solution of several tasks:

1. to determine the self-assessment of the level of development of information hygiene skills of young people;
2. to determine the real state of the level of development of young people's information hygiene skills;
3. to reveal the causes and consequences of the gap between self-assessment and the real state of the level of development of youth information hygiene skills.

5.2 Research Design

5.2.1 Participants

The research was conducted during 2022–2023 in institutions of higher education of the Sumy region: (Sumy National Agrarian University (74 students of 1 courses), Sumy Regional Institute of Postgraduate Education (28 students of 1 courses), Sumy State Pedagogical University named after A.S. Makarenko (82 students of 1–3 courses), Sumy Vocational College of Economics and Trade (86 students of 1–2 courses), Putivl S. V. Rudniev Pedagogical College (42 students of 2 courses), at Borys Grichnenko Kyiv Metropolitan University (26 students of 2 courses) and Donbas State Pedagogical University (21 students of 2–3 courses). The total number of participants is 359 students).

5.2.2 Methods

To solve the first task, a survey was used as an empirical method of scientific knowledge. The survey (Self-assessment 1) was conducted using the author's questionnaire, which helped determine the self-assessment of the level of information hygiene skills of each participant. The questionnaire passed an expert evaluation

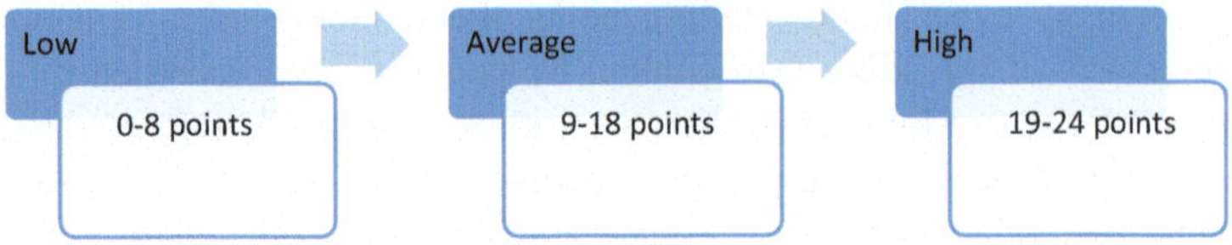

Fig. 5.1 Levels of development of youth information hygiene skills (questionnaire for self-assessment). (Source: Own research)

and was approved for use. Answers to the questions included ratings of 2 (yes), 1 (sometimes), and 0 (no) (Likert scale). The maximum number of points for the first questionnaire is 24. Questionnaires (two for each information hygiene skill) are presented in Table 5.2, which follows the consideration of the concept of information hygiene and the analysis of methods of self-assessment in Sect. 5.3.3.

Thus constructed questionnaire allowed to determine the level of development of the respondent's information hygiene skills in general (the levels of development of information hygiene skills of young people are shown in Fig. 5.1) and to specify the answers according to the information hygiene skills that were the subject of the study.

The testing method was used to solve the second task of the research. The set of tasks (Table 5.1) allowed us to determine the actual level of youth information hygiene skills (two for each skill). Note that this study partially used the tasks proposed by the authors in the works (Rudenko et al., 2022, 2023). The list of tasks also received a favourable rating after the examination. The specifics of the task assessment were similar: 0 points—wrong answer; 1 point—partially correct; 2 points—completely correct answer. The distribution of levels was similar (Fig. 5.1).

The testing allowed a real assessment of the formed skills and characterized the real level of development of youth information hygiene skills.

After analysing the test results, we again conducted a survey (Self-assessment 2) with the same questions (Table 5.2). The roadmap of the research of information hygiene skills is presented in Fig. 5.2.

To confirm the gap between self-assessment and the actual level of development of information hygiene skills, we used a statistical analysis based on the comparison of averages (Student's criterion) and implemented it with MS Excel tools. Statistical hypotheses ("H_0 – the averages are statistically the same (or the gap is insignificant)" and "H_a – the averages are statistically different (the gap is statistically significant)") made it possible to compare the average grades and determine/disprove their statistical similarity for each of the skills (para. 2.1).

We conducted a statistical analysis of the first and third survey (comparing Self-assessment-1 with Self-assessment-2) using a non-parametric sign criterion. Statistical hypotheses ("H_0 – changes in answers are insignificant" and "H_a – changes in answers are significant") allowed seeing whether students' self-assessment changes after the objective state of development of their information hygiene skills are identified.

Table 5.1 Tasks for determining the actual level of youth information hygiene skills (testing)

<table>
<tr><th>The skill of information hygiene</th><th>Tasks and answer options (two correct answers out of several)</th></tr>
<tr><td rowspan="2">The skill to distinguish fact from judgement.</td><td>1. Fact vs judgement. Choose which, in your opinion, is a fact. The question has two correct answers (0 points—no correct answer; 1 point—one correct answer; 2 points—two correct answers).
Information that contains a personal assessment.
An event that contains an emotional evaluation.
Information that is affirmed or denied.
A real event that can be tied to a specific time or place.
Evidenced information.
The expression of personal thoughts of a person.
Exaggeration, generalization, or one-eyed description of a fact.</td></tr>
<tr><td>2. Fact vs judgement. Choose which, in your opinion, is a fact. The question has two correct answers (0 points—no correct answer; 1 point—one correct answer; 2 points—two correct answers).
The war will be protracted
The world will pay a high price for Russia's war against Ukraine
On February 28, 2022, the president signed the application for Ukraine's membership in the EU
All information is verified on the Internet!
Volunteers will definitely go into politics after the war
Zelenskyi won the 2019 presidential election.</td></tr>
<tr><td>The skill to detect manipulative content.</td><td>1. Match the manipulative message to influence the audience (the correct answer is considered the correctly matched pair). Each correctly set logical pair is valued at 1 point.
<table>
<tr><td>1</td><td>“Shock!!!!”
“You won't believe it!”</td><td>A</td><td>To shape public opinion in a direction beneficial to the author. Disorientation, intimidation, propaganda.</td></tr>
<tr><td>2</td><td>“Poland's army plans to invade Ukraine at the end of July”</td><td>B</td><td>An appeal to the reader's emotions to reduce the degree of critical analysis</td></tr>
<tr><td>3</td><td>“Everyone says...” “You can't be silent...”</td><td>C</td><td>Formation of favourable stereotypes in the mind.</td></tr>
<tr><td>4</td><td>“More troops are mobilized from the western regions than from others”</td><td>D</td><td>Increase the number of views (clickbait).</td></tr>
</table></td></tr>
<tr><td>The skill to identify fake information</td><td>1. What do you think is effective in verifying the authenticity of information from the Internet? The question has two correct answers (0 points—no correct answer; 1 point—one correct answer; 2 points—two correct answers).
It is impossible to verify the photo, so it is not necessary to pay attention to it.
It is worth checking whether the title corresponds to the main part of the message.
The author with the specified name and photo is a real person, so such information should be trusted.
It is necessary to check the presence of emotional influence in the message.
The presence of errors in the text are only typographical errors, so you should not pay attention to them.</td></tr>
</table>

(continued)

Table 5.1 (continued)

The skill of information hygiene	Tasks and answer options (two correct answers out of several)
	2. Which of the following statements do you think are correct? The question has two correct answers (0 points—no correct answer; 1 point—one correct answer; 2 points—two correct answers). Fake news often has flashy titles written in capital letters with exclamation marks. Fake sites never make mistakes in spelling and punctuation. Copies of well-known pages or sites are often created to promote fake information. Fake news contains real photos and videos.
The skill to check the reliability and authority of information sources/ site	1. By what signs can you identify a garbage site? The question has two correct answers (0 points—no correct answer; 1 point—one correct answer; 2 points—two correct answers). A variety of content is used to attract the audience. Contains only humour, entertainment, or life hacks. Do not disguise themselves as mass media. Create information noise to mask the real problem. They contain real contacts of the editorial office, the names of the editor-in-chief and journalists.
	2. Which of the following statements is not an example of a violation of journalistic standards? The question has two correct answers (0 points—no correct answer; 1 point—one correct answer; 2 points—two correct answers). Inseparability/unity of the journalist's opinion and the opinions of experts/interviewees. Marking of all advertising materials. Coverage of only one point of view, which the journalist considers more objective. A balance of opinions and points of view on controversial issues.
The skill to protect personal data	1. Which of the following statements characterize, in your opinion, safe online behaviour? The question has two correct answers (0 points—no correct answer; 1 point—one correct answer; 2 points—two correct answers). I have the same logins and passwords for all my accounts on the network. A password that contains numbers, upper- and lower-case letters and symbols is the most reliable. Personal data should not be disclosed. I open all letters that come to my e-mail. I trust information from any sites.
	2. Choose only those situations in which safe actions are described. The question has two correct answers (0 points—no correct answer; 1 point—one correct answer; 2 points—two correct answers). "We need to collect a million votes so that Viber or Facebook are not paid!" You should not "like" such messages. The ad offers to see yourself at 50+ years old and download an application that requires access to the camera or photos. It's worth a try. It's always worth participating in quizzes and games posted on social networks. "Sent a message to 10 friends!". Never send "happiness letters".

(continued)

Table 5.1 (continued)

The skill of information hygiene	Tasks and answer options (two correct answers out of several)
The skill of managing time spent online	1. Which of the following statements do you think are correct? The question has two correct answers (0 points—no correct answer; 1 point—one correct answer; 2 points—two correct answers). Long-term scrolling and "clicking" in social networks form a psychological addiction. Installing the "parental control" function on a child's smartphone is a violation of his/her personal space. Falling asleep with a smartphone does not affect the quality of sleep. It is useful to set the function of turning off notifications on all smartphone applications.
	2. What allows you to do the "screen time" function configured in the smartphone? Control the duration of the application that spends the most time. Control how many photos have been taken by the phone's camera. Control how much money is spent per day. Control how many times the screen is unlocked during the day.

Source: Own research

Table 5.2 Questionnaire for self-assessment of the level of development of youth information hygiene skills

The skill of information hygiene	№	Question
Skill to distinguish fact from judgement	1	I can tell the difference between a fact or an unsubstantiated fact from a judgement.
	7	I can detect misinformation in the message.
Skill to detect manipulative content	2	I can understand by the title of the article that the article is manipulative.
	8	I understand the reasons for misrepresentation.
Skill to identify fake information	3	I can tell the difference between fake and fact.
	9	I know how to use fact-checking services.
Skill to check the reliability and authority of sources of information on the site	4	I can tell the difference between a reliable site and a garbage site.
	10	I know what a media whitelist is.
Skill to protect personal data	5	I can protect personal data.
	11	I understand all the consequences of sharing personal data.
Skill of managing the time spent on the network	6	I can control the time I spend online.
	12	I can always stop and "get out" of the network.

Source: Own research

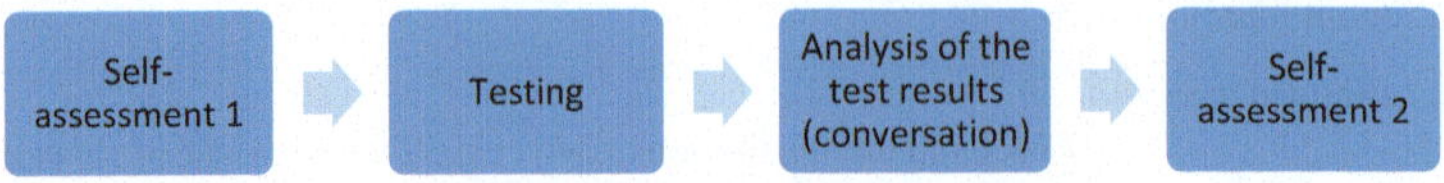

Fig. 5.2 The roadmap of the research of information hygiene skills. (Source: Own research)

To solve the third task of the research, we held conversations about the results of the research. At the meeting, we analysed the test results and asked a series of oral questions that helped us identify the reasons for the gap:

1. How often do you feel confident using digital technology and media?
2. How would you define what it means to adhere to the principles of information hygiene?
3. What aspects of information hygiene did you study at school or university?
4. Do you feel social pressure regarding your information hygiene skills?
5. From what sources do you most often get information about information hygiene?
6. Do you feel limited in your use of digital technology due to technical or financial reasons?
7. How do you most often use digital technology and media in your daily life?
8. How often do you question the information you receive from the media and the Internet?
9. How do you respond to situations where you feel less digital and media savvy?
10. How often are you faced with challenges that require the application of information hygiene skills?

5.3 The Results of the Study

5.3.1 Information Hygiene Skills

The importance of information hygiene began to be discussed at the end of the twentieth—beginning of the twenty-first century. Thus, in the work (Eremin, 1995), a new scientific direction is initiated—*information hygiene* as a field of knowledge that studies the regularities of the influence of information (information technologies) on the state of mental, physical, and social health of a specific person and society. The science of information hygiene in society explores the influence of environmental factors within the information environment on human health, performance, and life expectancy. It formulates standards, requirements, and sanitary measures aimed at enhancing the information quality of communities, living conditions, and human activities (Eryomin, 2020).

We cannot but agree with G. Nicholas that information hygiene is the conscious management of information habits and information behaviour (Nicholas, 2010). A valid perspective is offered by Z. Koval, who defines information hygiene as information immunity and as a means of ensuring information and psychological stability. The author pays attention to social information hygiene. "Without mechanisms, which should have a comprehensive nature, states risk losing their information sovereignty, and the population will be governed by misleading and staged information ghosts from the world of" crooked mirrors "of the opposite side," states his research paper (Koval, 2017). We agree with A. Dubrovyk-Rokhova, who considers information hygiene to be a guarantee against cyber fraud. In her understanding, information hygiene refers to a healthy way of interacting with information, it ensures effective perception, processing, and understanding of information sources (Dubrovyk-Rokhova, 2017).

The findings of V. Varynskyi et al. are intriguing, as they analyzed the state of the information hygiene segment of the Ukrainian Internet media in conditions of global threats (Varynskyi et al., 2021).

These findings are complemented by Loukas, G., Murugesan, S., and Andriole, S. J., who focus on countering misinformation in the news and mass media (Loukas et al., 2022), and R.K. Olsen, M.K. Solvoll and K.-A. Futsæter explore the importance of information hygiene for national security and information policy in general (Olsen et al., 2022). I. Kilimnik examines the legal framework supporting self-regulation in the media industry (Kilimnik, 2023).

A key point in the study is the perspective of Grimes D. R., who emphasizes the need to recognize the harmful effects of misinformation and take precautions to avoid the spread of misinformation (Grimes, 2020). Building on this, O. Kyiianytsia et al. offer a model aimed at managing informational hygiene in society through the elimination of risk factors through targeted intervention by social institutions and management subjects (Kyiianytsia et al., 2023).

L. Slusky and P. Parviz-Navin analyzed the methodological foundations of information security and personal hygiene (Slusky & Parviz-Navin, 2012). The authors model information hygiene as an open dynamic system of activity that includes individuals, groups, and society.

The findings of M. Ostroga, A. Yurchenko, and A. Korovai are interesting when they investigate the problem of detecting information noise for an average member of society through the prism of the need to observe information hygiene (Ostroga et al., 2023).

Along with the "information hygiene" phenomenon the "cyber hygiene" phenomenon is also studied as a multilayer system of rules and activities users must follow when working on the Internet. It is a distinctive feature. On the other hand, cyber hygiene means abilities and habits that help users and organizations mitigate the networks' potential vulnerability to cyber-attacks and enhance resilience against cyber-attack threats (Mednikarov et al., 2023).

We also note that all interpretations of the concept of information hygiene necessarily refer to a set of measures that ensure it. Z. Koval talks about measures aimed at ensuring the quality, truthfulness, and usefulness of the information received (Koval, 2017), and V. Halamendyk—about mechanisms for overcoming pathological dependencies related to the information space (Halamendyk, 2008).

Understanding the essence of information hygiene as a field of knowledge, scientists when mastering it point to media literacy skills, information habits and information behaviour; security and protection of information space; informational and psychological stability and resistance to negative influences; stable value orientations; psychological and physical health. We believe that some skills are related to media literacy and can be acquired through the development of critical thinking; some to digital well-being, they form a healthy psychological and physiological stay in the information space; part of the skills that characterize information literacy and involves digital protection and security skills; to information behaviour and develop through the formation of a healthy and responsible way of interaction and communication in the information space.

We agree with N. Prykhodkina's definition, according to which media literacy is the ability to critically select media information, the skills of conscious use of information technologies, and the production of one's own media product (Prykhodkina, 2020). As for information literacy, the UNESCO strategy interprets it as "the ability to determine and formulate one's own information needs, to find access to information, to have the skills to process information using IT, and to critically evaluate it" (UNESCO, 2020). Consequently, both concepts are equivalent. However, they have a different direction: information literacy—for access to information and its processing, media literacy—for analysis and evaluation of media messages. This opinion is held by M. Osyukhina, who claims that media and information literacy (MIL) is an umbrella term, combining the signs of information and media literacy (Osyukhina, 2016). According to M. Osyukhina, media information literacy includes the skills to gain access to information, analyse, evaluate, use, create, and distribute it by legal and ethical norms.

In their previous research, the authors singled out the ability to resist informational influences, which also refers to information hygiene. This is the ability to take a responsible attitude to the consumption of information content, the ability to protect one's own information space, the ability to counteract destructive influences based on emotional colouring, the ability to verify media products for their critical evaluation (Rudenko et al., 2023).

We emphasize the importance of digital security skills and consider the position of M. Dodel and G. Mesch to be valid, as they define them as basic abilities needed to work in digital systems, as well as skills needed to understand and use online content on the Internet (Dodel & Mesch, 2018). According to J. Dostal, X. Wang, W. Steingartner, and P. Nuangchalerm, we supplement these skills with the ability to detect cyber threats (e.g. hacking, fraud, malware), understand best practices, and use appropriate security tools to protect data (Dostal et al., 2017). Digital security is identified with the ability to use modern information technologies and software effectively, ethically, and safely, which includes privacy management, understanding the digital footprint, personal cyber security skills (countering online fraud, malware), the ability to prevent risks in communication (countering cyberbullying, grooming) (Drushlyak et al., 2022).

The generalization of the interpretation of the concepts of "media literacy", "information literacy", "media-information literacy", "ability to resist informational influences", and "digital security", as well as the author's own experience make it possible to highlight the leading skills of information hygiene:

1. the skill to distinguish a fact from a judgement;
2. the skill to detect manipulative content;
3. the skill to identify fake information;
4. the skill to check the reliability and authority of the site's information sources;
5. the skill to protect personal data;
6. the skill of managing time spent online.

5.3.2 *Self-Assessment*

In many situations, the real behaviour of people and their everyday life is influenced not so much by the level of certain knowledge and skills, but by their subjective assessment.

Self-assessment is often (implicitly or otherwise) conceptualized as a personal, unguided reflection on performance to create an individual summary of one's own level of knowledge, skills, and understanding in a particular domain (Eva & Regehr, 2008).

In educational practice, self-assessment is conventionally understood as an assessment practice that asks subjects of learning to express judgments about their own work, products, and learning processes (Panadero et al., 2016). At the same time, it is worth distinguishing between "self-assessment" and "reflection", since reflection is a broader concept that includes an understanding of the evaluated process itself, its causes, and consequences (Nieminen & Boud, 2024).

Atikuzzaman Md., Zabed Ahmed S.M. consider the problem of self-assessment of self-efficacy (Self-assessment of self-efficacy), emphasizing that this method has become quite popular among methods of assessing information literacy skills (Atikuzzaman & Zabed Ahmed, 2023). According to (Bandura, 1986), self-efficacy is people's judgement of their ability to organize and carry out the activities necessary to achieve specified types of performance.

S. Cassidy and P. Eachus define self-assessment as a belief in one's ability to successfully perform a certain task (Cassidy & Eachus, 2002) and define self-assessment as the level of confidence that a person can successfully perform a given task.

This approach is currently promising, and we agree with T. Rosman et al. (2014) because the subjective assessment of skill is considered "a basic belief that underlies human motivation, productivity, achievement, and emotional well-being" (Bandura, 2010), and therefore can have a positive effect on effort expenditure and persistence in completing tasks, especially in the presence of certain obstacles. This goes some way to explaining why subjects fail objective tests (e.g., due to low self-efficacy), shedding light on the motivational reasons for success and failure.

We align with the view that self-assessment allows learners to reflect on their abilities and strategies, monitor their learning progress, identify their strengths and weaknesses, and direct their actions to eliminate possible weaknesses (Lew et al., 2010).

The problem of studying the self-assessment of certain skills of an individual is quite relevant, which is confirmed by studies of the self-assessment of computer competence (Sieverding & Koch, 2008); digital and Infomedia literacy (Jones Procter, 2023); information literacy (Prihandoko, 2021; Inazu & Soyemi, 2022; Rosman et al., 2014); digital literacy (Cote & Milliner, 2017); digital competence (Nebot et al., 2021).

5.3.3 Methods of Measuring Self-Assessment of Information Hygiene Skills

To measure self-assessment, psychologists use special scales, for example, the scale of self-assessment of media literacy of primary school pupils (Chang et al., 2011). The scale contains 13 items (13-item Scale) to assess the cognitive response of each subject to the use of media technologies in education and their behavioural attitude to the use of media technologies. All items were rated on a 5-point Likert scale that ranged from "strongly disagree" to "strongly agree".

The Information Literacy Self-Assessment Scale developed by (Kurbanoglu et al., 2006) is considered the most widely used self-assessment tool, as reported by Mahmood K. in a review article (Mahmood, 2017).

A. Pandian, S. B. Baboo, and L. J. Yi propose a 10-domain self-assessment tool for determining Digital Media Literacy, which is oriented towards Operational, Information Navigation, Social, Mobile, Creative, Critical Understanding, Digital Citizenship, Safety, Regulation, and Problem Solving. The Likert-type format was used in the self-assessment section. Students were asked to indicate their confidence levels using the scales and scores as follows: 1 for 'Don't know', 2 for 'Know a little', 3 for 'Fair', 4 for 'Competent', and 5 for 'Very competent' (Pandian et al., 2020).

To determine digital competence, O. Ovcharuk and I. Ivaniuk used five blocks of questions about competence in using digital tools and instruments for distance learning and self-education: information and digital literacy, communication and cooperation, digital content creation, security, and problem-solving (Ovcharuk & Ivaniuk, 2021). Their methodology was based on international approaches, including the Digital Competence Framework for Citizens: Eight Skills with Examples of Use (DigComp 2.1: Digital Competence Framework for Citizens) (Redecker & Punie, 2017; Carretero et al., 2017). DigComp 2.1 includes the following levels: basic user, independent user, and professional user. It outlines five areas of this competence: information and digital literacy, communication and collaboration, digital content creation, security, and problem-solving.

T. Rosman, A.-K. Mayer, and G. Krampen, in the context of the study of the level of information literacy based on the results of the analysis of scientific literature, note that three fundamentally different concepts of information literacy assessment are observed: achievement tests, information search tasks, and self-assessment. Unfortunately, these methods are mostly used independently of each other, with a lack of comparative studies on their objectivity, reliability, and validity (Rosman et al., 2014).

Since information hygiene skills include media literacy skills, information literacy skills, and digital literacy skills, when creating the author's questionnaire (Table 5.2), we partially relied on and modified the questions of the scales, the validity and effectiveness of which have been experimentally confirmed.

Self-assessment is a rather controversial method of measuring skills. It is worth noting that the accuracy of self-assessment depends on the individual (Leach, 2012): people with a lower level of ability tend to overestimate their own ability because they lack experience in the relevant field and do not have the relevant knowledge to recognize the correctness of the judgement.

It is worth noting that according to the results of research (Ackerman & Wolman, 2007), the accuracy of self-assessment increases if this diagnosis is carried out after the completion of the corresponding verification of the actual level of skills. The authors explain this by the fact that the level of effort required by the test is more adequately perceived "after" and by "internal feedback".

Empirical data on the relationship between self-assessment and actual assessment of skills is rather ambiguous. For example, (Freund & Kasten, 2012) did not see a correlation when determining IQ scores, and (Boud et al., 2013) showed that the relationship is significant. Therefore, self-assessment cannot replace the actual assessment of the level of significance, but it can help in the interpretation of research results. For example, overestimated skills can lead to a decrease in motivation, causing feelings of disappointment and failure. Freund and Kasten (2012), while low self-assessment can impair outcome expectations and therefore also lead to reduced motivation (Bandura, 1994). Scientists record cases when research results indicate a discrepancy between self-assessment and the real level of knowledge and skills, for example, when studying the level of development of digital competence of future teachers (Tsankov & Damyanov, 2019). This fact is confirmed by research (Rosman et al., 2014), which shows that the self-estimated information literacy level exceeds the actual estimate.

5.3.4 Comparison of the Results of the Self-Assessment 1 Survey and Testing

Based on the results of the surveys, we received the following empirical data (Table 5.3 and Fig. 5.3).

Statistical analysis of the results revealed a statistical significance of the gap for all skills except "Skill to protect personal data". The analysis of the answers confirmed our expectations regarding the trustworthiness of young people to

Table 5.3 Results of the "Self-assessment 1"[a] survey and testing[b]

The skill of information hygiene	Survey "Self-assessment1"	Testing	Gap	Accepted hypothesis
(1) The skill to distinguish fact from judgement	88%	52%	36%	H_a (P(T < =t) = 7,92E-17)
(2) The skill of detecting manipulative content	84%	53%	31%	H_a (P(T < =t) = 4,57E-13)
(3) The skill of identifying fake information	87%	59%	28%	H_a (P(T < =t) = 1,41E-11)
(4) The skill to check the reliability and authority of sources of information on the site	53%	40%	13%	H_a (P(T < =t) = 0,034)
(5) The skill to protect personal data	83%	72%	11%	H_0 (P(T < =t) = 0,118)
(6) The skill of managing time spent online	86%	42%	44%	H_a (P(T < =t) = 4,75E-19)

[a]Answers "Yes" and "Sometimes" were considered
[b]Scores 2.4 were taken into account and 0 or 1 were not taken into account

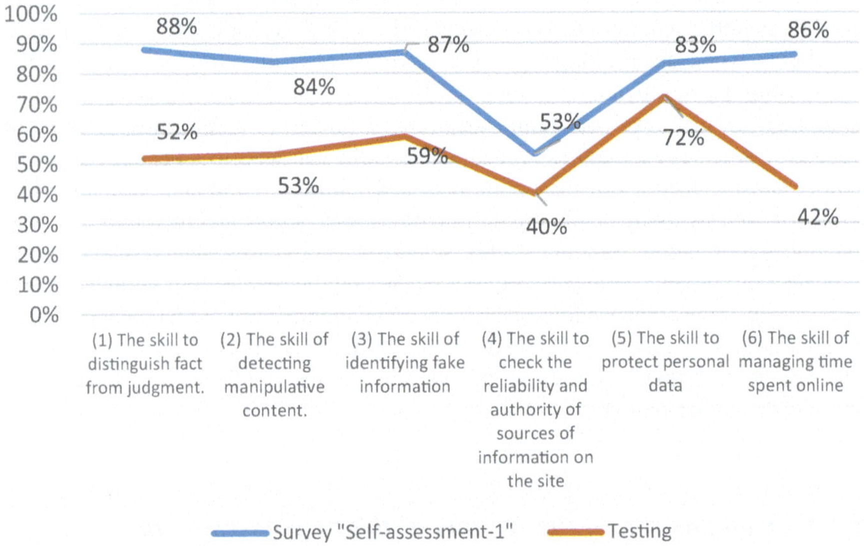

Fig. 5.3 Results of comparison of empirical data. (Source: Own research)

information content and their lack of ideas about the possible negative impact of information technologies, computer applications, social networks, and services on their health.

It should be noted that 5% of students, due to their age-related psychological characteristics, find it difficult to recognize the fact characterised by the words

"I can't", and "I don't know", especially in questions related to IT. There is a certain intolerance to one's own real abilities and a desire to embellish reality (especially in boys).

5.3.5 *Statistical Analysis of Surveys (Self-Assessment-1 and Self-Assessment-2)*

Statistical analysis of the results of two surveys (Self-assessment-1 and Self-assessment-2) using the non-parametric sign criterion revealed that 149 answers remained unchanged, while 210 students changed their own results: 36 students' self-assessment increased, and 174 students' self-assessment decreased. According to the hypothesis acceptance rule, we have: T(a = 0.05) = 104.92, n−T (a = 0.05) = 105.075 for n = 210; T(empirical) = 36 < 104.92.

Since T(empirical) did not fall into the interval (104.92; 105.075), the null hypothesis should be rejected and the alternative accepted: the changes in the answers are significant, and since the pairs with the "−" sign are more than the pairs with the "+" sign, it is fixed statistically a significant trend toward a decrease in student self-assessment after testing analysis and corresponding to a narrowing of the gap. Visualisation of the results (Fig. 5.4) shows the levelling of self-assessment.

Based on the results of the conversation and summarising the students' answers, we formulated the reasons for the recorded gap.

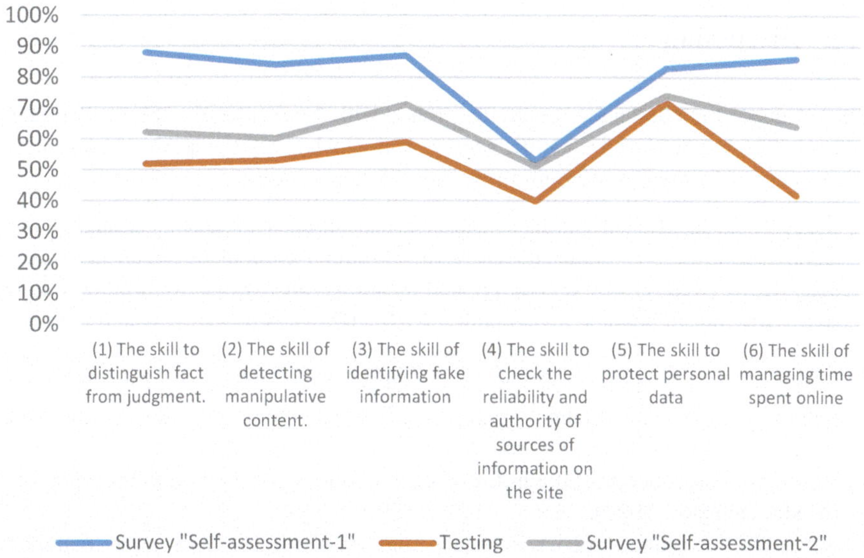

Fig. 5.4 Distribution of empirical data for the results of Self-assessment 1, testing, and Self-assessment-2. (Source: Own research)

1. Exaggeration of one's own capabilities regarding the use of digital technologies and social networks.
2. Insufficient education in the field of information hygiene (media education, information literacy), which leads to the fact that young people with a low level of knowledge in this field cannot objectively assess their level and tend to overestimate their knowledge and skills.
3. The influence of the social environment, that is, the desire to be on a par with peers, the desire to appear competent in front of others, as well as the fear of admitting one's own incompetence can encourage young people to exaggerate their capabilities. The surrounding environment creates the opinion that young people are more knowledgeable and literate in the field of media and information technologies than it is.
4. Orientation of young people to unreliable sources of information, which affects their ability to critically evaluate information and media content.
5. The digital divide, which means that different access to modern technologies and the Internet in different regions of the country and among different segments of the population can lead to uneven development of information hygiene skills.
6. Inappropriate use of digital technologies and media by young people, in most cases for entertainment, does not contribute to the development of information hygiene skills and leads to an underestimation of the complexity and depth of problems related to information hygiene. The lack of practical application of information hygiene skills in real situations can lead to the fact that young people are not aware of their weaknesses and cannot objectively assess their skills.

5.4 Discussion

The results of the conducted research correlate with the results of analytical studies presented in statistical and scientific sources and which allow comparing key aspects of the state of formation of youth information hygiene skills:

1. Study of the level of ML skills of Ukrainian social network users. Features of the sample: volume—15 thousand users of Ukrainian content of social networks, age—from 18 to 35 years, geographical location—all regions of Ukraine. Initiative "How not to become a vegetable" 2020–2022, 4 waves of research (Moroz, 2023).
2. Index of media literacy of Ukrainians 2020–2023. Features of the sample: volume—1200 respondents, aged 18–25 years, geographical location—all regions of Ukraine. Special project of the NGO "Media Detector", the goal is to validate the concept of the index of media literacy for the Ukrainian audience (Media Detector, 2023).
3. Research "Ukrainian media, attitudes and trust" in 2023. Sample features: volume—21,000 respondents, age 18–35 years, geographical location—all

regions of Ukraine (International organisation Internews, initiator of the "Media Program in Ukraine" (USAID, 2023)).
4. Survey "Fakes and facts. Do Ukrainian youth distinguish between them?" Features of the sample: 4237 respondents, age 18–35, geographical location—all regions of Ukraine (UNICEF, 2023).
5. Report on the interim study on the awareness of target audiences about the main aspects of cyber security, 2022. Features of the sample: volume—1244 respondents, age 17–25 years, geographical location—all regions of Ukraine. CRDF Global. National Cyber Security Cluster, Research supported by the US Department of State's Office of the Coordinator for US Assistance to Europe and Eurasia (CRDF Global, 2021).

Data analysis and comparison strengthened the existing results. In particular:

1. Young people aged 18–25 are the most intensive users of the network (99.5%), in 2023, 97% used the Internet almost every day, which is 18% more than in 2017. The purpose of the application: searching for information, getting to know the news, communicating on social networks, watching videos, shopping, online payments, Internet television, creating and posting one's own content, and training. This finding correlates with the results we obtained in the "Skill of managing time spent online" section, as only 42% of respondents know how to control the time they spend online, with the largest gap at 44%.
2. During 2017–2023, the structure of youth media activity changed. There is a systematic reduction in the consumption of television, radio, press, and Internet sites and an increase in the use of social networks from 45% to 74% (USAID, 2023).
3. There is a gradual reduction in activity in the use of print media, radio, and television. Instead, social networks are gaining popularity every year. In 2023, Telegram became the main social network for communication and news consumption. This finding is consistent with the results of our study on the definition of "Skill to check the reliability and authority of sources of information on the site." According to actual indicators, it corresponds to the lowest indicators—only 40%, as young people abandon authoritative media and increasingly trust social networks, even for the purpose of obtaining news information, and not knowing how to check their reliability. In addition, this skill is associated among young people with learning, mental, and pre-research activities and is not considered necessary and everyday significant, she self-assessment low scores (53%) and confirmed them.
4. Along with the growth of media activity in social networks, there is an increase in posts and distribution of IPSO, as well as publications that violate the signs of information hygiene (according to the research approach (Moroz, 2023), such publications with links to garbage sites, garbage Facebook pages, links to materials containing manipulative information, fakes identified by fact-checking companies, tests, flash mobs, games, etc., in which a person must provide access to his data or which is a data collection tool). In 2022, such publications among young people were recorded at 57%, and the spread of IPSO at 49% (Moroz, 2023). This

result echoes the result of our research on "Skill to distinguish fact from judgement" (gap 36%), "Skill to detect manipulative content" (gap 31%), and "Skill to identify fake information" (gap 28%). We identified problems with understanding the essence of the concepts of fact-judgement, manipulation, fake-not-fake. A superficial, unconscious attitude to these concepts was partly the cause of such a gap. We note that getting acquainted with the results and demonstrating the gap between self-assessment and actual indicators made students think and more reflexively approach the assessment of their own skills. In turn, after explanations, demonstration of comparative results, and conducting a certain "reading", the results of the repeated testing of "Self-assessment 2" allowed students to understand the essence of skills and lower their ambitious self-estimates.

5. Despite the increase in the percentage of people who are aware of the possibilities of media manipulation, the self-esteem t of media literacy differs significantly among young people with its actual results (UNICEF, 2023). The estimates in our study are generally similar and testify to the similarity of behavioural patterns, worldviews, and practical skills of young people to information hygiene both on a national scale and in higher education institutions of a separate local region.
6. There has been an increase in the experience of encountering cyber threats, including the hacking of accounts on social networks (from 28% in 2021 to 41% in 2023) and the hacking of gaming accounts in computer games (from 19% to 38%). 29% of young people have a general understanding of the concept of "Digital Security" in 2023 compared to 18% in 2021. Self-reported knowledge and application of cyber hygiene rules was recorded in 63% of young people (2023) versus 33% in 2021. The actual data of the test of skills in the application of cyber security rules in 2023 showed that 49% of young people adhere to the rules of using two-factor authentication; able to distinguish all types of cyber threats 34% (CRDF Global, 2021). This finding correlates with our result on "Skill to protect personal data". Two-factor authentication, complex passwords, the use of a fingerprint and other methods of protection are used by young people confidently, which is why they are rated highly, but adequately. This is the only indicator for which the statistical analysis of the results did not reveal a statistically significant gap.

In general, according to the results of analytical reports presented in official sources, the situation shows that young generation, who grew up during the time of rapid development of digital technologies, uses the Internet almost 100% every day, has a diversified consumption of media mostly from social networks, and confidently possesses IT skills of any complexity, quickly learns and relearns, self-confident in own information hygiene, in particular, in digital literacy, media competence, and cyber security skills. At the same time, self-assessment does not coincide with the actual results of measurements of many indicators (the ability to distinguish a fake from the truth, to identify real cyber threats), the difference between which in 2022 is 40–60% on average. A self-confident attitude to the digital space gives rise to an almost defiant behaviour in the information space, which manifests itself in ignoring informational dangers, unlimited consumption of information of dubious quality, lack of critical analysis and evaluation, trusting attitude to manipulative techniques and fake media content.

The conducted research is unanimous in its conclusions regarding the insufficient level of information hygiene skills: against the background of increased information threats and an increase in the time spent in the information space, imaginary and self-confident observance of information hygiene skills is observed among young people. Observation of the contingent of students showed that the majority of them do not always distinguish a fact from judgement, tend to believe misinformation and spread fakes; do not know how and do not have the habit of checking information for reliability; have superficial ideas about the operation of search algorithms, the functioning of social networks; characterised by insufficient self-control and psychological stability during their stay in the information space; have a low threshold of awareness of the importance of personal data protection; may unknowingly indulge in criminal activities and share personal data without realising the possible consequences.

The consequences of such a gap are almost defiant behaviour in the cyber environment, which is manifested in ignoring information dangers, irresponsible consumption of information of dubious quality, dissemination and further dissemination of false data, biased assessment of facts, trusting attitude to manipulative techniques, and others.

We consider information hygiene as an interdisciplinary science at the intersection of informatics, psychology, information security, hygiene, ethics, physiology, biology, mathematics, physical education, and other sciences, therefore the content of training should be implemented comprehensively. To overcome the gap between self-assessment and the actual assessment of the level of development of information hygiene skills, we propose the integration of information hygiene elements into general education and professional disciplines; the integration of information hygiene courses into the curriculum (the content of such a special course is proposed by the authors in (Rudenko et al., 2022)) at the university level; creating campaigns to raise awareness among young people about the importance of critical thinking as extracurricular and research activities (the club of fact-checkers, project-research activities, creative studio, training); as independent work; as additional informal education (the analysis of open educational courses was carried out in (Rudenko et al., 2023)); through the implementation of media practices and media traditions; through the popularization of physical and mental health skills in the digital space.

5.5 Conclusions

The modern world is full of information threats (disinformation and fake news, cyberbullying, phishing and online fraud, hacker attacks, and privacy violations). Therefore, young people who face such threats every day should develop their own information hygiene skills and be aware of information threats, be able to effectively interact with information in cyberspace, protecting themselves from information influences and their data from potential threats.

Observation of the contingent of students showed that the majority of them do not always distinguish facts from judgements, tend to believe misinformation and spread fakes; do not know how and do not have the habit of checking information for reliability; have superficial ideas about the operation of search algorithms, the functioning of social networks; characterized by insufficient self-control and psychological stability during their stay in the information space; have a low threshold of awareness of the importance of personal data protection; may unknowingly indulge in criminal activities and share personal data without realizing the possible consequences.

The conducted research proved the existence of a problem of a gap between self-assessment and the actual assessment of the level of information hygiene skills. The statistical analysis of the results revealed the statistical significance of the gap for the indicators "Skill to distinguish fact from judgement", "Skill to detect manipulative content", "Skill to identify fake information", "Skill to check the reliability and authority of website information sources", "Skill for managing time spent online" except "Skill for protecting personal data". After analysing the test results, we again surveyed students' self-assessments. A statistically significant trend towards a decrease in students' self-assessment was recorded after the analysis of the test results and the explanation of the main errors. We believe that this allowed students to realize their own gaps in knowledge and skills and lower their ambitious self-assessment.

We believe that the reasons for the recorded gap between self-assessment and the actual assessment of the level of information hygiene skills are gullibility characteristic of the younger generation, lack of life experience, orientation towards unreliable sources of information, an insufficiently formed value outlook, exaggeration of one's own capabilities regarding the use of digital technologies and social networks, certain educational gaps in the field of information hygiene, exaggerated expectations of society from young people in the field of media and information technologies, use of digital technologies and media only for entertainment. Such characteristics create the foundation for the formation of prejudices, unfounded and often harmful stereotypes, and hostile narratives in the minds of young men and women. As a result, students can easily be manipulated, believe misinformation, spread fakes, unknowingly indulge in criminal activities, and share personal data without realizing the possible consequences.

References

Ackerman, P. L., & Wolman, S. D. (2007). Determinants and validity of self-estimates of abilities and self-concept measures. *Journal of Experimental Psychology: Applied, 13*(2), 57–78. https://doi.org/10.1037/1076-898X.13.2.57

Atikuzzaman, M., & Zabed Ahmed, S. M. (2023). Information literacy self-efficacy scale: Validating the translated version of the scale for use among Bangla-speaking population. *The Journal of Academic Librarianship, 49*(1), 102623. https://doi.org/10.1016/j.acalib.2022.102623

Bandura, A. (1994). *Self-efficacy*. Wiley.
Bandura, A. (2010). Self-efficacy. In I. B. Weiner & W. E. Craighead (Eds.), *The Corsini encyclopedia of psychology* (pp. 1534–1536). Wiley.
Bandura, A., & National Inst of Mental Health. (1986). *Social foundations of thought and action: A social cognitive theory*. Prentice-Hall, Inc.
Boud, D., Lawson, R., & Thompson, D. G. (2013). Does student engagement in self-assessment calibrate their judgement over time? *Assessment & Evaluation in Higher Education, 38*(8), 941–956. https://doi.org/10.1080/02602938.2013.769198
Carretero, S., Vuorikari, R., & Punie, Y. (2017). *DigComp 2.1: The digital competence framework for citizens with eight proficiency levels and examples of use*. Publications Office of the European Union. https://doi.org/10.2760/38842
Cassidy, S., & Eachus, P. (2002). Developing the computer user self-efficacy (Cuse) scale: Investigating the relationship between computer self-efficacy, gender and experience with computers. *Journal of Educational Computing Research, 26*(2), 133–153. https://doi.org/10.2190/JGJR-0KVL-HRF7-GCNV
Chang, C., Liu, E. Z., Lee, C., Chen, N., Hu, D., & Lin, C. (2011). Developing and validating a media literacy self-evaluation scale (MLSS) for elementary school students. *The Turkish Online Journal of Educational Technology, 10*(2), 63–71.
Cote, T. J., & Milliner, B. (2017). Preparing Japanese students' digital literacy for study abroad: Is more training needed? *The JALT Call Journal, 13*(3), 187–197. https://doi.org/10.29140/jaltcall.v13n3.218
CRDF Global. (2021). *Baseline research report on target audiences awareness of key aspects of cybersecurity*. Retrieved from http://surl.li/mrsldl
Dodel, M., & Mesch, G. (2018). Inequality in digital skills and the adoption of online safety behaviors. *Information, Communication & Society, 21*(5), 712–728. https://doi.org/10.1080/1369118X.2018.1428652
Dostal, J., Wang, X., Steingartner, W., & Nuangchalerm, P. (2017). Digital intelligence – New concept in context of future school of education. In *Proceedings of ICERI2017 conference 16th–18th Nov. 2017, Seville, Spain*. Retrieved from https://ssrn.com/abstract=3255366
Drushlyak, M. G., Semenog, O. M., Hrona, N. V., Ponomarenko, N. P., & Semenikhina, O. V. (2022). Typology of Internet resources for the development of youth's Infomedia literacy. *Information Technologies and Learning Tools, 88*(2), 1–22. https://doi.org/10.33407/itlt.v88i2.4786
Dubrovyk-Rokhova, A. (2017). What do you need to know about "fitting", "vishing" and "cache trapping"? Day. 18–19. 11. Retrieved from http://surl.li/bqzvee
Eremin, A. L. (1995). On the issue of developing a new direction – Information ecology. In *Abstracts of the 1st international conference* (pp. 238–239). MANEB Center. Retrieved from http://surl.li/ewjrqx
Eryomin, A. L. (2020). Information hygiene: Modern approaches to hygienic evaluation of content and physical signals of information carriers. *Hygiene and Sanitation, 99*(4), 351–355. https://doi.org/10.33029/0016-9900-2020-99-4-351-355
Eva, K. W., & Regehr, G. (2008). "I'll never play professional football" and other fallacies of self-assessment. *The Journal of Continuing Education in the Health Professions, 28*, 14–19. https://doi.org/10.1002/chp.150
Freund, P. A., & Kasten, N. (2012). How smart do you think you are? A meta-analysis on the validity of self-estimates of cognitive ability. *Psychological Bulletin, 138*(2), 296–321. https://doi.org/10.1037/a0026556
Grimes, D. R. (2020). Health disinformation & social media. The crucial role of information hygiene in mitigating conspiracy theory and infodemics. *EMBO REPORTS, 21*(11), 51819. https://doi.org/10.15252/embr.202051819
Halamendyk, V. B. (2008). Information hygiene as a factor in maintaining mental health of a person. *Humanitarian Bulletin of Zaporizhzhya State Engineering Academy, 35*, 83–91. Retrieved from http://surl.li/kholwl

Inazu, I., & Soyemi, O. D. (2022). Self-assessment of information literacy skills by students of federal universities in South-West, Nigeria. *Jewel Journal of Librarianship, 17*(1), 13–25. Retrieved from http://surl.li/seqied

Jones, S.-L., & Procter, R. (2023). Young peoples' perceptions of digital, media and information literacies across Europe: Gender differences, and the gaps between attitudes and abilities. *Technology, Pedagogy and Education, 32*(4), 435–456. https://doi.org/10.1080/1475939X.2023.2210152

Kilimnik, I. (2023). Information society and information security. New challenges and new ways to overcome information threats. *Uzhhorod National University Herald. Series: Law, 2*, 53–57. https://doi.org/10.24144/2307-3322.2022.76.2.8

Kolot, A. M., & Herasymenko, O. A. (2020). Generation Z and socio-labor platform "Work 4.0": Interaction imperatives. *Demography and Social Economy, 2*(40), 103–138. https://doi.org/10.15407/dse2020.02.103

Koval, Z. (2017). Introduction of complex mechanisms of information hygiene as a means of ensuring the information and psychological stability of a person, society, and the state. *Actual problems of state administration, 2*(70), 62–68. http://lib.rada.gov.ua/static/about/text/Zbirnuk-2%2870%29-2017.pdf

Kurbanoglu, S., Akkoyunlu, B., & Umay, A. (2006). Developing the information literacy self-efficacy scale. *Journal of Documentation, 62*(6), 730–743. https://doi.org/10.1108/00220410610714949

Kyiianytsia, O., Holik, O., Kurban, D., Yatsiuk, Y., & Shkurov, D. (2023). Information hygiene in the media space as a factor of environmental consciousness of society. *Ad Alta: Journal of Interdisciplinary Research, 13*(2), 73–79. Retrieved from http://surl.li/xnyywx

Leach, L. (2012). Optional self-assessment: Some tensions and dilemmas. *Assessment & Evaluation in Higher Education, 37*(2), 137–147.

Lew, M. D., Alwis, W. A. M., & Schmidt, H. G. (2010). Accuracy of students' selfassessment and their beliefs about its utility. *Assessment & Evaluation in Higher Education, 35*(2), 135–156.

Loukas, G., Murugesan, S., & Andriole, S. J. (2022). Information hygiene: The fight against the misinformation "Infodemic". *IT Professional, 24*(2), 16–18. https://doi.org/10.1109/MITP.2022.3163007

Mahmood, K. (2017). Reliability and validity of self-efficacy scales assessing students' information literacy skills: A systematic review. *The Electronic Library, 35*(5), 1035–1051.

Mednikarov, B., Tsonev, Y., Nikolov, B., & Lazarov, A. (2023). Characteristics and Components of the Cyber Hygiene as a Subclass of Cyber Security in Military Environment and Educational Issues. *Strategies for Policy in Science and Education, 31*(2), 154–169. https://doi.org/10.53656/str2023-2-3-cha

Melnychuk, S. K. (2017). Psychological features of the formation of types of self-confidence in youth. *Science and Education, 9*, 42–47. https://doi.org/10.24195/2414-4665-2017-9-7

Moroz, O. (2023). Research on the level of infohygiene of Ukrainian Facebook users (4th wave, January 2023). Retrieved from http://surl.li/hftggu

Nebot, M. Á. L., Cosentino, V. V., Esteve-Mon, F. M., & Segura, J. A. (2021). Diagnostic and educational self-assessment of the digital competence of university teachers. *Nordic Journal of Digital Literacy, 16*(3–4), 115–131. https://doi.org/10.18261/issn.1891-943x-2021-03-04-03

Neha, S., Kevin, B., & Rani, T. (2021). Technological perspectives of education for Generation Z and Generation Alpha. *Utkal Historical Research Journal, 34*(III), 153. https://ssrn.com/abstract=4416901

Nicholas, G. (2010). *The shallows: What the Internet is doing to our brains*. W. W. Norton & Company.

Nieminen, J. H., & Boud, D. (2024). Placing authenticity at the heart of student self-assessment: an integrative review. *Teaching in Higher Education*, 1–23. https://doi.org/10.1080/13562517.2024.2367669

Olsen, R. K., Solvoll, M. K., & Futsæter, K.-A. (2022). Gatekeepers as safekeepers-mapping audiences' attitudes towards news media's editorial oversight functions during the COVID-19 crisis. *Journalism and Media, 3*, 182–197. https://doi.org/10.3390/journalmedia3010014

Ostroga, M., Yurchenko, A., & Korovai, A. (2023). *Information hygiene and information noise* (p. 22). Academic Visions. https://doi.org/10.5281/zenodo.8252019

Osyukhina, M. (2016). Ways of developing media information literacy in Ukraine (on the example of the School of Journalism of Sumy State University). In *XII International science and practice conference "Journalism education in Sumy Oblast: Global professional standards"* (pp. 12–15). Sumy State University.

Ovcharuk, O., & Ivaniuk, I. (2021). A self-assessment tool of the level of digital competence of Ukrainian teachers in the context of lifelong learning: The results of an online survey 2021. *Educational Dimension, 5*, 75–88. https://doi.org/10.31812/educdim.4719

Panadero, E., Brown, G. T., & Strijbos, J. W. (2016). The future of student self-assessment: A review of known unknowns and potential directions. *Educational Psychology Review, 28*, 803–830. https://doi.org/10.1007/s10648-015-9350-2

Pandian, A., Baboo, S. B., & Yi, L. J. (2020). Reading the 21st century world: Self-assessment of digital media literacy among secondary school students in Malaysia. *International Journal of Asian Social Science, 10*(7), 350–359. https://doi.org/10.18488/journal.1.2020.107.350.359

Posnick-Goodwin, S. (2010). *Meet generation Z*. Retrieved from http://surl.li/djyzcg

Prihandoko, L. (2021). Self-assessment on information literacy competency of students in Indonesia-Papua New Guinea border area during the COVID-19 Epidemic. In *Proceedings of the 2nd international conference on progressive education, ICOPE 2020*. https://doi.org/10.4108/eai.16-10-2020.2305235.

Prykhodkina, N. O. (2020). *Media education of students in schools in English-speaking countries: Monograph*. Krok. https://lib.iitta.gov.ua/727838/1

Public Association «Media Detector». (2023). Media Literacy Index of Ukrainians: 2020–2022. Retrieved from http://surl.li/hknga

Redecker, C., & Punie, Y. (2017). *European framework for the digital competence of educators: DigCompEdu*. Publications Office of the European Union. https://doi.org/10.2760/178382

Rosman, T., Mayer, A.-K., & Krampen, G. (2014). Combining self-assessments and achievement tests in information literacy assessment: Empirical results and recommendations for practice. *Assessment & Evaluation in Higher Education, 40*, 740–754. https://doi.org/10.1080/02602938.2014.950554

Rudenko, Yu., Drushlyak, M., Naboka, O., Proshkin, V. & Semenikhina, O. (2022). Special course on information hygiene as a tool for developing youth's ability to resist informational influences. Smyrnova-Trybulska E. (ed.). *E-learning in the Transformation of Education in Digital Society, 14*, 268–287. https://doi.org/10.34916/el.2022.14.20

Rudenko, Y. O., Drushlyak, M. G., Shamonia, V. G., Ostroha, M. M., & Semenikhina, O. V. (2023). Development of student's ability to resist information influences. *Information Technologies and Learning Tools, 94*(2), 54–71. https://doi.org/10.33407/itlt.v94i2.5162

Sieverding, M., & Koch, S. C. (2008). (Self-)evaluation of computer competence: How gender matters. *Computers & Education, 52*(3), 696–701. https://doi.org/10.1016/j.compedu.2008.11.016

Slusky, L., & Parviz-Navin, P. (2012). Student information security practices and awareness. *Journal of Information Privacy and Security, 8*(4), 3–26.

Tsankov, N., & Damyanov, I. (2019). The digital competence of future teachers: Self-assessment in the context of their development. *International Journal of Interactive Mobile Technologies, 13*(12), 4–18. https://doi.org/10.3991/ijim.v13i12.11068

UNESCO. (2020). *Seoul Declaration on Media and Information Literacy for Everyone and by Everyone: a Defence against Disinfodemics*. Retrieved from http://surl.li/nnpaom

UNICEF. (2023). *U-Report*. Retrieved from http://surl.li/ksmlw

USAID. (2023). *Ukrainian media, attitude and trust in 2023*. Retrieved from http://surl.li/guqgc

Usta-Azizova, D., Akhmedova, M., Muhammadiev, N., & Khudoiberdiev, A. (2022). Innovative mechanisms for protecting youth from global information attacks. *International Journal of Early Childhood Special Education (INT-JECSE), 14*(01), 2842–2844. https://doi.org/10.9756/INT-JECSE/V14I1.339

Varynskyi, V., Varynska, A., Kostytsky, M., & Kushakova-Kostytska, N. (2021). Information security and information hygiene on internet media. *Nexo Revista Científica, 34*(01), 120–128. https://doi.org/10.5377/nexo.v34i01.11291

Chapter 6
The Use of Educational Applications in Primary and Lower Secondary School Teaching and Their Impact on the Development of Learners' Key Competences

Magdalena Zavodna, Tatiana Havlaskova, Tomas Javorcik, and Katerina Kostolanyova

Abstract Digital technologies have become an integral part of the educational process, which has fundamentally changed the way we teach. Educational applications offer innovative approaches to learning and have the potential to significantly influence the development of students' key competences. Many of them have become an important tool for self-study or online learning and now also for face-to-face learning, so they need to be given sufficient attention. This paper focuses on the current state of the use of applications in teaching at the lower secondary schools in the Czech Republic, their impact on the development of key competences and presents a possible categorization of educational applications with regard to their use. The research was conducted through a questionnaire survey among lower secondary school teachers across the Czech Republic. Based on the collected data, an analysis of the most well-known teaching applications for lower secondary schools was conducted. The results show that the frequency of use of apps in teaching exceeded 80% regardless of the length of experience of the teachers and the subject of teaching. The most frequently used applications include Kahoot, Google Classroom and Wordwall. Depending on the task, these applications can develop communicative, learning and problem-solving or digital competences.

Keywords Lower secondary school · Teaching · Learning applications · Learners · Teachers · Key competences

M. Zavodna (✉) · T. Havlaskova · T. Javorcik · K. Kostolanyova
University of Ostrava, Faculty of Education, Ostrava, Czech Republic
e-mail: Magdalena.Zavodna@osu.cz; Tatiana.Havlaskova@osu.cz; Tomas.Javorcik@osu.cz; Katerina.Kostolanyova@osu.cz

E. Smyrnova-Trybulska et al. (eds.), *E-Learning and Enhancing Soft Skills*,
https://doi.org/10.1007/978-3-031-82243-8_6

6.1 Introduction

In recent years, with the rapid development of digital technologies and the impact of the Covid-19 pandemic, the ways in which teaching in lower secondary schools are being implemented are changing. Significant innovations include the integration of learning applications into the educational process used initially only in online learning, such as MS Teams and Google Classroom (Zavodna et al., 2024). These applications offer various interactive forms of learning that can assist in the development of students' key competencies. In the current educational environment, where the emphasis is on individual approach and the promotion of creative and critical thinking, learning apps are becoming an indispensable tool.

Instructional apps can support not only the development of subject knowledge but also the development of key competencies or soft skills such as collaboration, critical thinking, problem solving and creativity (Broadbent et al., 2020). While traditional teaching methods are mostly based on the passive reception of information, digital technologies and various learning applications enable students to actively engage in the learning process, thereby increasing their motivation and interest in learning (Deutsch et al., 2022; Nasution & Saragih, 2017).

Although there are a number of studies that focus on technological innovations in education, such as microlearning, mobile-learning or game-based learning, especially in higher education settings, e.g. Yang (2023); Javorcik et al. (2023); Quispe et al. (2018); Sunarto et al. (2020); Osman et al. (2023), there is still little research that specifically explores the use of learning apps in primary or lower secondary schools (such as Nasution and Saragih (2017), or Deutsch et al., 2022), their possible classification and their impact on the development of students' key competences in lower secondary schools. Therefore, this article focuses on the analysis of the use of learning apps in the context of lower secondary schools and their impact on the development of pupils' key competences. It examines how these technologies are used in the classroom. Whether they serve to innovate teaching and what potential they have to support pupils' motivation, knowledge and key competences. The aim is to provide a comprehensive view of current trends in the use of learning applications in primary and lower secondary schools and to offer a classification and recommendations for the effective use of these tools in educational practice.

In order to clarify the terms used in this paper, we need to describe the functioning of the school system in the Czech Republic. The term primary school covers pupils aged 6 to 15 years and the education is divided into first and second grades. The first grade therefore corresponds to the international term primary school, but for the second grade it was necessary to use the international term lower secondary school. Our research was conducted at both grades of primary schools, i.e. primary and lower secondary schools. At the primary school grade, teachers are not differentiated by subject area, so in all tables and graphs they are referred to only by the term first grade. At the lower secondary school grade, teachers' specialization is already differentiated according to their teaching fields.

6.2 Learning Applications

Teaching apps are a modern tool that allows teachers to enrich and diversify their teaching through interactive elements and multimedia content. These applications include a wide range of types and types differing in their focus and way of working. These can include a variety of quiz games and competitions (Kahoot), independent work based on adaptive learning (Duolingo), collaborative platforms (Google Classroom), or just interactive presentations (Nearpod) and infographics (H5P) that facilitate the understanding of complex concepts and promote active learners engagement in the learning process.

In addition to the learning content itself, learning apps can also assist in assessing and monitoring learners progress. Teachers can use these apps to easily collect data on learners' performance, identify areas for improvement and adapt their teaching methods to the needs and level of the learners. Apps can also provide immediate feedback to learners, which is crucial for their motivation and self-management. Some apps can also contribute to the aforementioned development of key competencies or soft skills (Broadbent et al., 2020), such as collaboration and communication between learners (Nasution & Saragih, 2017).

As evidenced by a number of studies, learning apps can lead to greater learner motivation as well as better learning outcomes or understanding of the material (Deutsch et al., 2022; Nasution & Saragih, 2017). For example, in a study by Atin et al. (2022), they concluded that gamification in a mobile-learning app for mathematics increased learners' interest by 35%, motivation by 33%, and understanding of mathematics by 42%. Furthermore, the use of Google apps also leads not only to improved knowledge of the curriculum, but also to better technology skills, development of creative and critical thinking skills, and positive impact on group activities. Google applications were rated as user-friendly and easy to access, which facilitated access to materials and participation in the classroom through a collaborative environment (Osman et al., 2023). Our research focuses on educational applications in general, i.e., not only mobile applications, but also applications used on desktop computers and other technologies.

On the contrary, too many animations and motivational elements can be distracting for learners, so the design of learning apps should help learners focus on the key data and the given tasks. Concentration is a key factor in the use of learning apps, which positively influences the intention to continue using mobile-learning apps, while fun elements had no effect on this (Yang, 2023).

6.3 Key Competences and Soft Skills

The terms 'soft skills' and 'key competences' are closely related, although not identical. Soft skills include personal attributes, communication and interpersonal skills that enable individuals to work and interact effectively with others. These include communication skills, teamwork, empathy, time management, and problem-solving skills (Sadiki & Azalmad, 2024), and emotional intelligence, which includes social awareness, relationship management, and self-management skills (Bradberry & Greaves, 2016). As reported by Ellah and Azmi (2024), in some states, the development of soft skills is not formally incorporated into educational policy and anchored in school curriculum.

However, in the Czech Republic, soft skills are already included in the school curriculum under the term key competences. These represent a broader concept that includes not only soft skills, but also other skills and knowledge that are considered necessary for life in contemporary society (Obst, 2017; Skalková, 1999; Vališová & Kovaříková, 2021). Key competences are precisely defined and defined in the National Framework Education Programme (FEP) issued by the Ministry of Education of the Czech Republic. "Key competences represent a set of knowledge, skills, abilities, attitudes and values important for the personal development and fulfilment of every member of society. Their selection and conception is based on values generally accepted in society and on generally shared ideas about which competences contribute to an individual's education, to a happy and successful life and to strengthening the functions of civil society (RVP, 2023, p.10)." And these competences are the main goal of primary and lower secondary education in the Czech Republic. After the 2021 revision, when a new competence was added to the document based on the European DigComp document, the FEP defines a total of seven key competences:

1. **Learning competences**—The learner chooses appropriate methods and strategies for effective learning, plans and manages his/her own learning, and is willing to be a lifelong learner. Seeks, classifies and uses information effectively for learning, creative activity and practical life. Uses general terms and symbols, integrates knowledge from different areas and develops a comprehensive view of different phenomena. Independently observes, experiments, compares results and draws conclusions for future use. Understands the meaning and purpose of learning, relates positively to it, evaluates their progress, identifies barriers and plans how to improve their learning (RVP, 2023).
2. **Problem solving competences**—The learner recognises and understands problem situations, reflects on their causes and plans solutions using their own judgement and experience. He/she searches for the information needed to solve them, recognizes their similarities and differences, and perseveres to find different solutions despite possible setbacks. Independently solves problems using logical, mathematical, and empirical procedures, verifies the correctness of solutions, and applies best practices to new situations. Thinks critically, makes considered decisions, defends them and evaluates the results of his/her actions with an awareness of responsibility (RVP, 2023).
3. **Communicative competences**—Pupils formulate their ideas and opinions logically and concisely in written and oral expression. Listens to others, understands them, responds appropriately, engages in discussion and argues. Understands, reflects on, responds to and makes creative use of different types of texts, records

and communication media. Uses information and communication technologies to communicate effectively with the outside world. Uses acquired communication skills to build relationships and collaborate with others (RVP, 2023).

4. **Social and personal competences**—The pupil works effectively in a group, co-creates rules of work with teachers and positively influences the quality of work by accepting a new role. Contributes to a pleasant atmosphere and good interpersonal relationships by being considerate and respectful, providing or asking for help when needed. Actively participates in discussions, understands the need for collaboration, values the experience of others and respects different points of view. Builds a positive self-image that promotes self-confidence and self-development and manages his/her actions to achieve self-satisfaction and self-esteem (RVP, 2023).
5. **Civic competences**—The student respects the beliefs of others, appreciates their values, empathizes with the situations of others, rejects oppression, and recognizes the obligation to oppose violence. Understands the principles of laws and social norms and is aware of his/her rights and responsibilities. Makes responsible choices and provides help in crisis situations. Respects and appreciates traditions, cultural and historical heritage, shows a positive attitude towards art and culture, and actively participates in cultural and sporting events. Understands ecological contexts, respects the requirements for a quality environment and makes decisions in favour of health and sustainable development (RVP, 2023).
6. **Working competences**—The learner uses materials, tools and equipment safely and effectively, follows rules, performs duties and adapts to new working conditions. Evaluates work outcomes in terms of quality, functionality, economy, health, environmental and social values. Uses knowledge and experience to develop and prepare for the future, making decisions about further training and career direction. Is oriented in entrepreneurial activities, understands the nature, purpose and risks of entrepreneurship and develops entrepreneurial thinking (RVP, 2023).
7. **Digital competences**—The learner is proficient with digital devices, applications and services, uses them in learning, school and society, and makes independent decisions about which technologies to use. Critically assesses, manages and shares data and digital content, choosing appropriate practices and resources. Creates and edits digital content, combines different formats and expresses themselves using digital means. Uses technology to facilitate work, automate activities, streamline processes and improve outcomes. Understands the importance of digital technologies, learns about new technologies, evaluates the benefits and reflects on the risks. Prevents safety threats and looks after their own and others' health and acts ethically when collaborating and sharing information (RVP, 2023; Vuorikari et al., 2022).

However, there are problems that although teachers are aware of the importance of developing soft skills in learners, or their development is directly part of the school curriculum, they rarely help their learners to develop these skills, for many reasons (Ellah & Azmi, 2024). There are also significant differences in prospective teachers' perceptions of soft skills depending on their major and years of experience (Sadiki & Azalmad, 2024). Teachers need to be imaginative, resourceful, and willing to try new methods in order to sufficiently develop soft skills in learners (Ellah & Azmi, 2024).

The development of soft skills and key competencies is an essential element for personal and professional success in modern society (Ellah & Azmi, 2024; Obst, 2017; Skalková, 1999; Vališová & Kovaříková, 2021). In recent years, learning applications have become an important tool that can contribute significantly to the development of these skills. These apps offer an interactive and engaging environment that can effectively support learning in different areas of knowledge, skills and key competencies. The aim of our research is to present a list of the most commonly used apps in primary and lower secondary schools in the Czech Republic and a proposal for their possible use in the development of selected key competences.

6.4 Methods

The main aim of our research was to find out what applications are currently used by primary and lower secondary school teachers and whether they use them at all. Subsequently, we wanted to examine these apps in detail, to determine their purpose and impact on the development of pupils' key competences, and finally to categorise them based on their use.

For our research, we set the following four research questions:

Q1: Does the length of teaching experience influence teachers' use of apps in the classroom?
Q2: Does teachers' educational field of study influence the use of apps in teaching?
Q3: Which applications do teachers use most often in their teaching?
Q4: How can these applications be categorized?

Based on the first two research questions, we established two hypotheses. The last two research questions are descriptive, so we could not set any hypotheses for them.

H1: Teachers' length of experience has an impact on whether teachers use apps in their teaching.
H2: Teachers' field of training has an effect on whether teachers use apps in teaching.

For our research we used a questionnaire survey. The questionnaire was created and sent electronically to primary school principals with a request to distribute it among the teachers of the school. These schools were selected randomly from each region of the Czech Republic. In total, we received responses from 218 respondents. The research sample consisted of primary school teachers across all subject areas taught, see Fig. 6.1 for the specific composition in terms of percentage by subject area. In terms of length of experience, the range of respondents is wide, consisting of both novice teachers and teachers with over 40 years of experience. Our main requirement

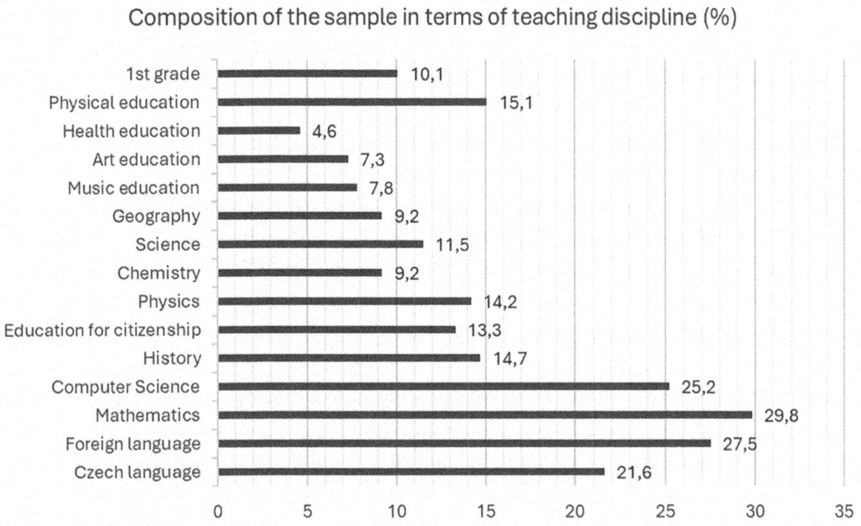

Fig. 6.1 Composition of the research sample in terms of teaching discipline (N = 218). (Source: Own research)

was that the respondents teach pupils aged between 8 and 15, which corresponds to grades 3 to 9 of primary school, according to the Czech education system.

The questionnaire consisted of nine questions. Of these, two questions were open-ended and seven were closed-ended. One question was dichotomous, the others were multiple choice and respondents could indicate more than one answer (e.g. the question on field of study, see Fig. 6.1). Four questions were used to obtain the necessary information about the respondents or their schools needed to compare the data. Five questions were thematically focused according to the aim of our research on the use of apps in teaching.

The research survey was conducted during the school year 2023/2024. PAST4 was used to process the statistical data from the contingency tables.

6.5 Results

The results of the survey show that currently the majority of teachers (87.2%) in primary and lower secondary schools use apps in their teaching, only 12.8% of teachers do not use teaching apps. Of those respondents who do not use apps, 25% gave the reason that they do not know any apps. In the same number (10.7%), respondents gave three different reasons, namely lack of experience, lack of equipment and lack of internet connection. A further 7.1% cited fear of using them as a reason, and the same number also cited fear of a negative impact on pupils. Only

3.6% of respondents did not know where to find them, the same number could not integrate them into their teaching and 21.4% did not give a reason.

We were also interested in whether pupils have access to the internet at school. A total of 94.5% of respondents said that pupils have access to the Internet, while only 3.2% said that pupils do not have access to the Internet. The remaining respondents did not comment on this question.

In Table 6.1, you can see a specific comparison of the length of experience and whether respondents use apps in their teaching. These data have not been converted to percentages for clarity. Based on this contingency table (Table 6.1), a Chi-Square test was conducted (Table 6.2). H0: There is no statistically significant association between the length of experience and the use of apps in teaching. Therefore, our H1 was rejected.

See Table 6.3 for the correlation between the field of study and the use of apps in teaching. Due to the fact that some respondents taught more than one subject, the data presented here are not percentages and the numbers of disciplines exceed the number of respondents. It can be seen from the data that the largest difference between the use and non-use of apps by subject area is in computer science, mathematics, and both native and foreign language. Based on this contingency table (Table 6.3), the Chi-square test was again conducted (Table 6.4). Here again, H0 was accepted: there is no statistically significant relationship between the subject taught and the use of apps in the classroom. Our H2 was therefore rejected.

Table 6.1 Use of applications in teaching in terms of length of experience

length of practice	Yes	No
Up to 2 years	15	1
3 to 6 years	33	4
7 to 10 years	23	1
11 to 19 years	30	6
20 to 29 years	52	5
30 to 39 years	33	9
Over 40 years	4	2
Total	190	28

Source: Own research

Table 6.2 Chi-square test for H1

Chi squared			
Rows, columns:	7,2	Degrees freedom:	6
Chi2:	8,701	p (no assoc.):	0,191
Monte Carlo p:	0,191		
Fisher's exact			
Not available			
Other statistics			
Cramer's V:	0,199	Contingency C:	0,195

Source: Own research

Table 6.3 Use of applications in teaching in terms of teaching discipline

Teaching field	Yes	No
Czech language	42	5
Foreign language	54	6
Mathematics	54	11
Computer science	52	3
History	29	3
Education for citizenship	27	2
Physics	29	2
Chemistry	18	2
Science	24	1
Geography	17	3
Music education	14	3
Art education	14	2
Health education	10	0
Physical education	28	5
1st grade	20	2
Total	432	50

Source: Own research

Table 6.4 Chi-square test for H2

Chi squared			
Rows, columns:	15,2	Degrees freedom:	14
Chi2:	9,974	p (no assoc.):	0,764
Monte Carlo p:	0,767		
Fisher's exact			
Not available			
Other statistics			
Cramer's V:	0,143	Contingency C:	0,142

Source: Own research

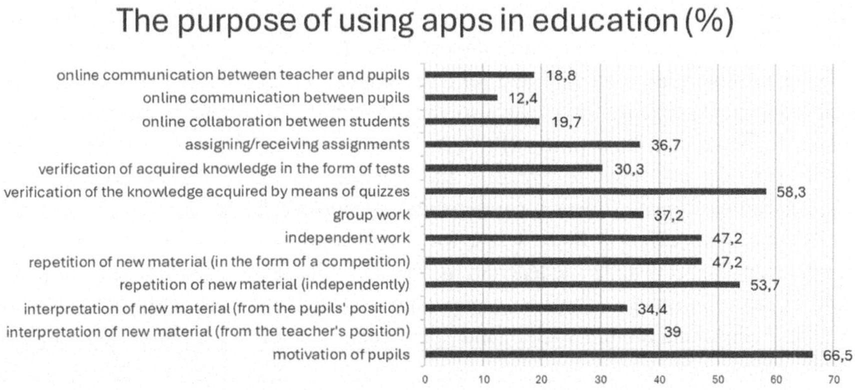

Fig. 6.2 Purpose of applications used in education at primary and lower secondary schools (N = 218). (Source: Own research)

Findings on the purpose of teachers' use of learning apps in their teaching (Fig. 6.2) show that the most common reason for their use is to increase learners' motivation, which is also related to the second most frequently cited reason, namely, to test knowledge through playful quizzes. On the other hand, online communication, either between learners and the teacher, or learners with each other, or their collaboration is not very common.

6.5.1 Educational Applications and Their Categorization

Based on the questionnaire survey, we obtained a wide range of many applications (see below) that teachers use simultaneously in their teaching. The applications utilized in the educational process can be classified in a number of ways. The first method of classification is based on the original purpose of the apps in question. This entails differentiating between those that were created with the specific intention of being used in an educational context and those whose initial purpose did not include such a use. The second possible method of categorization aligns with the primary phases of the learning process, i.e. the objective of the part of the lesson, which include motivation (Mentimeter, YouTube…), exposure (PowerPoint, Nearpod…), fixation (Kahoot, Blooket…), application (Scratch, GeoGebra…), and diagnostic (Google Forms, Umimeto…) phases (Malach, 2003; Skalková, 1999). The third method of categorization is based on the function of the application in an educational context. This results in three fundamental categories, which may be subject to some degree of overlap:

1. **Applications used to organize teaching**

This category includes applications that the teacher can directly integrate into the classroom. This means that these are not applications used for class register, grading and absence reporting. This includes applications such as Google Classroom (Meet), MS Teams, Zoom, Skype, etc. The learners work directly with these applications, so the teacher can use them to organise the teaching. These applications can be used not only for online video calls but also for assignments. Learners can work directly on their own or in groups. This can also include applications that serve as bulletin boards for student collaboration (Padlet), but also bulletin boards that the teacher simply projects and uses to organise the teaching indirectly (ClassroomScreen), e.g. for learners draws, timers for independent/group work, etc.

This category offers a huge scope for the development of many different key competences and it is up to the teacher, his/her formulation of the assignment and the topic of the assignment. The first competences offered here are learning, problem solving, communicative, social and personal competences, but also digital competences.

2. **Applications for the interpretation of the curriculum**

In this category we can include applications that can be used by the teacher in any way to explain new curriculum content. They can vary in their degree of interactivity from regular presentations (PowerPoint) and teaching materials in different formats (Canva, Word, YouTube) to interactive materials (Nearpod, H5P) that allow self-study.

However, it does not have to be only materials created by the teacher who shares them with the learners, but a reverse way of teaching, where learners themselves (independently, in pairs or in groups) search for new information and create these materials. In this way, they are not just passive listeners, but actively acquire the necessary knowledge along with other skills and key competences.

In this case, the development of digital competences would primarily focus on digital content creation and information retrieval. Learners would also develop their learning and problem-solving competences and, in the case of group work, communicative, social and personal competences.

3. **Applications used to practice the curriculum**

These apps have achieved great popularity and spread in education thanks to the Covid-19 pandemic. They are very often used because of their strong motivational effect on learners. In addition, some of them can facilitate the teachers' work in preparing practice tasks, as they already contain ready-made tasks and questions (Umimeto), unlike, for example, Liveworksheet, where shared worksheets can be used, but do not always match the needs of the learners in the class and the teacher's ideas. In the Czech environment, Kahoot has become the most popular application of this type, but other similar applications such as Blooket are now beginning to proliferate.

This type of application encourages learners' competitiveness and leads to the development of communication competences as well as social and personal competences. However, applications such as Scratch or GeoGebra also fall into this category. These, on the other hand, are not based on the principle of quizzes, but also serve to practise the subject matter through independent or group work. For example, if a learner creates a program in Scratch according to the teacher's instructions and practises the principle of 'repeat with condition', he or she will also be developing his or her digital competence or learning and problem-solving competence.

See Table 6.5 for the 20 most commonly used applications, along with their possible categorisation by learning phase. Some of these applications can be used across all disciplines (e.g. Kahoot, Umimeto), while other applications can only be used in one discipline (e.g., GeoGebra, Scratch).

Table 6.5 Most frequently used applications

Application name	Quantity (%)	Category
Kahoot	40,8	3
Google	38,1	1
MS teams	26,1	1
Wordwall	17,4	3
Learning apps	9,6	3
Umimeto	7,8	3
YouTube	4,1	2
Liveworksheets	4,1	3
MS Office	3,7	2
Quizlet	3,7	3
Mentimeter	3,2	2
Quizizz	2,8	3
Nearpod	2,8	2
Blooket	2,3	3
Scratch	2,3	3
Flippity	1,8	3
Socrativ	1,8	3
Geogebra	1,8	3
Vividbooks	1,8	2
Canva	1,4	2

Source: Own research

6.6 Conclusions

It is clear that the use of apps in primary and lower secondary school teaching has become an integral part of modern educational practice. Our research reveals that neither the length of experience of teachers nor their field of teaching has any influence on whether these teachers use educational apps in their teaching. A surprising finding might be that not only novice teachers from the "new modern generation" use teaching apps, but also older and more experienced teachers with many years of experience who did not grow up with digital technologies.

In this digital era, where technology is permeating every aspect of our lives, apps in schools represent a significant step forward. They can not only contribute to the development of subject knowledge and key competences, but also facilitate the work of teachers, and some of them have a significant motivational effect. The data from our research clearly show the relatively large spread of these apps in primary and lower secondary school teaching in the Czech Republic (Table 6.1) and the diversity of their use (Fig. 6.2). We can see that apps are not only used in teaching as a motivational element, but they also become an important tool in explaining the curriculum or organizing learners' work. As evidenced by our previous research (Zavodna et al., 2024), even after the Covid-19 pandemic, MS Teams and Google Classroom applications are still widely used and have become an integral part of some classes, not only in computer science. Their purpose has shifted from online learning to supporting face-to-face teaching and self-study, as mentioned above, collectively their use can be termed as supporting the organisation of learning.

Different apps serve different purposes in the classroom, allowing teachers and learners to use them flexibly according to their specific needs. Apps aimed at explaining the curriculum can provide an interactive and multimedia environment that can increase learners engagement, facilitate understanding of complex topics, and also significantly motivate learners through game elements, as also demonstrated by, e.g., Osman et al. (2023); Yadav and Oyelere (2020); Atin et al. (2022); and Deutsch et al. (2022). An important benefit of these applications is also to support the development of key competences such as digital, communicative, social, personal, learning and problem-solving competences, thus preparing learners for future challenges in their personal and professional lives. However, they can also contribute to the development of creativity, critical thinking, collaboration, competitiveness and motivation to learn. It all depends on the way the teacher integrates these applications into the classroom and the degree of creativity in designing the assignments.

Given these advantages, it is clear that apps are a valuable tool for improving the quality of teaching in primary and lower secondary schools. It is therefore important to continue to integrate and develop them so that they can best serve the educational goals and needs of learners in today's modern age.

References

Atin, S., Syakuran, R. A., & Afrianto, I. (2022). Implementation of gamification in mathematics m-learning application to creating student engagement. *International Journal of Advanced Computer Science and Applications, 13*(7), 542–556.

Bradberry, T., & Greaves, J. (2016). *Emotional Intelligence 2.0*. TalentSmart.

Broadbent, J., Panadero, E., & Fuller-Tyszkiewicz, M. (2020). Effects of mobile-app learning diaries vs online training on specific self-regulated learning components. *Educational Technology Research and Development, 68*(5), 2351–2372.

Deutsch, T., Hinterhölzl, S., & Groher, I. (2022). A game-based learning app for primary school children with German as a second language. In *Proceedings of the 14th International Conference on Computer Supported Education* (pp. 556–561). SCITEPRESS. https://doi.org/10.5220/0011126300003182.

Ellah, H. R., & Azmi, N. (2024). Soft skills in language education: From perception to frequency of instruction. *International Journal of Language and Literary Studies, 6*(2), 533–544.

Javorcik, T., Havlaskova, T., & Kostolanyova, K. (2023). Microlearning in the education of future teachers: Monitoring and evaluating students' activity in a microlearning course. *The Electronic Journal of e-Learning, 21*(1), 13–25.

Malach, J. (2003). *Základy didaktiky*. Ostravská univerzita v Ostravě, Pedagogická fakulta.

Nasution, M. I. P., & Saragih, A. H. (2017). Fun Mobile-based teaching media for primary school. In *Advances in social science, education and humanities research* (pp. 128–131). Atlantis Press. https://doi.org/10.2991/aisteel-17.2017.28

Obst, O. (2017). *Obecná didaktika*. Univerzita Palackého v Olomouci.

Osman, N., Mohd Noor, S. S., Che Hat, N., Mohmad Rouyan, N., & Mat Saad, K. N. (2023). Students' acceptance of flipped learning through Google application. *Online Journal of Communication and Media Technologies, 13*(4), e202360. https://doi.org/10.30935/ojcmt/13786

Quispe, B. M., Casas, L. A., Oviedo, O. A., Quispe, J. H., Chavez, C. V., Pucho, R. V., Choquehuayta, A., & Joaquín, B. D. (2018). Influence of the use of educational apps on smart mobile devices in the academic performance of students. In *The 10th International Conference on Education Technology and Computers* (pp. 341–346). Association for Computing Machinery. https://doi.org/10.1145/3290511.3290541.

Rámcový vzdělávací program pro základní vzdělávání (2023).

Sadiki, H., & Azalmad, N.-E. (2024). An investigation into the soft skills of prospective teachers. *Journal of Language and Linguistics in Society, 4*(3), 43–53.

Skalková, J. (1999). *Obecná didaktika*. ISV nakladatelství.

Sunarto, M. J. D., Hariadi, B., Sagirani, T., Amelia, T., & Lemantara, J. (2020). MoLearn, a web-and android-based learning application as an alternative for teaching-learning process in high schools. *International Journal of Instruction, 13*(1), 53–70.

Vališová, A., & Kovaříková, M. (2021). *Obecná didaktika a její širší pedagogické souvislosti v úkolech a cvičeních*. Grada.

Vuorikari, R., Kluzer, S., & Punie, Y. (2022). *DigComp 2.2: The digital competence framework for citizens – With new examples of knowledge, skills and attitudes*. Publications Office of the European Union.

Yadav, A. K., & Oyelere, S. S. (2020). Contextualized mobile game-based learning application for computing education. In *Education and information technologies* (pp. 2539–2562). Springer. https://doi.org/10.1007/s10639-020-10373-3

Yang, X. (2023). Mobile learning application characteristics and learners' continuance intentions: The role of flow experience. In *Education and information technologies* (pp. 2259–2275). Springer. https://doi.org/10.1007/s10639-023-11910-6

Zavodna, M., Kilianova, K., Kockova, P., & Kostolanyova, K. (2024). Comparing online and face-to-face learning through the eyes of primary and secondary school teachers. In *INTED2024 Proceedings* (pp. 4722–4729). iated.

Chapter 7
Programming in Technically Focused Education

Miroslav Hrubý

Abstract The main goal of the article is to describe the role of programming in the contemporary technical civilization and some consequences for the technically focused education. The author attempts to express his own opinions on the questions such as, what importance the programming teaching and learning bring for the students of technically focused branches, what tools could be used, how effectively distance learning and e-learning could be used and the possible approach to programming exams in the time of availability of AI. The importance of connections between the students' attributes such as, previous education, previous work experience, personality and hobbies, and programming teaching and learning, is discussed. Programming can be seen as a specific field of art. A painter uses colours, a programmer uses data structures and statements which are available in a selected programming language. The article may serve as a starting point for discussions of both conference participants and future readers.

Keywords Programming · Technical education · Teaching · Learning · Exams

7.1 Introduction

Due to the technical development of the civilization digital competences are generally required. Knowledge of any programming language can be an advantage in wider and wider scope of professions. Programming in technical focused education can be recognized as a necessary component of study programs but required outcomes should be carefully selected according to the specifics of every study field. The author works at the military technical faculty where both military and civilian students are educated. The approach to programming teaching, learning and testing used at his work place is discussed.

M. Hrubý (✉)
University of Defence, Brno, Czech Republic
e-mail: miroslav.hruby@unob.cz

E. Smyrnova-Trybulska et al. (eds.), *E-Learning and Enhancing Soft Skills*,
https://doi.org/10.1007/978-3-031-82243-8_7

7.2 Programming Teaching

In most cases the programming language selection is done in accreditation materials. It can be JavaScript, C++, C#, Python, Java, etc. The author means that the basics of programming is an irreplaceable topic for students of all technical focused branches. The teacher's first task is to meet the target group attributes such as

- previous secondary education (multi-year general, 4-year general, technical, military, other)
- previous university (uncompleted) education
- previous work experience (in the field of study, other)
- current programming knowledge and skills from the own point of view
- personal vocational focus
- study and work preferences
- hobbies.

The author can recommend carrying out the relevant questionnaire survey in the first lesson of the semester. Study and work preferences comprise very important data for the teacher. The student can prefer

- national language (Czech) or English
- printed study materials or electronic online sources
- independent study or cooperative study.

The target group attributes should be strictly respected in selected methods of teaching and recommendation of suitable study materials. Programming knowledge is not a priority for all technically focused education but in the contemporary computerized world it is a necessary complement for all technical branches. The teacher must decide how deep inside to go is useful for students of every technical branch. On the other hand, the general structure of programming teaching should contain

- terminology
- data types
- data structures
- constants
- variables
- statements
  - input/output
  - branching
  - cycles
- functions, methods
- graphical user interface.

The core of programming teaching and learning should be solving of problems in a specified high level programming language, practical work on own students' notebooks. Students' approaches which are expressed in their source codes, should be discussed and compared. The teacher should keep in mind that the initial knowledge and skills of individual students can be quite different. There may also be different suitable methods of study for individual students. Some were used to using their memory more from their previous studies, while others preferred logical thinking, structuring knowledge, and derivation. Some people are used to studying independently, others are more comfortable with group work and discussions. Students who are able to successfully study from English study resources have a certain advantage, as most quality study materials are available on the Internet in English. These are not only English manuals, but also videos narrated in English by native speakers, which are sometimes additionally provided with English subtitles, Banas (2024), freeCodeCamp.org (2024), TutorialGenius.com (2024), W3Schools (2024). For studying the C# programming language, it is possible to recommend videos by two Czech authors. David Čápka Hartinger and Petr Voborník make their work freely available on the Internet, Hartinger (2024a, b, c, d, e), Voborník (2024a, b).

For students of technical fields for whom programming is not a key skill, the JavaScript programming language can be recommended for a basic introduction to programming. Its advantage is that students do not need to install anything on their notebooks. The source code can be written in the Notepad application that is part of the Windows operating system. JavaScript source codes can be run in any web browser (Edge, Chrome, Firefox, etc.). These students can use Hrubý (2023a, b).

Programming languages such as C++, C#, Python, Java, etc. should be required by accredited study programs for students whose focus is Information Technologies, Communication Technologies, Robotics, etc. The concrete programming language and the depth of its mastery should be in compliance with accredited materials.

7.3 Programming Teaching at the Technical Faculty of the University of Defence in Brno

University of Defence (UoD), Faculty of Military Technology (FMT), which is the only technically focused faculty of the UoD, has for almost all their military students the subject Information Technology in the Armed Forces (ITAF) in the scope of 56 teaching hours, which is guaranteed by the Department of Informatics and Cyber Operations FMT UoD. The structure of the subject is presented in Table 7.1.

The subject ITAF is included in the first semester of three accredited study programs of the FMT UoD. These are Military Technologies—Electrical, Military Technologies—Mechanical and Military Geography and Meteorology. The history of the subject ITAF is influenced by the results of the students' evaluation of the course, which are available after the end of the semester. Students prefer increasing the number of exercises and laboratory exercises at the expense of lectures.

Table 7.1 Structure of the subject ITAF in the academic year 2023–2024

1.	Lectures Exercises	2 2	Introduction to the subject, basic terms
2.	Lectures	4	Cyber security
3.	Lectures	2	Information systems of the CAF
4.	Lectures	2	Viruses in IT—malware
5.	Lectures	2	Protection of classified information
6.	Lectures	4	Command and control systems
7.	Lectures	2	Introduction to algorithmization
8.	Lectures Exercises	2 2	Data types and data structures
9.	Lectures Exercises	2 2	Basic commands of a higher programming language
10.	Lab. exercises	6	Basic programming techniques
11.	Lectures	4	Introduction to computer networks, history, models of network architecture
12.	Lectures	6	Link and network layer protocols
13.	Lectures	4	Routing, application layer protocols
14.	Lectures Lab. exercises	2 2	Basic configuration of the network with respect to security
15.	Lab. exercises	4	Design and realization of simple networks

Source: Own work

The ITAF course contains 16 lessons of the basics of programming. The JavaScript programming language seems to be a good solution for the programming part of the course.

An assignment for students in programming (Topic 10 in Table 7.1) can be as follows.

Example 1 Generate random integers from the interval $<-9;-1>$ into a vector named snake that has 10 components. Find out and print the indexes of folders that do not contain either of the two numbers entered by the user. Print the data on the web page according to the pattern below.

```
Numbers stored in vector snake:
??
??
...
??

Tested numbers: ?? and ??

Indexes found:
??, ??, ..., ??
```

The solution of the problem in Example 1 can be divided in two files (File 1, File 2) because we want the JavaScript code to be separated. The student can run the snake.html file in any web browser available at his/her notebook.

File 1: snake.html

```
<!DOCTYPE html>
<html>
 <head>
 <title>Example SNAKE</title>
 </head>
 <body>
 <script src="snake.js"></script>
 </body>
</html>
```

File 2: snake.js

```
var snake=[],i,test_num1,test_num2,first=false;
document.write("Numbers stored in vector snake:<br>");
for(i=0;i<10;i++){
 snake[i]=Math.round(Math.random()*8)-9;
 document.write(snake[i]+"<br>");
}
test_num1=Number(prompt("Enter 1. tested number",-5));
test_num2=Number(prompt("Enter 2. tested number",-2));

document.write("<br>Tested numbers: "+test_num1+" and
"+test_num2+"<br><br> Indexes found:<br>");
for(i=0;i<10;i++){
 if((snake[i]!==test_num1)&&(snake[i]!==test_num2)){
 if(!first){
 document.write(i);
 first=true;
 }else{
 document.write(", "+i);
 }
 }
}
```

General programming knowledge and skills from the first semester when the JavaScript programming language is used, are further developed at the FMT UoD according to the accredited study materials of various study modules such as Information Technologies, Military Robotics, etc. For instance, students of the Information Technologies or Military Robotics module develop their programming knowledge and skills in the C# programming language in the sixth semester and later also in the Python language.

7.4 Programming Testing

It can be a difficult question to find out the universal proper rules for programming testing which cover all necessary knowledge and skills required for credits, classified credits or exams. The author means that a suitable approach can contain the elements as follows:

- individual task which is solved by a student during a semester
- written test
- final discussion.

The individual tasks should be formulated according to the student's personal focus, i.e. with respect to the student's field of study, current capabilities, his/her hobbies. The main goal is a systematic work of a student during a couple of weeks. A necessary coordination by a teacher, consulting with schoolmates and friends can be recommended. Students could prepare a few solutions and compare them according to some criterions. Students should defend their written final protocols in front of other students and their teacher.

Written tests should cover all important parts of the subject and be based on short source codes. The students can use only pencil and eraser. For the sake of objectivity no other aids can be permitted. Tested students should express understanding of the two aspects. Firstly, what is contained in variables and how is this content of variables step by step changed. Secondly, what outcomes are visible on the screen. The assignment for students is: *Draw the situation in the memory and display on the computer screen after executing the codes below.*

The concrete example of written test focused on the C# programming language which consists of six tasks, can be prepared as follows:

Task 1

```
int pavel = 13;
while (pavel >= 11)
{
    Console.WriteLine("Pavel: " + pavel);
    pavel--;
}
Console.WriteLine(pavel);
Console.ReadKey();
```

It is a very easy task to start. The student should demonstrate an understanding of the while loop, forming a simple output on the display and gain self-confidence for the next tasks.

Task 2

```
sbyte pat = 0, mat = -2;
Console.WriteLine(pat - mat);
int[] bob = new int[5] { 10, 20, 30, 40, 50 };
for (pat = 1; pat <= 3; pat++)
{
    bob[pat] = mat * pat;
    Console.Write($"{pat}: {bob[pat]},  ");
}
Console.WriteLine();
Console.WriteLine("{0,4}", bob[4] + bob[1]);
Console.ReadKey();
```

The student should demonstrate an understanding of the for loop, work with a vector and form an output on the display.

Task 3

```
int[] s = new int[8] { 1, 2, 2, 3, 1, 3, 2, 2 };
int i, p1 = 0, p2 = 0, p3 = 0;
for (i = 2; i < s.Length; i++)
{
    switch (s[i])
    {
        case 1:
            p1++;
            break;
        case 2:
            p2++;
            break;
        case 3:
            p3++;
            break;
    }
}
Console.WriteLine("{0,2},{1,2},{2,2}", p1, p2, p3);
Console.ReadKey();
```

The student should demonstrate an understanding of the switch statement.

Task 4

```
FileStream ds = new FileStream(@"d:\test.bin",
FileMode.Create, FileAccess.ReadWrite);
BinaryWriter w = new BinaryWriter(ds);
byte b = 3;
w.Write(b);
b = 1;
w.Write(b);
b = 0;
w.Write(b);
w.Write(b);
BinaryReader r = new BinaryReader(ds);
r.BaseStream.Position = 0;
int i;
i = r.ReadInt32();
Console.WriteLine(i);
w.Close();
ds.Close();
Console.ReadKey();
```

The student should demonstrate an understanding of the binary file processing.

Task 5

```
string[,] pole = {
    { "Jan", "Josef", "Zoro" },
    { "Milivoj", "Klara", "Viktor" },
    { "Erik", "Miloš", "Leon" },
    { "Pepino", "Karel", "Petr" }
};
Console.WriteLine(pole.Length);
Console.WriteLine(pole.GetLength(0));
Console.WriteLine(pole.GetLength(1));
int suma = 0;
foreach (string j in pole)
{
    for (int i = 0; i < j.Length; i++)
    {
        if (j[i] == 'a')
        {
            suma++;
            Console.WriteLine("{0}", j);
        }
    }
}
Console.WriteLine(suma);
Console.ReadKey();
```

The student should demonstrate an understanding of the 2D array (string matrix) processing and foreach statement.

Task 6

```
short bingo = 0x_f;
int dingo = ~bingo;
Console.WriteLine(bingo);
Console.WriteLine(dingo);
Console.WriteLine($"{Convert.ToString(bingo, toBase:
2).PadLeft(16, '0')}");
Console.WriteLine("{0}", Convert.ToString(dingo, toBase:
2));
Console.WriteLine("{0}", Convert.ToString(bingo + dingo,
toBase: 2));
Console.WriteLine("{0}", Convert.ToString(bingo + dingo));
Console.ReadKey();
```

The student should demonstrate an understanding of the various input and output forms.

Every written test should be prepared by the teacher as an original. The proper outcomes on the screen which are in Figs. 7.1, 7.2, 7.3, 7.4, 7.5, 7.6 and 7.7, could serve for final discussion between teacher and student.

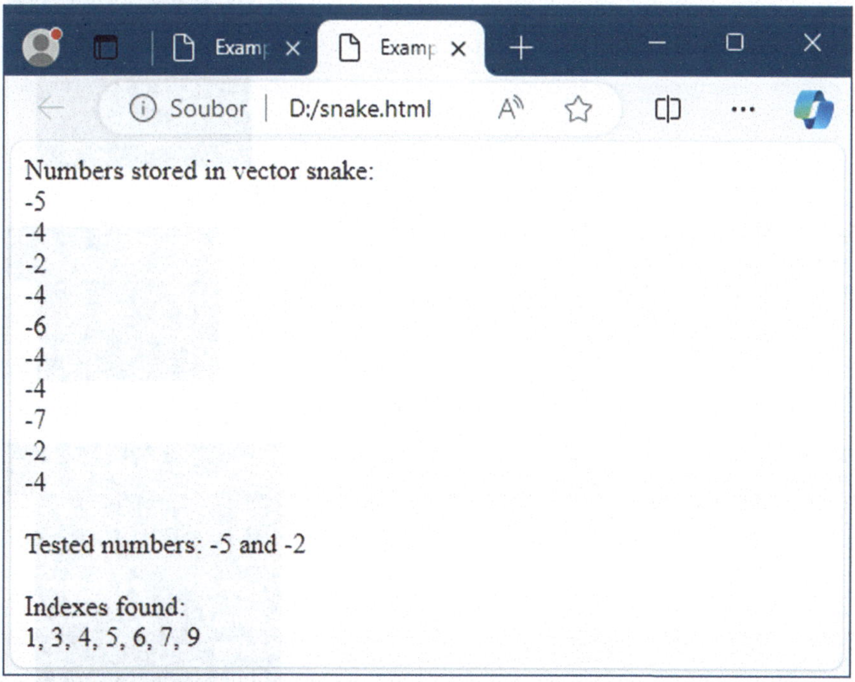

Fig. 7.1 Outcome on the screen of the Example SNAKE which was run in MS Edge web browser. (Source: Own work)

Fig. 7.2 Outcome on the screen of the source code Task 1. (Source: Own work)

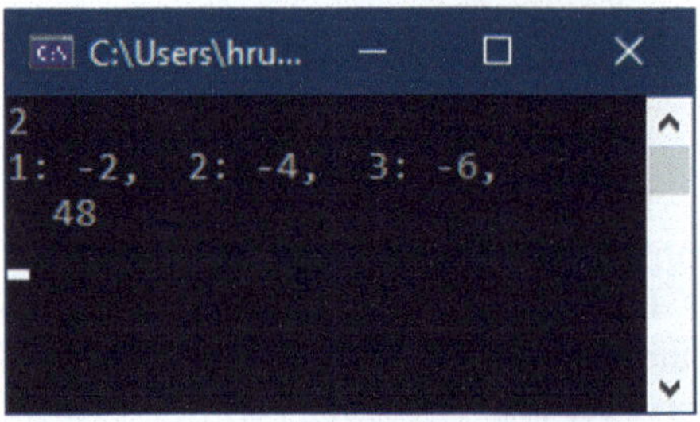

Fig. 7.3 Outcome on the screen of the source code Task 2. (Source: Own work)

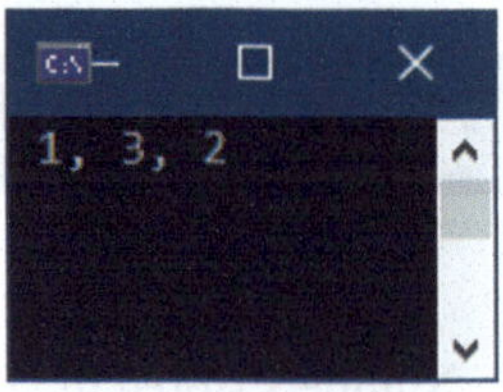

Fig. 7.4 Outcome on the screen of the source code Task 3. (Source: Own work)

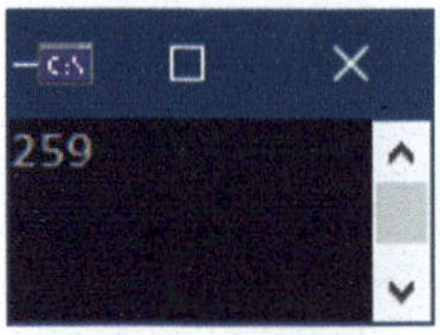

Fig. 7.5 Outcome on the screen of the source code Task 4. (Source: Own work)

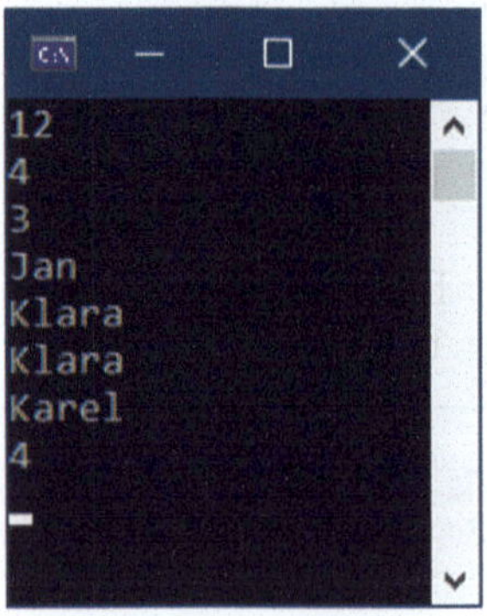

Fig. 7.6 Outcome on the screen of the source code Task 5. (Source: Own work)

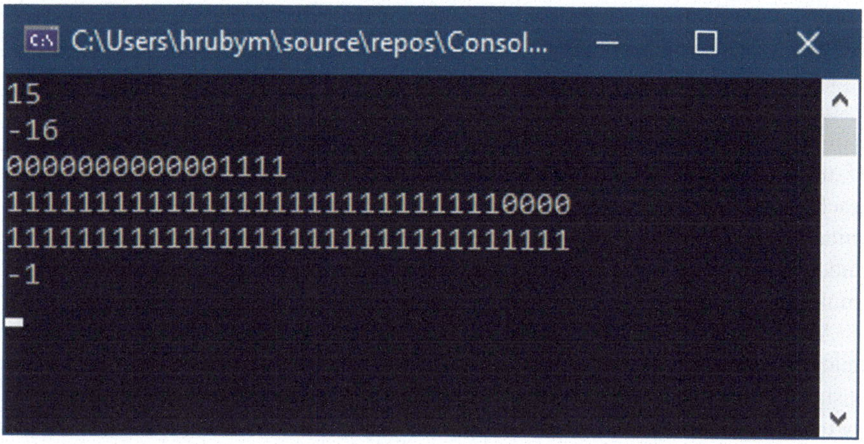

Fig. 7.7 Outcome on the screen of the source code Task 6. (Source: Own work)

Exam results in summer semester 2023–2024 when 6 tasks described above were used, unveiled simply and reliably each student's understanding of key concepts and operations in the C# language.

Final discussion should be especially focused on the individual task, written test, terminology. If the teacher means that a student's exam results are worse than his/her capability, he/she should recommend the student to repeat the exam with promising better assessment.

7.5 Conclusion

Military and civilian technical practice requires similar knowledge and understanding of similar mechanisms in the field of programming. Programming, like mathematics, contributes to the development of students' technical thinking. For students of various technically oriented fields of study, it is currently necessary to become familiar with basics of IT terminology, data types, data structures and commands of a higher programming language. Debugging solutions to at least a few simple problems provides students with an understanding of how to think when writing source code in a higher-level programming language. This experience can significantly help them in their possible future cooperation with IT specialists.

Currently, there is a wide range of study resources for programming that are freely available on the Internet. These sources can be of different quality. The teacher should provide students with an up-to-date overview of these study resources at the introductory lesson of the subject and recommend their use by students from his point of view. The evaluation of the usability of these materials for the fulfillment of study goals can be different from the point of view of the teacher and from the point of view of students of different technical fields and different initial knowledge.

Assessment of not only Internet study sources from the students' point of view can be an interesting topic for research.

E-learning comprises many different approaches and many different study resources. The students should have a choice what way to the required competency will be used in every single case. Recommendation of e-learning resources and usage of them should be the student's right. On the other hand, AI in programming teaching and learning should be a subject of research and testing, searching for suitable solutions. Programming is also about a style of thinking which should be under individual development. AI should help the development but not to force to think the only way.

A very debatable problem is the verification of students' academic results in the field of programming. If the test was carried out on a computer, it is very difficult to ensure correctness. One of the possible solutions is an individual assignment solved by the student during the semester and its defense, a written exam without the use of any electronic devices and a final interview between the examiner and the student.

Acknowledgments The work presented in this paper has been supported by the Projects for the Development of the Organization: *Cyber forces and assets.*

References

Banas, D. (2024). *C# Tutorial.* https://www.youtube.com/watch?v=0p0JLFZj2C8

freeCodeCamp.org. (2024). *C# Programming for Absolute Beginners | FREE 12 hour course.* https://www.youtube.com/watch?v=ALXpVlPYqYE

Hartinger, D. Č. (2024a). *Základní konstrukce jazyka C# .NET – Online kurz.* https://www.itnetwork.cz/csharp/zaklady

Hartinger, D. Č. (2024b). *Objektově orientované programování v C# .NET – Online kurz.* https://www.itnetwork.cz/csharp/oop

Hartinger, D. Č. (2024c). *Soubory v C# .NET – Online kurz.* https://www.itnetwork.cz/csharp/soubory-a-sit

Hartinger, D. Č. (2024d). *WPF – Okenní aplikace v C# .NET – Online kurz.* https://www.itnetwork.cz/csharp/wpf

Hartinger, D. Č. (2024e). *Windows Forms – Okenní aplikace v C# .NET – Online kurz.* https://www.itnetwork.cz/csharp/winforms

Hrubý, M. (2023a). *Information technology. Algorithmization and programming* (study texts). University of Defence, 100 pp. ISBN 978-80-7582-273-4.

Hrubý, M. (2023b). *Information technology. A collection of solved "study" examples from the JavaScript language* (study texts). University of Defence, 102 pp. ISBN 978-80-7582-272-7.

TutorialGenius.com. (2024). *C# full course: C# tutorial for beginners.* https://www.youtube.com/watch?v=ALXpVlPYqYE

Voborník, P. (2024a). *Základy programování v C#.* https://www.youtube.com/playlist?list=PLxTqV9i8bnb8tanSMrno74A4tvOF7eIqT

Voborník, P. (2024b). *OOP v C#.* https://www.youtube.com/watch?v=fubIDH9v4ZA&list=PLxTqV9i8bnb83MI7zkib02OoJKnKyUH7V

W3Schools. (2024). *C# Tutorial.* https://www.w3schools.com/cs/index.php

Chapter 8
Examining M-Learning Trends Among University Students—Mobile Devices in Students' Study and Life

Eugenia Smyrnova-Trybulska, Małgorzata Przybyła-Kasperek, Marcin Gierczyk, and Dagmara Dobosz

Abstract The study investigates the use of mobile devices for educational purposes among university students, focusing on 51 individuals from computer science and pedagogy specializations at the University of Silesia in Katowice, Poland. The research explores the frequency and purposes of mobile device usage, such as participation in e-courses, MOOCs, and educational apps like Duolingo. It also examines the impact of the COVID-19 pandemic on these practices, identifying key features and perceived drawbacks of mobile learning. Five research questions guide the inquiry, addressing usage frequency, priority purposes, pandemic effects, important m-learning features, and perceived negative aspects. Namely: RQ1: What is the frequency of use of mobile devices (smartphones, tablets, notebooks)? RQ2: What are the highest priority purposes for the use of mobile devices—educational purposes, e.g. to participate in e-courses and MOOCs, to use mobile learning applications, i.e. Duolingo? RQ3: What is the students' self-assessment of how the pandemic has affected the frequency of use and which mobile devices? RQ4: What features of teaching m-learning are most important to students? RQ5: What negative features of teaching m-learning do students perceive? Statistical inference and descriptive analysis are employed to test hypotheses and compare results. The

The original version of the chapter has been revised. A correction to this chapter can be found at https://doi.org/10.1007/978-3-031-82243-8_20

E. Smyrnova-Trybulska
Faculty of Arts and Educational Sciences, University of Silesia in Katowice, Cieszyn, Poland
e-mail: esmyrnova@us.edu.pl

M. Przybyła-Kasperek (✉)
Faculty of Science and Technology, University of Silesia in Katowice, Sosnowiec, Poland
e-mail: malgorzata.przybyla-kasperek@us.edu.pl

M. Gierczyk · D. Dobosz
Faculty of Social Science, University of Silesia in Katowice, Katowice, Poland
e-mail: marcin.gierczyk@us.edu.pl, dagmara.dobosz@us.edu.pl

E. Smyrnova-Trybulska et al. (eds.), *E-Learning and Enhancing Soft Skills*,
https://doi.org/10.1007/978-3-031-82243-8_8

findings provide insights that can inform further research across different faculties and international contexts.

Keywords M-learning · Trends · Students · Mobile devices · Questionnaire · Post-hoc Dunn Bonferroni test

8.1 Introduction

The development of technology, its wide availability, and the improvement of internet infrastructure have contributed to the increase in the use of mobile technologies in various spheres of life and led to the adoption of mobile learning (Moca & Badulescu, 2023). This has resulted in an increase in the quality of the learning process and an improvement in academic performance (Chao & Chen, 2023). Mobile technologies have significantly changed the way users search, store, and share information, eliminating time and location constraints (Barfi & Du Preez, 2024). The phenomenon of mobile technology is evident in today's rapidly changing educational environment, where it has been noted that the introduction of mobile technology in universities proves to create a more effective learning environment compared to traditional classroom lectures (Alsswey & Al-Samarraie, 2019). Crompton (2013, p. 3) emphasizes that in order to fully understand the nature and significance of m-learning, it is necessary to view it in the context of a wide range of historical events that have gradually prepared the ground for contemporary approaches to mobile education: "A consideration of all the various historical and cultural events that have led to mobile learning (m-learning) would trace back through history far beyond the invention of Gutenberg's printing press and the influence of the Industrial Revolution".

Also Traxler (2005, p. 262) **provides** an early definition: 'any educational provision where the sole or dominant technologies are handheld or palmtop devices.' However, Alsswey and Al-Samarraie (2019, p. 3653) highlight differences between m-learning and e-learning in the context of their study, defining m-learning as:

> …an extension of e-learning in which learners use their mobile devices to make learning more accessible and more convenient than we are used to in the existing e-learning environments

It is noteworthy that the transition from e-learning to m-learning is considered a key element in the implementation of advanced technologies in higher education (Seliaman & Al-Turki, 2012). However, it is important to note that this integration process may involve overcoming significant barriers that could impede this transition, such as social and cultural aspects, data protection and privacy, licensing and intellectual property issues, lack of strategy and support from institutions, resistance to change, and lack of appropriate equipment (Alsswey & Al-Samarraie, 2019).

Mobile learning enables learning anywhere, anytime, and provides the opportunity to adapt the pace and style of learning to the individual user's needs and preferences. More broadly, it provides the opportunity to empower all sections of society through education and training. However, as Pedro et al. (2018) note, there are also those who question the benefits of using mobile devices in the classroom, mainly because of the as yet unexplored effects resulting from their integration in higher education.

Universities around the world are increasingly paying attention to the potential of m-learning to set new directions for education (Naveed et al., 2023). Research on m-learning in universities not only identifies which devices and apps are popular with students (Kumar & Sharma, 2020), but also explores how these technologies affect learning performance, motivation, and time organization (Moya & Camacho, 2021). Furthermore, the m-learning phenomenon is not limited to lecture theatres; mobile devices are becoming an integral part of students' daily lives, influencing the way they communicate, acquire knowledge, and entertain themselves. After the Covid-19 pandemic, which accelerated the development of m-learning in universities, the use of smartphones for learning has increased significantly. Students and lecturers alike began to recognize the benefits of m-learning, which changed their attitudes towards mobile technologies (Gull et al., 2023). A meta-analysis by Qazi et al. (2024) found that the quality of the e-content, the quality of the platform, internal management at the university, and the behavior of lecturers influence the level of use of m-learning platforms.

Given the growing implementation of m-learning as a new and rapidly developing technology in our educational systems, researchers are becoming increasingly interested in exploring and analyzing its impact on training processes and lifelong learning (Qazi et al., 2024). As technology gains prominence in educational contexts, it is provoking a deeper understanding of how it can transform traditional teaching methods and support the development of lifelong skills and knowledge. For this reason, more and more attention is being paid to researching the potential of m-learning in the context of long-term educational and professional development.

8.2 Literature Review

Students use technology daily and find mobile devices both attractive and easy to use. As these devices are an integral part of their lives, students increasingly expect technology to be integrated into education. The effective use of technology in teaching not only fosters innovation but also facilitates the development of new skills, enabling students to study and exchange information more efficiently. The use of mobile learning (m-learning) is generally evaluated positively by students, with noted benefits including the ability to control individual progress, the blending of formal and informal education, and increased engagement and motivation. Furthermore, students find that completing exercises and participating in discussions via mobile devices are more rewarding compared to traditional classroom instruction (Dias & Victor, 2022). However, Wu et al. (2012) highlight in their meta-analysis that mobile phones and personal digital assistants (PDAs) may be supplanted by emerging technologies such as tablets. The potential of these newer technologies suggests they may become the preferred tools in m-learning, offering users more advanced features and enhanced learning experiences.

A study by Ataş and Çelik (2019) found that students spend an average of 5 h per day using their smartphones and check their devices approximately 28 times daily.

The highest levels of activity occur in the evenings and at night, leading the researchers to suggest that emails or teaching materials should be sent during these times. Smartphones are primarily used by students for leisure activities, particularly when they are bored, lonely, or waiting for someone. The most common uses of smartphones include texting, chatting, browsing social media, and searching for information on the internet.

Students use mobile devices such as phones and tablets because of their convenience, portability, and ease of use. The most frequently cited activities related to the educational use of smartphones include taking notes, reading or downloading learning materials, checking deadlines or grades, and communicating. Nearly one in two students believes that accessing learning materials on a smartphone significantly helps their academic performance. Negative aspects associated with using mobile apps for learning often include problems with internet connectivity, lack of confidence in the reliability of information sources, hardware limitations such as small screen size, distractions, or eye fatigue (Nikolopoulou, 2022).

Individuals who have a strong goal orientation and view learning as central to their objectives are more likely to use smartphones for learning purposes rather than for distracting activities during self-directed learning. The motivation to learn may protect against distracting smartphone use by reducing cognitive load (Fu et al., 2021).

The results of a study by Jeong (2022) indicate that the use of mobile apps in language learning can increase student motivation and make learning more sustainable and enjoyable compared to traditional methods. Participants in the study noted that the main benefits of mobile English learning were easy access to learning materials, tool portability, flexibility and self-learning opportunities, better interaction, and an improved sense of efficiency in learning. Similarly, research by Chen (2022) found that the ease of use of mobile learning (M-learning) apps had a significant impact on the perceived usefulness of these apps. That is, the easier these apps were to use, the more students perceived them as useful and effective learning tools. Furthermore, ease of use had a significant positive impact on students' attitudes and behavior towards using M-learning apps. In practice, this means that students who found these apps convenient, intuitive, and offering a wide range of possibilities were more likely to use them regularly. A study by Shishakly et al. (2023) showed the same relationship. In addition, their findings showed that trust in the app and its quality positively impacted students' attitudes towards the adoption of M-learning apps.

In a study conducted by Nikolopoulou (2022), more than two-thirds of students reported that their academic activities changed after the pandemic. Students began to use their mobile phones more frequently for various activities, leading to better communication with lecturers and peers and greater utilization of available technological tools. As a result, the use of these devices, according to those surveyed, became easier and more efficient.

Research of Gikas and Grant (2013, p. 23) in the context of class-activities such as a lecture, shows: "Traditional college-aged students reported that mobile devices could sometimes be distracting, particularly due to the temptation of social networking applications unrelated to class. Conversely, older students emphasized that they did not find the devices distracting. Additionally, although there isn't overwhelming data to confirm this, some evidence suggests that students blurred the lines between their personal identity and their mobile devices, feeling a strong need to stay constantly near them".

The study reveals that the availability of course-related contents, the role of authority, and the limitations of mobile devices are among the main determinant factors to the adoption of mobile learning in high schools (Hamzah & Muchlis, 2018). The research provides an overview of a study seeking to predict adoption attributes for mobile learning (m-learning) using the Theory of Planned Behavior (TPB) framework (Seyal et al., 2017). Data analysis via SPSS demonstrated that out of the three attributes investigated, only attitude remains a significant determinant in adopting m-learning technology.

Experts from different countries in their research based on *Unified Theory of Technology Acceptance (UTAUT)* provided and described different approaches to m-learning. The study (Kuciapski, 2016) investigated how students perceive the use of mobile technologies during the studying process Therefore, proper study was based on the UTAUT to explain determinants impacting students' intention to use mobile devices and software for studying (Kuciapski, 2016). Acceptance of mobile learning by higher educational institutions in Sri Lanka: An UTAUT2 approach analysed by Nawaz and Mohamed (2020). The study aims to investigate the determinants of mobile learning (m-learning) acceptance of Korean university students based on the UTAUT2 developed by Venkatesh et al. (2012) and Kang et al. (2015). This research aimed at examining factors influencing college students to adopt mobile learning (m-learning) services. An integrated model combining the information systems success (ISS) and UTAUT2, was developed to identify m-learning determinants (Tarhini et al., 2024). Bharati and Srikanth (2018) published their research on a modified UTAUT2 model for m-learning among students in India.

In this study Mobile learning readiness of the students was at moderate level. Also the results indicated positive and significant effect of attitude and behavioral control on the intention of students to accept m-learning (Azizi & Khatony, 2019). The TPB-based model was a suitable model for identifying psychological factors that affect the intention of students of medical sciences to adopt m-leaning. In order to increase the students' acceptance of mobile learning, we suggest that other psychological, behavioral, social, and cultural factors affecting the acceptance of m-learning should be identified (Azizi & Khatony, 2019).

Stepwise regression results indicate that performance expectancy, social influence, facilitating conditions, hedonic motivation and habit significantly affect behavioral intention to use m-learning among Korean university students (Kang et al., 2015). This study identifies and analyzes the factors that influence students' attitude and readiness towards mobile technology use in education, providing valuable insights into improving the adoption of new technologies and to evaluate the sustainability of m-learning in economic education (Moca & Badulescu, 2023).

While price value (PV), effort expectancy (EE), performance expectancy (PE), and privacy (PR) were not significant predictors of BI, the results of the neural network model ranked the predictive power of the factors in the following order: information quality, habit (HB), system quality (SYQ), hedonic motivation (HM), facilitating condition (FC), and social influence (SI), positively influenced m-learning adoption (Tarhini et al., 2024). In the research (Qamar et al., 2024) the data were quantitatively analysed using reliability analysis, confirmatory factor analysis, and multiple regression analysis. The results show that out of three first-order m-learning variables, only two (system and service quality items) have a positive impact on students' learning satisfaction in the Indian context (Qamar et al., 2024).

Research revealed electronic learning (e-learning) as the most common validated mode of delivery, followed by mobile learning (m-learning), learning management system (LMS), personal learning environment (PLE), and massive open online course (MOOC), along with different supportive facilitating technologies used in education such as social media platforms, teaching assistant robots, simulators, as well as virtual reality (VR) and augmented reality (AR) technologies (Granić, 2023; Chromiński et al., 2023).

A total of 200 higher education students voluntarily participated in the research (Alowayr, 2022). The research outcomes are significant for educational institutions and teachers to implement m-learning effectively. Many recommendations can be suggested to help enhance learners' willingness to adopt m-learning technology: Teachers need to design their courses in an interactive way and provide several different activities to ensure that learners obtain real benefits in their learning outcomes. When learners have positive perceptions about the use of this technology in the learning process, such perceptions should be supported by providing beneficial courses (Alowayr, 2022).

An Empirical Study on Factors Affecting Mobile Learning Acceptance in Higher Education were conducted by Abdallah et al. (2021). Research on Mobile learning acceptance among secondary school students in Indonesia presented in their own research Pratama (2021). Al-Shihi, Sharma, and Sarrab analysed a Neural network approach to predict mobile learning acceptance (2018). M-learning as a next generation in the context of the hotel training system presented in their research (Kim & Kizildag, 2011).

Heard (2020) conducted an empirical analysis of mobile learning acceptance in Puerto Rico's higher learning institutions. An integrated model on an empirical examination of continuous intention to use m-learning was described by Al-Emran et al. (2020a). Exploring students' awareness and perceptions in the context of influencing factors and individual differences driving m-learning adoption was presented by Sabah (2016). A sequential mediation analysis and the moderating role of personal innovativeness in information technology in the area of adoption of m-learning apps provided by Chen (2022). Aloqaily et al. (2019) presented a study on a neural network analytical model for predicting determinants of mobile learning acceptance. The potentials and dangers of mobile gaming for language learning exams were discussed by Robison (2012) and Naveed et al. (2020) explored and described an empirical study on a structural equation modeling for mobile learning acceptance by university students.

Towards a conceptual model for examining the impact of knowledge management factors on mobile learning acceptance was presented by Al-Emran et al. (2020b). Baghcheghi et al. (2020) analysed factors affecting mobile learning adoption in healthcare professional students based on technology acceptance model. Research on behavioral intention towards mobile learning in South Asia countries as Taiwan, China, Indonesia, and Vietnam was? conducted and described Lin et al. (2020). Lisana and Suciadi (2021) analysed and described the Acceptance of Mobile Learning in a case study of 3D simulation Android app for learning physics.

Sophea et al. (2022) explored factors influencing students' behavioral intention on using mobile learning (M-learning) in the tourism and hospitality major in Phnom Penh, Cambodia. Research on investigating the determinants and age and gender differences in the acceptance of mobile learning was provided by Wang et al. (2009). Darren Pullen et al. (2015) explored different indicators of pre-service teachers' acceptance and use of mobile learning in Malaysia.

8.3 Research Design

The article considers the following research questions:

RQ1: What is the frequency of use of mobile devices (smartphones, tablets, notebooks)?

RQ2: What are the highest priority purposes for the use of mobile devices—educational purposes, e.g. to participate in e-courses and MOOCs, to use mobile learning applications, i.e. Duolingo?

RQ3: What is the students' self-assessment of how the pandemic has affected the frequency of use and which mobile devices?

RQ4: What features of teaching m-learning are most important to students?

RQ5: What negative features of teaching m-learning do students perceive?

The article considers the following research hypotheses:

H1: Smartphones are the most commonly used mobile device tool.

H2: Students most often use mobile devices for entertainment, use in everyday life, use for communication (Messengers WhatsApp, Viber, Telegram) or use to search for information.

H3: Students believe that the pandemic has increased the frequency of mobile device use in general.

H4: The most important features of teaching m-learning to students is accessibility at any time.

H5: The features: limitation of material in mobile courses, inability to perform advanced tasks (for example, limited computing power) and unfilled level of quality of graphics and video is not recognized as negative aspects of teaching via m-learning to students.

8.4 The Results of the Study

This section will discuss the questionnaire that was prepared, as well as analyse the results obtained.

8.4.1 Questionnaire

Our focus was on exploring students' opinion on m-learning trends and mobile devices in their study and life. The survey was con-ducted in June 2024. Students of the University of Silesia in Katowice, Poland two faculties—Faculty of Science and Technology, Faculty of Arts and Educational Science—were asked to respond. They were mainly students of two specializations—Computer Science and Pedagogy. A total of 51 responses were received. Respondents were randomly selected. Emails were sent to students in different years of studies and faculty asking them to fill out the survey. Taking part in the survey was voluntary. The questionnaire was anonymous and the students agreed to use their responses for research.

In order to understand the structure and characteristics of the students who took part in the questionnaire, the first part of the questionnaire included sociological questions. The metric included questions on the respondent's field of study, gender, age, year of study and place of residence. A description of students who took part in the questionnaire in relation to the above-mentioned characteristics is presented in Table 8.1.

As can be seen, the respondents are mostly male, aged 19–22. Thus, first, second or third year university students live either in a large city or in a village.

8.4.2 Frequency of the Use of Mobile Devices (Smartphones, Tablets, Notebooks)

Now we proceed to the analysis of the first research question and the first hypothesis on the frequency of use of mobile devices. The main questions in the questionnaire that addressed this issue were as follows:

Table 8.1 Descriptive statistics on responses to sociological questions

Age	19–22	23–25	26–35		
Quantity/percentage	32/63%	11/22%	8/15%		
Year of study	1st	2nd	3rd	4th	5th
Quantity/percentage	6/12%	23/45%	5/10%	16/31%	1/2%
Gender	Female	Male			
Quantity/percentage	15/29%	36/71%			
Place of residence	Village	Small city	Medium city	Large city	
Quantity/percentage	20/39%	3/6%	10/20%	18/35%	

Source: Own research

Q1. Please specify how many hours per day do you spend using mobile devices?

- Smartphone
- Tablet
- Notebook

The students' responses to the above question are presented in Fig. 8.1. The data shows that tablets are the least used device among students, with 45 students reporting they use a tablet for less than an hour a day. Smartphones are the most frequently used, with many students spending more than 2, 3, 4, or even 5 h per day on these devices. A significant number of students rarely use notebooks, although a small group (6 students) use them extensively, exceeding 5 h of use per day.

The analysis of the first research question focused on the frequency of mobile device use, with the main goal of comparing usage patterns for smartphones, tablets, and notebooks. To do this, we collected responses from 51 students for each of the three devices, creating three dependent samples (smartphone, tablet, and notebook usage). Given that the measurement scale used in the questionnaire was ordinal—reflecting ordered categories rather than continuous numerical data—we selected non-parametric statistical methods for the analysis. The Friedman test is a non-parametric alternative to the repeated measures ANOVA. It is designed to test for differences in the median values across multiple related groups when the data does not meet the assumptions of normality and homogeneity of variances that parametric tests like ANOVA require.

Each student provided responses for all three devices, so the data is paired (dependent samples). The Friedman test is designed to handle this type of data by considering the ranks of each student's response for smartphone, tablet, and notebook usage. Our null hypothesis (H0) was that there was no statistically significant

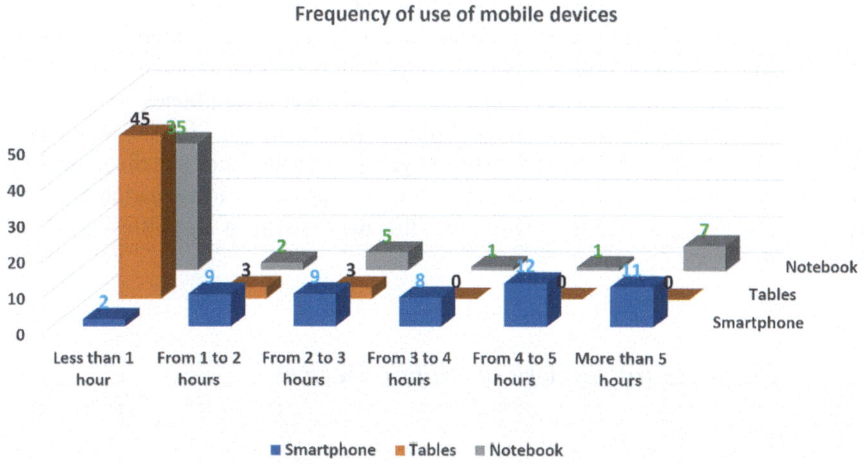

Fig. 8.1 Report on answers obtained for question Q1. (Source: Own research)

Table 8.2 p-value of the post-hoc Dunn Bonferroni test for frequency of use smartphon, tablet, notebook

	Smartphone	Tablet	Notebook
Smartphone		0.000001	0.000008
Tablet	0.000001		
Notebook	0.000008	0.08815	

Source: Own research

difference in the median usage across the three devices. The Friedman test allowed us to evaluate whether the observed differences were likely due to chance.

The test confirmed a significant difference (we can reject the null hypothesis) in medians, the results obtained: $H(2,51) = 62.696$; p-value $= 0.000001$.

After performing the Friedman test, which indicated that there were significant differences in median usage across smartphones, tablets, and notebooks, we needed to determine which pairs of devices were responsible for these differences. The Friedman test itself only tells us that at least one group differs but does not specify where the differences lie. To address this, we applied the Dunn-Bonferroni post-hoc test to perform pairwise comparisons between the three devices.

A post-hoc test (Dunn-Bonferroni) was then performed, the results are shown in Table 8.2. The test found no significant median difference between the time students spent using tablets and notebooks. This indicates that these two devices are used similarly by students, suggesting that the frequency of their use is comparable. The test detected a significant median difference between smartphone and tablet use, with smartphones being used more frequently. This means that students, on average, spend significantly more time using smartphones compared to tablets. Similarly, there was a significant median difference between smartphone and notebook use, with smartphones again being used more frequently than notebooks.

These results indicate that smartphones are used much more often by students than either tablets or notebooks, while the usage patterns of tablets and notebooks are relatively similar. The pairwise comparisons clarify that smartphones stand out as the most frequently used device, confirming that there is indeed a significant difference in device usage within the student group. This indicates that the frequency of tablet and notebook use is similar within the group, while smartphones are used significantly more often. Therefore, the **first research hypothesis, H1, is confirmed**.

8.4.3 Purposes for the Use of Mobile Devices

The next issue analyzed was to determine the primary purposes for which students use mobile devices. To investigate this, the following questions were posed:

Q2: Please specify for which purposes you use your mobile devices most often? Where 1—very rarely, 5—very often

- for educational purposes e.g. MS Teams, Google Meet or to participate in e-courses for example on the Moodle platform,
- for educational purposes, including MOOCs (Massive Open Online Courses, for example, Navoica, Khan, Udacity, Coursera, Udemy, EdeX, etc.)
- the use of mobile learning applications, i.e. Duolingo
- for entertainment purposes
- use in everyday life (online shopping, bank transfers, orders, etc.)
- use to communicate (Messengers WhatsApp, Viber, Telegram)
- use for information retrieval

The students' responses to the above question are presented in Fig. 8.2. It is evident that students most frequently use mobile devices for entertainment, daily activities, communication, and information search. They are much less likely to use these devices for educational purposes, MOOCs, and mobile learning applications. However, within the realm of education, students most often use mobile devices for accessing platforms such as MSTeams.

As before, to confirm the statistical significance of the median differences in subgroups defined by the purposes of mobile device use, seven dependent samples were defined, each with 51 observations. Given the ordinal scale of the data, the significance of median differences within the groups was verified using the Friedman test. The test confirmed the statistical significance of the median differences across the groups, obtaining the following results: H(6,51) = 158.90; p-value = 0.000001.

Also post-hoc test (Dunn Bonferroni) was then performed, the results are shown in Table 8.3. Significant differences were observed between pairs of purposes where one is from the {Educational, MOOCs, Mobile learning} group and the other is from the {Entertainment, Everyday life, Communication, Information} group. No

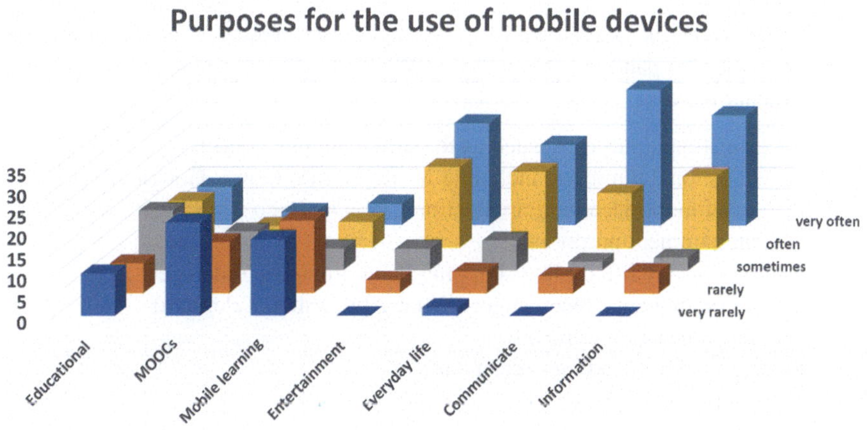

Fig. 8.2 Report on answers obtained for question Q2. (Source: Own research)

Table 8.3 p-value of the post-hoc Dunn Bonferroni test for purposes for the use of mobile devices

	Educational	MOOCs	Mobile learning	Entertainment	Everyday life	Communicate	Information
Educational		0.108676	0.694312	0.000204	0.022017	0.000004	0.000576
MOOCs	0.108676		1	0.000001	0.000001	0.000001	0.000001
Mobile learning	0.694312	1		0.000001	0.000001	0.000001	0.000001
Entertainment	0.000204	0.000001	0.000001		1	1	1
Everyday life	0.022017	0.000001	0.000001	1		1	1
Communicate	0.000004	0.000001	0.000001	1	1		1
Information	0.000576	0.000001	0.000001	1	1	1	

Source: Own research

significant median differences were found within each of these groups of purposes. Therefore, we can conclude that the frequency of mobile device use for Entertainment, Everyday life, Communication, and Information is homogeneous and significantly higher than for Educational, MOOCs, and Mobile learning purposes. Thus, **hypothesis H2 is confirmed.**

8.4.4 How the Pandemic Has Affected the Frequency of Use and which Mobile Devices

Another issue under investigation is how the pandemic has affected the frequency of use and which mobile devices. For this purpose, the following question was asked:

Q3: What is your self-assessment of the change in the frequency of use of mobile devices after the pandemic by purpose? Where 1—Significant decrease and 5—Significant increase

- for educational purposes, such as to participate in e-courses
- for educational purposes, including MOOCs, MS Teams, Google Meet
- the use of mobile learning applications, i.e. Duolingo
- for entertainment purposes
- use in everyday life (online shopping, bank transfers, orders, etc.)
- use to communicate (Messengers WhatsApp, Viber, Telegram)
- use for information retrieval
- to keep track of deadlines (calendar)
- for group work

The students' responses to the above question are presented in Fig. 8.3. The results show that it was rare for students to report a decrease in mobile device usage due to

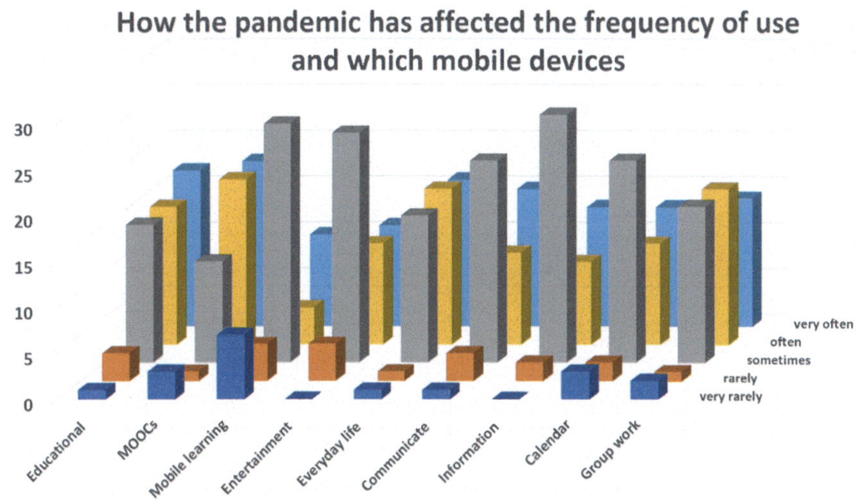

Fig. 8.3 Report on answers obtained for question Q3. (Source: Own research)

the pandemic. A large number of students indicated that the pandemic did not change their frequency of mobile device use. However, a notable number of students reported an increase in usage during the pandemic. The highest increase in frequency was observed for MOOC-related purposes.

As before, a statistical test was performed to confirm significant differences in medians. Nine dependent samples each with 51 observations were created. The scale is ordinal, so we again perform the Friedman test to confirm the significance of median differences across samples. The test confirmed a significant difference in medians, and the following results were obtained: $H(8,51) = 33.413$; $p\text{-value} = 0.000052$.

Also a post-hoc test (Dunn Bonferroni) was then performed, the results are shown in Table 8.4. The post-hoc test only confirmed a statistically significant difference in median of the frequency of use of mobile devices for the following pairs: (Mobile learning, MOOC) and (Mobile learning, Everyday life). Based on the results, it can be concluded that the pandemic has broadly contributed to changes in the frequency of mobile device use across various purposes, generally leading to an overall increase in usage. Thus, **hypothesis H3 is confirmed**.

8.4.5 Features of Teaching M-Learning Are Most Important to Students

Now, we investigated what feature of using mobile devices is most important to students, the one that makes their lives the easiest. The questions in the survey that addressed this issue were as follows:

Table 8.4 p-value of the post-hoc Dunn Bonferroni test for how the pandemic has affected the frequency of use

	Educational	MOOCs	Mobile learning	Entertainment	Everyday life	Communicate	Information	Calendar	Group work
Educational		1	0.056	1	1	1	1	1	1
MOOCs	1		0.017	1	1	1	1	1	1
Mobile learning	0.056	0.017		1	0.008	0.204	1	1	0.097
Entertainment	1	1	1		1	1	1	1	1
Everyday life	1	1	0.008	1		1	1	1	1
Communicate	1	1	0.204	1	1		1	1	1
Information	1	1	1	1	1	1		1	1
Calendar	1	1	1	1	1	1	1		1
Group work	1	1	0.097	1	1	1	1	1	

Source: Own research

Q4: What, in your opinion, are the positive features of m-learning:

- availability at any time
- short portions of material and thus short learning times
- alerts / reminders
- easy and quick management of activities
- easy and quick contact with other users (and Chatbots?)
- wide range of learning opportunities
- other

The students' responses to the above question are presented in Fig. 8.4. Availability at any time in courses is considered the most important feature of m-learning by students. Forty-nine students responded positively when asked if this was the most crucial feature. Regarding other features, the student responses were evenly split, with half considering them important and the other half not. Additional benefits of m-learning courses mentioned by students include:

- The consistency of the learning experience
- The ability to spend more time with family by avoiding the time spent commuting, which often takes 1–2 h
- The possibility learning anywhere
- The ability to learn at one's own pace, whether faster or slower

Statistical tests were performed to identify the significance of differences in the distributions of the responses given by the students in groups. Six dependent samples were created, each with 51 observations. As the measurement scale is nominal (Yes, No) the Cochran test was performed to verify the significance of differences in the

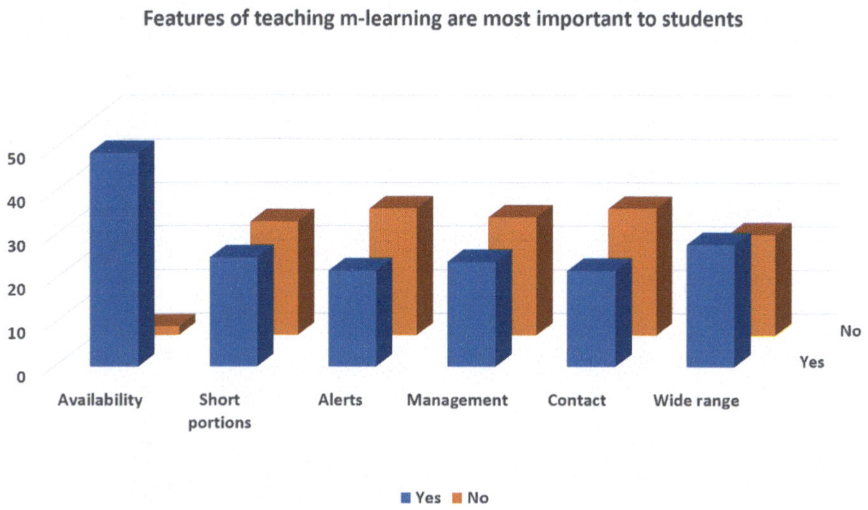

Fig. 8.4 Report on answers obtained for question Q4. (Source: Own research)

distribution of values. The test confirmed the statistical significance of the differences of the distributions in the subgroups, and test statistic values of Q = 47.135 were obtained, with 5 degrees of freedom and a p-value of 0.000001.

Subsequently, the Dunn post-hoc test was performed to identify pairs of groups for which there are significant differences, the results are shown in Table 8.5. As can be seen, the distribution of values for availability at any time is significantly different from all other characteristics of m-learning courses. Hypothesis H4, that the most important features of teaching m-learning to students is accessibility at any time, is thus confirmed.

8.4.6 *Negative Features of Teaching M-Learning Students Perceive*

The final aspect examined in the questionnaire was the question of negative features of m-learning students perceive. The question that addressed this issue was as follows:

Q5: What, in your opinion, are the negative features of m-learning:

- limitation of material in mobile courses
- limited media and online contact with the trainer
- inability to perform advanced tasks (for example, limited computing power)
- unfilled level of quality of graphics and video
- small screen, text not always legible

Table 8.5 p-value of the post-hoc Dunn test for the most important features of teaching m-learning for students

	Avail ability	Short portio ns	Alert s	Mana geme nt	Conta ct	Wide range
Availability		0.000008	0.000001	0.000002	0.000001	0.000164
Short portions	0.000008		1	1	1	1
Alerts	0.000001	1		1	1	1
Management	0.000002	1	1		1	1
Contact	0.000001	1	1	1		1
Wide range	0.000164	1	1	1	1	

Source: Own research

- not always provided responsiveness (ability to change scale/resolution of website screen/application window, course, etc.)
- crashes at important moments of learning
- unstable Internet connection
- technical problems
- other

The students' responses to the above question are presented in Fig. 8.5. The data shows that the most definitive responses were given to the question regarding the limitation of material in mobile courses. A significant majority of students (47) did not consider this to be a disadvantage of m-learning. This suggests that the course material is generally perceived to be at an adequate level. When asked about the negative aspects of m-learning courses, students most frequently mentioned the small screen size, text legibility issues, and unstable internet connections.

For nine dependent samples, each with 51 observations and a nominal scale dependent variable, the Cochran test was performed to verify significant differences in student responses for various negative aspects of m-learning courses. The test confirmed the statistical significance of the differences of the distributions in the subgroups, and test statistic values of $Q = 38.766$ were obtained, with 8 degrees of freedom and a p-value of 0.000005.

The Dunn post-hoc test was performed to identify pairs of groups for which there are significant differences, the results are shown in Table 8.6. The statistical analysis revealed significant differences between the responses regarding the limitation of material in mobile courses and other negative characteristics, except for the inability to perform advanced tasks (e.g., limited computing power) and inadequate quality of graphics and video. The test identified two homogeneous groups:

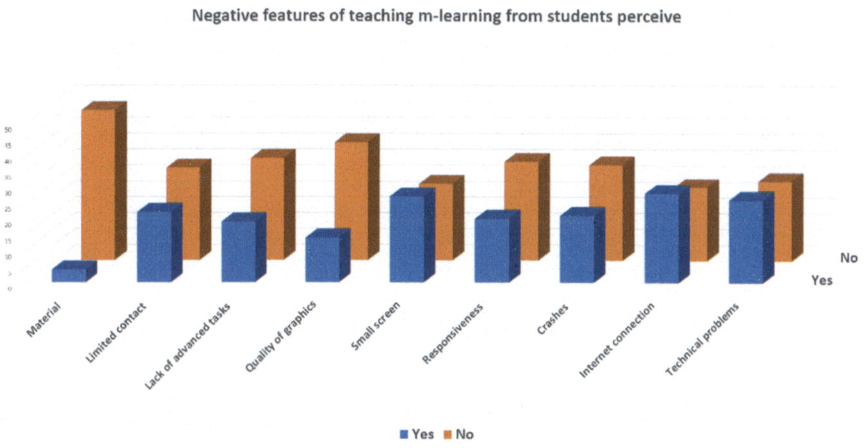

Fig. 8.5 Report on answers obtained for question Q5. (Source: Own research)

{Material, Lack of advanced tasks, Quality of graphics}
{Limited contact, Small screen, Responsiveness, Crashes, Internet connection, Technical problems}

It can be concluded that the first group of characteristics is not perceived negatively by students at all. As for the remaining features, opinions were divided, making it impossible to definitively confirm any negative aspect of m-learning. Hypothesis **H5 is thus confirmed**.

8.5 Discussion

8.5.1 Discussion of the Results

The study aimed to understand the perception and use of mobile devices by students at the University of Silesia in Katowice in the context of m-learning. The results indicate several important trends related to the frequency of using mobile devices, the purposes for which they are used, and the impact of pandemics on patterns of technology use.

8.5.2 Frequency of Using Mobile Devices

Analysis of the frequency of the use of the three types of mobile devices (smartphones, tablets, notebooks) showed significant differences between them. The results clearly indicated higher average smartphone use values than the other

Table 8.6 p-value of the post-hoc Dunn test for negative features of teaching m-learning

	Material	Limited contact	Lack of advanced tasks	Quality of graphics	Small screen	Responsiveness	Crashes	Internet connection	Technical problems
Material		0.0064	0.064	1	0.0001	0.031	0.0143	0.0001	0.0001
Limited contact	0.0064		1	1	1	1	1	1	1
Lack of advanced tasks	0.064	1		1	1	1	1	1	1
Quality of graphics	1	1	1		0.244	1	1	0.128	0.448
Small screen	0.0001	1	1	0.244		1	1	1	1
Responsiveness	0.031	1	1	1	1		1	1	1
Crashes	0.0143	1	1	1	1	1		1	1
Internet connection	0.0001	1	1	0.128	1	1	1		1
Technical problems	0.0001	1	1	0.448	1	1	1	1	

Source: Own research

devices. Interestingly, in a similar study conducted among Polish students in 2016, the most popular device was a notebook (80% of students), followed only by respondents who reported using a smartphone or tablet (Lorencowicz et al., 2016). Research conducted in 2024 by the National Media Institute (https://kim.gov.pl/wyniki-badan/) shows that in the 16–29 age group, a smartphone is used by 96% of people; similarly, among those using the internet, a smartphone is the most common device used for this purpose, followed by a computer/laptop and the tablet the least common. Smartphones have revolutionised how we communicate, moving it to the online space. This multifunctional device is ideally suited to young people’s needs, making it their most frequently used device. Thanks to the mobility of smartphones, students can use them anywhere and anytime, increasing their functionality.

8.5.3 Goals of Using Mobile Devices

The results on the purposes of using mobile devices indicate that students use them most often for entertainment, daily activities and communication. Using them for educational purposes, including online courses (MOOCs) and m-learning applications, is much less frequent, confirming hypothesis H2. Students prefer to use mobile devices for non-academic purposes, which is in line with the results of other studies conducted in this area, which show that most students use smartphones for social interaction or entertainment (Andersson, 2022; Gerosa et al., 2022; Arumugam et al., 2020). In a study by Kocak and Goktas (2021), an activity-monitoring application was installed on students' smartphones to record their device use weekly. The results showed that students used their smartphones for an average of 4 h and 50 min daily, launching the same or different apps approximately 380 times daily. Social media and instant messaging appeared to be the most frequently used apps. These results are relevant to the implementation of mobile technology in higher education. This is because they show that although students have extensive access to mobile devices, they mainly use them in a private rather than academic context. In a similar study carried out by Ng and co-authors (2017), it was found that in terms of academic use, students were most likely to use their smartphones to check reference materials, communicate with others via SMS and record lectures. In contrast, they were least likely to use smartphones to send emails regarding assignments, watch course-related videos and record class presentations. Importantly, as Green's (2019) study showed, in situations where students felt motivated in class, they used mobile devices to actively participate in discussions, access learning materials and take notes. However, when lecturers failed to present content in an engaging or comprehensible way, students often reached for their smartphones, viewing them as sources of distraction. These observations suggest that teachers should view smartphone use as a reflection of teaching methods and course atmosphere rather than as universal signs of disinterest or distraction. These findings may provide guidance for designers of m-learning courses to engage students more by integrating entertainment elements into the educational content, which could increase interest in this form of learning.

8.5.4 Impact of the Pandemic on the Use of Mobile Devices

The study of the impact of the pandemic on the frequency of use of mobile devices indicated that students rarely reported a decrease in the use of these devices. On the contrary, an increase in frequency of use was observed for most purposes, such as taking online courses (MOOCs) and using m-learning applications. Faced with the constraints of the pandemic, many educational institutions have moved to a remote learning mode. This has forced students and teachers to use mobile devices such as smartphones and tablets to participate in online classes, use learning platforms and communicate with others. It can be concluded that the pandemic has contributed to the increased adoption of digital technologies in higher education, which is in line with global trends (Matsieli & Mutula, 2024). In a study by Guppy and co-authors (2022), lecturers, students and professionals attempted to make predictions about the future of digital learning after the pandemic. Although the majority of respondents agreed that there would be a further expansion of blended-hybrid learning after the pandemic and a slight increase in fully online courses. Interestingly, students were the most sceptical group towards the anticipated changes in teaching, which may affect their engagement and adaptation to new educational methods. The research also showed doubts about teaching and participation in courses that are fully or partially online.

8.5.5 Positive and Negative Features of M-Learning

Results of the study on students' preferences for m-learning features indicate that the most important advantage of this form of learning is the availability of materials at any time, suggesting that time flexibility is a key factor influencing students' choice of m-learning. In terms of the negative features of m-learning, material limitations and technical problems such as unstable internet connection were identified as the most significant barriers. Kang (2024) analysed the benefits and challenges of m-learning, in which he showed that the greatest advantages of mobile learning included flexibility and accessibility, increasing learning effectiveness, developing intrinsic motivation, and increasing student engagement. In contrast, the negatives included a lack of opportunities to achieve learning and teaching outcomes, user distraction and possible visual impairment. However, many researchers emphasise that m-learning positively impacts student satisfaction (Togaibayeva et al., 2022; Hamidi & Jahanshaheefard, 2019).

Presented in this article research results could be considered as a continuing of wider complementary research in the area of using the ICT tools, Internet services and apps by students and young learners. Some previous results were published in Smyrnova-Trybulska et al. (2022, 2023) and Przybyła-Kasperek et al. (2023) and other. Future step is a wider of sample of research, involving students of different specializations and courses, verification of new hypotheses and deep analysis of received results.

8.6 Conclusions

In the article Author presented a review and description of literature and research on m-learning in different contexts, fulfilled by experts from different disciplines and countries, which shows that, e. g. that mobile learning enables learning anywhere, anytime, and provides the opportunity to adapt the pace and style of learning to the individual user's needs and preferences. Experts from different countries in their research based on *Unified Theory of Technology Acceptance (UTAUT)* provided and described different approaches to m-learning. More and more attention is being paid to researching the potential of m-learning in the context of long-term educational and professional development.

Although many questions have been answered, many aspects remain unexplored. The article provided answers to research questions and verified hypotheses 1–5. All hypotheses were confirmed, including: H1: Smartphones are the most commonly used mobile device tool; H2: Students most often use mobile devices for entertainment, use in everyday life, use for communication (Messengers WhatsApp, Viber, Telegram) or use them to search for information; H3: Students believe that the pandemic has increased the frequency of mobile device use in general; H4: The most important features of teaching m-learning to students is accessibility at any time; H5: The features: limitation of material in mobile courses, inability to perform advanced tasks (for example, limited computing power) and unfilled level of quality of graphics and video is not recognized as negative aspects of teaching m-learning to students.

In the near future, the Authors plan to conduct research on a larger scale and a larger group of respondents and verify further hypotheses regarding various didactic, ethical, technical aspects, social, professional use of mobile devices by students—future teachers, computer scientists as well as artists and engineers. The results may be helpful for academic teachers in planning their teaching procedures, developing tasks, activities, and ways of communicating, which may positively condition the level of quality of education.

References

Abdallah, N., Abdallah, O., & Bohra, O. (2021). Factors affecting mobile learning acceptance in higher education: An empirical study. *International Journal of Advanced Computer Science and Applications, 12*(4), 664–671.

Al-Emran, M., Arpaci, I., & Salloum, S. A. (2020a). An empirical examination of continuous intention to use m-learning: An integrated model. *Education and Information Technologies, 25*(4), 2899–2918. https://doi.org/10.1007/s10639-019-10094-2

Al-Emran, M., Mezhuyev, V., & Kamaludin, A. (2020b). Towards a conceptual model for examining the impact of knowledge management factors on mobile learning acceptance. *Technology in Society, 61*, 101247. https://doi.org/10.1016/j.techsoc.2020.101247

Aloqaily, A., Al-Nawayseh, M. K., Baarah, A. H., Salah, Z., Al-Hassan, M., & Al-Ghuwairi, A.-R. (2019). A neural network analytical model for predicting determinants of mobile learning acceptance. *International Journal of Computer Applications in Technology, 60*(1), 73–85. https://doi.org/10.1504/IJCAT.2019.099502

Al-Shihi, H., Sharma, S. K., & Sarrab, M. (2018). Neural network approach to predict mobile learning acceptance. *Education and Information Technologies, 23*, 1805–1824. https://doi.org/10.1007/s10639-018-9691-9

Alowayr, A. (2022). Determinants of mobile learning adoption: Extending the unified theory of acceptance and use of technology (UTAUT). *The International Journal of Information and Learning Technology, 39*(1), 1–12.

Alsswey, A., & Al-Samarraie, H. (2019). M-learning adoption in the Arab gulf countries: A systematic review of factors and challenges. *Education and Information Technologies, 24*(5), 3163–3176. https://doi.org/10.1007/s10639-019-09923-1

Andersson, C. (2022). Smartphones and online search: Shifting frames in the everyday life of young people. *Information and learning sciences, 123*(7/8), 351–370. https://doi.org/10.1108/ils-03-2022-0025

Arumugam, N., Selvanayagam, S., & Sathiyasenan, S. T. (2020). The effects of smartphone usage on university students. *International Journal of Academic Research in Progressive Education and Development, 9*(3), 170–183. https://doi.org/10.6007/IJARPED/v9-i3/7960

Ataş, A. H., & Çelik, B. (2019). Smartphone use of university students: Patterns, purposes, and situations. *Malaysian Online Journal of Educational Technology, 7*(2), 59–70.

Azizi, S. M., & Khatony, A. (2019). Investigating factors affecting on medical sciences students' intention to adopt mobile learning. *BMC Medical Education, 19*, 381. https://doi.org/10.1186/s12909-019-1831-4

Baghcheghi, N., Koohestani, H. R., Karimy, M., & Alizadeh, S. (2020). Factors affecting mobile learning adoption in healthcare professional students based on technology acceptance model [Faktori koji utiču na učenje putem mobilnog telefona kod studenata zdravstvene nege na osnovu modela prihvatanja tehnologije]. *Acta Facultatis Medicae Naissensis, 37*(2), 191–200. https://doi.org/10.5937/afmnai2002191b

Barfi, K. A., & Du Preez, M. (2024). University of Cape Coast doctoral students' use of mobile devices to seek information. *New Review of Academic Librarianship*. https://doi.org/10.1080/13614533.2024.2381521

Bharati, V. J., & Srikanth, R. (2018). Modified UTAUT2 model for m-learning among students in India. *International Journal of Learning and Change, 10*(1), 5–20. https://dpi.org/10.1504/IJLC.2018.089532

Chao, C.-M., & Chen, W.-R. (2023). Factors determining behavioral intention of nursing students to use mobile learning: An application and extension of the UTAUT model. *International Journal of Mobile Communications, 21*(4), 402–421. https://doi.org/10.1504/IJMC.2023.131160

Chen, J. (2022). Adoption of M-learning apps: A sequential mediation analysis and the moderating role of personal innovativeness in information technology. *Computers in Human Behavior Reports, 8*, 100237. https://dpi.org/10.1016/j.chbr.2022.100237

Chromiński, K., Przybyła-Kasperek, M., Smyrnova-Trybulska, E., Bazarbayeva, A., & Morze, N. (2023). Comparative international research in area of educational platform and MOOCs in opinion of IT students using data mining analysing. In E. Smyrnova-Trybulska (Ed.), *E-learning & Artificial Intelligence* (E-learning Series) (Vol. 15, pp. 15–30). ISSN 2451-3644 (print edition), ISSN 2451-3652 (digital edition), ISBN 978-83-66055-39-1). STUDIO NOA for University of Silesia. https://doi.org/10.34916/el.2023.15.02

Crompton, H. (2013). A historical overview of M-learning: Toward learner-centered education. In L. Y. M. Z. L. Berge (Ed.), *Handbook of mobile learning* (pp. 3–14). Routledge.

Darren Pullen, J.-F., Swabey, K., Abadooz, M., & Ranjit Sing, T. K. (2015). Pre-service teachers' acceptance and use of mobile learning in Malaysia. *Australian Educational Computing, 30*(1).

Dias, L., & Victor, A. (2022). Teaching and learning with Mobile devices in the 21st century digital world: Benefits and challenges. *European Journal of Multidisciplinary Studies, 7*(1), 26–34. https://doi.org/10.26417/ejms.v5i1.p339-344

Fu, E., Gao, Q., Wei, C., Chen, Q., & Liu, Y. (2021). Understanding student simultaneous smartphone use in learning settings: A conceptual framework. *Journal of Computer Assisted Learning, 37*, 91–108. https://doi.org/10.1111/jcal.12471

Gerosa, T., Gui, M., & Büchi, M. (2022). Smartphone use and academic performance: A pervasiveness approach beyond addiction. *Social Science Computer Review, 40*(6), 1542–1561. https://doi.org/10.1177/08944393211018969

Gikas, J., & Grant, M. M. (2013). Mobile computing devices in higher education: Student perspectives on learning with cellphones, smartphones & social media. *The Internet and Higher Education, 19*, 18–26.

Granić, A. (2023). Technology acceptance and adoption in education. In O. Zawacki-Richter & I. Jung (Eds.), *Handbook of open, distance and digital education*. Springer. https://doi.org/10.1007/978-981-19-2080-6_11

Green, M. (2019). Smartphones, distraction narratives, and flexible pedagogies: Students' mobile technology practices in networked writing classrooms. *Computers and Composition, 52*, 91–106. https://doi.org/10.1016/j.compcom.2019.01.009

Gull, M., Ahmed, A., Hassan, S., Nazir, A., & Nasir, A. (2023). Breaking barriers in higher education: How students are embracing mobile learning in the post-COVID era. *Pakistan Journal of Multidisciplinary Research (PJMR), 4*(2).

Guppy, N., Verpoorten, D., Boud, D., Lin, L., Tai, J., & Bartolic, S. (2022). The post-COVID-19 future of digital learning in higher education: Views from educators, students, and other professionals in six countries. *British Journal of Educational Technology, 53*(2), 678–692. https://doi.org/10.1111/bjet.13212

Hamidi, H., & Jahanshaheefard, M. (2019). Essential factors for the application of education information system using mobile learning: A case study of students of the university of technology. *Telematics and Informatics, 38*, 207–224. https://doi.org/10.1016/j.tele.2018.12.006

Hamzah, A., & Muchlis, N. F. (2018). The exploration through the factors affecting students' adoption on m-learning technologies. In *AIP Conference Proceedings*. https://doi.org/10.1063/1.5042879.

Heard, I. R. (2020). An empirical analysis of mobile learning acceptance in Puerto Rico's higher learning institutions. In *Mobile devices in education: Breakthroughs in research and practice* (pp. 325–335). https://doi.org/10.4018/978-1-7998-1757-4.ch020, https://kim.gov.pl/wyniki-badan/

Jeong, K.-O. (2022). Facilitating sustainable self-directed learning experience with the use of Mobile-assisted language learning. *Sustainability, 14*, 2894.

Kang, J. (2024). Benefits and challenges of Mobile-learning brought to student learning outcomes in higher education: A systematic review from 2014–2023. *International Journal of Academic Research in Progressive Education and Development, 13*(1), 2107–2123.

Kang, M., Liew, B. Y. T., Lim, H., Jang, J., & Lee, S. (2015). Investigating the determinants of mobile learning acceptance in Korea using UTAUT2. In *Emerging issues in smart learning* (pp. 209–216). Springer.

Kim, J. S., & Kizildag, M. (2011). M-learning: Next generation hotel training system. *Journal of Hospitality and Tourism Technology, 2*(1), 6–33. https://doi.org/10.1108/17579881111112395

Kocak, O., & Goktas, Y. (2021). Are we online or in class? Students' smartphone usage habits. *Journal of Computer and Education Research, 8*(2), 31–45. https://doi.org/10.18009/jcer.886748

Kuciapski, M. (2016). Students acceptance of m-learning for higher education–UTAUT model validation. In *Information systems: Development, research, applications, education: 9th SIGSAND/PLAIS EuroSymposium 2016, Gdansk, Poland, September 29, 2016, Proceedings 9* (pp. 155–166). Springer International Publishing.

Kumar, B. A., & Sharma, B. (2020). Context aware mobile learning application development: A systematic literature review. *Education and Information Technologies, 25*, 2221–2239.

Lin, S. H., Lee, H.-C., Chang, C.-T., & James, F. C. (2020). Behavioral intention towards mobile learning in Taiwan, China, Indonesia, and Vietnam. *Technology in Society, 63*, 101387. https://doi.org/10.1016/j.techsoc.2020.101387

Lisana, L., & Suciadi, M. F. (2021). The acceptance of mobile learning: A case study of 3D simulation android app for learning physics. *International Journal of Interactive Mobile Technologies, 15*(17), 205–214. https://doi.org/10.3991/IJIM.V15I17.23731

Lorencowicz, E., Koszel, M., Kocira, S., & Uziak, J. (2016). Student use of mobile devices in university classes. *Education – Technology – Informatics, 2*(16), 27–34. https://doi.org/10.15584/eti.2016.2.27

Matsieli, M., & Mutula, S. (2024). COVID-19 and digital transformation in higher education institutions: Towards inclusive and equitable access to quality education. *Education Sciences, 14*(8), 819. https://doi.org/10.3390/educsci14080819

Moca, M., & Badulescu, A. (2023). Determinants of economical high school students' attitudes toward mobile devices use. *Sustainability, 15*(12), 9331. https://doi.org/10.3390/su15129331

Moya, S., & Camacho, M. (2021). Identifying the key success factors for the adoption of mobile learning. *Education and Information Technologies, 26*, 3917–3945.

Naveed, Q. N., Alam, M. M., & Tairan, N. (2020). Structural equation modeling for mobile learning acceptance by university students: An empirical study. *Sustainability (Switzerland), 12*(20)., art. no. 8618, 1–20. https://doi.org/10.3390/su12208618

Naveed, Q. N., Choudhary, H., Ahmad, N., Alqahtani, J., & Qahmash, A. I. (2023). Mobile learning in higher education: A systematic literature review. *Sustainability, 15*(18), 13566. https://doi.org/10.3390/su151813566

Nawaz, S. S., & Mohamed, R. (2020). Acceptance of mobile learning by higher educational institutions in Sri Lanka: An UTAUT2 approach. *Journal of Critical Reviews, 7*(12), 1036–1049. https://doi.org/10.31838/jcr.07.12.183

Ng, S. F., Che Hassan, N. S. I., Mohammad Nor, N. H., & Abdul Malek, N. A. (2017). The relationship between smartphone use and academic performance: A case of students in a Malaysian tertiary institution. *Malaysian Online Journal of Educational Technology, 5*(4).

Nikolopoulou, K. (2022). Students' Mobile phone practices for academic purposes: Strengthening post-pandemic university digitalization. *Sustainability, 14*(22), 14958. https://doi.org/10.3390/su142214958

Pedro, L. F. M. G., Barbosa, C. M. M. d. O., & Santos, C. M. d. N. (2018). A critical review of mobile learning integration in formal educational contexts. *International Journal of Educational Technology in Higher Education, 15*(10). https://doi.org/10.1186/s41239-018-0091-4

Pratama, A. R. (2021). Fun first, useful later: Mobile learning acceptance among secondary school students in Indonesia. *Education and Information Technologies, 26*, 1737–1753. https://doi.org/10.1007/s10639-020-10334-w

Przybyła-Kasperek, M., Chromiński, K., Smyrnova-Trybulska, E., Morze, N., & Bazarbayeva, A. (2023). Importance, popularity and elements of educational platforms—A study of the opinions of students from Poland, Ukraine and Kazakhstan. *International Journal of Research in E-Learning*, 1–30. https://doi.org/10.31261/IJREL.2023.9.2.08

Qamar, M. T., Ajmal, M., Malik, A., Ahmad, M. J., & Yasmeen, J. (2024). Mobile learning determinants that influence Indian university students' learning satisfaction during the COVID-19 pandemic. *International Journal of Continuing Engineering Education and Life Long Learning, 15*(2–3), 123–145.

Qazi, A., Qazi, J., Naseer, K., Hasan, N., Hardaker, G., & Bao, D. (2024). M-learning in education during COVID-19: A systematic review of sentiment, challenges, and opportunities. *Heliyon, 10*, e32638. https://doi.org/10.1016/j.heliyon.2024.e32638

Robison, D. (2012). Learning on location with ami: The potentials and dangers of mobile gaming for language learning. In *Left to my own devices: Learner autonomy and mobile-assisted language learning* (Vol. 6, pp. 67–88). https://doi.org/10.1108/S2041-272X(2012)0000006007

Sabah, N. M. (2016). Exploring students' awareness and perceptions: Influencing factors and individual differences driving m-learning adoption. *Computers in Human Behavior, 65*, 522–533. https://doi.org/10.1016/j.chb.2016.09.009

Seliaman, M. E., & Al-Turki, M. (2012). Mobile learning adoption in Saudi Arabia. *World Academy of Science, Engineering and Technology, 69*(9), 391–293.

Seyal, A. H., Abd Rahman, M. N., Sy, M. Y., & Siau, N. Z. H. (2017). Examining M-learning in higher education: An application of the theory of planned behavior. In *Theory of planned behavior: New research* (pp. 17–21). Nova Science Publishers, ISBN: 978-1-53611-096-8.

Shishakly, R., Almaiah, M. A., Al-Otaibi, S., Lutfi, A., Alrawad, M., & Almulhem, A. (2023). A new technological model on investigating the utilization of mobile learning applications: Extending the TAM. *Multimodal Technologies and Interaction, 7*(9), 92. https://doi.org/10.3390/mti7090092

Smyrnova-Trybulska, E., Morze, N., & Varchenko-Trotsenko, L. (2022). Adaptive learning in university students' opinions: Cross-border research. *Education and Information Technologies*. eISSN 1573-7608, Print ISSN 1360-2357. https://doi.org/10.1007/s10639-021-10830-7

Smyrnova-Trybulska, E., Przybyła-Kasperek, M., & Kommers, P. (2023). Educational, social and ethical aspects of AI in the opinion of IT and pedagogy students – A case study. In P. Kommers, I. A. Sanches, & P. Isaias (Eds.), *E-Society. Conference proceedings, international conference e-learning, March 11–13, 2019* (pp. 171–178). IADIS Press, ISBN: 978-989-8704-47-4.

Sophea, D., Sungsuwan, T., & Viriyasuebphong, P. (2022). Factors influencing students' behavioral intention on using mobile learning (M-learning) in tourism and hospitality major in Phnom Penh, Cambodia. *Current Applied Science and Technology, 22*(2).

Tarhini, A., AlHinai, M., Al-Busaidi, A. S., Govindaluri, S. M., & Shaqsi, J. A. (2024). What drives the adoption of mobile learning services among college students: An application of SEM-neural network modeling? *International Journal of Information Management Data Insights, 4*, 100235.

Togaibayeva, A., Ramazanova, D., Yessengulova, M., Yergazina, A., Nurlin, A., & Shokanov, R. (2022). Effect of mobile learning on students' satisfaction, perceived usefulness, and academic performance when learning a foreign language. *Frontiers in Education, 7*, 946102. https://doi.org/10.3389/feduc.2022.946102

Traxler, J. (2005, June). *Defining mobile learning*. Paper presented at the IADIS international conference mobile learning 2005, Qawra, Malta.

Venkatesh, V., Thong, J. Y., & Xu, X. (2012). Consumer acceptance and use of information technology: Extending the unified theory of acceptance and use of technology. *MIS Quarterly*, 157–178.

Wang, Y.-S., Wu, M.-C., & Wang, H.-Y. (2009). Investigating the determinants and age and gender differences in the acceptance of mobile learning. *British Journal of Educational Technology, 40*(1), 92–118. https://doi.org/10.1111/j.1467-8535.2007.00809.x

Wu, W. H., Wu, Y. C. J., Chen, C. Y., Kao, H. Y., Lin, C. H., & Huang, S. H. (2012). Review of trends from mobile learning studies: A meta-analysis. *Computers & Education, 59*(2), 817–827. https://doi.org/10.1016/j.compedu.2012.03.016

Chapter 9
E-learning in Developing Interaction with the Environment as Soft Skills in 6-Year-Old Children (on the Example of Preschool Educational Institutions in Ukraine and France)

Liubov Lokhvytska, Nataliia Martovytska, Olena Kolomiiets, and Tetiana Sheleg

Abstract The article provides a substantiation for the use of e-learning in developing interaction with the environment as soft skills in six-year-old children within preschool educational institutions. The essence of using e-learning in the organization of the educational process with preschool children and the specifics of developing their skills of interaction with others are revealed. The aim was to theoretically comprehend the problem through an empirical study of the use of e-learning in the real practice of preschool educational institutions (on the example of Ukraine and France) to develop interaction with the environment as soft skills in six-year-old children. The method used included the analysis of the educational process plans for six-year-old children and conducting a survey (based on the questionnaire, developed by the authors) of preschool teachers. The research group consisted of 28 educators, 17 of whom were from two preschool educational institutions of Ukraine and 11 ones from France, who work with six-year-old children. The analysis of the obtained results made it possible to analyse the state of real practice in the use of e-learning in developing interaction with others as soft skills in six-year-old children and to outline its feasibility for the system of preschool education. This formed the

L. Lokhvytska (✉)
Hryhorii Skovoroda University in Pereiaslav, Pereiaslav, Ukraine
e-mail: Liubov.Lokhvytska@uhsp.edu.ua

N. Martovytska · O. Kolomiiets
Bohdan Khmelnytsky Cherkasy National University, Cherkasy, Ukraine
e-mail: nv_mart@vu.cdu.edu.ua; kolomiietsolena@vu.cdu.edu.ua

T. Sheleg
University of Versailles Saint-Quentin-en-Yvelines, Bilingual Montessori Preschool in Vincennes, Versailles, France
e-mail: Tetiana.Sheleh@ac-creteil.fr

E. Smyrnova-Trybulska et al. (eds.), *E-Learning and Enhancing Soft Skills*,
https://doi.org/10.1007/978-3-031-82243-8_9

basis for the development of the conceptual framework of an experimental model for implementing e-learning in developing such soft skills as interaction with the environment in six-year-old children. Further research will involve specification of this model (selection of educational resources, didactic tools, diagnostic instruments, etc.) and testing it through the organization of the educational process in preschool institutions.

Keywords E-learning · Soft skills · 6-year-old children · Interaction with the environment · Teachers of preschool educational institutions · Ukraine · France

9.1 Introduction

Current transformational processes in the educational system directly affect its first component—early preschool education. In particular, innovative teaching methods, including electronic learning (hereinafter referred to as e-learning), deserve attention. E-learning is defined as a type of education based on the use of electronic resources via the Internet, various information and communication and digital technologies and multimedia. It is generally viewed as a teaching methodology that implies digital devices (computers, smartphones, tablets, and other gadgets) (Bylieva & Nordmann, 2022).

Researchers (Behnamnia et al., 2023; Hashem, 2022; Olisah Kingsley & Mohamed Ismail, 2015) have demonstrated the effectiveness of e-learning in preschool educational institutions (hereafter referred to as PEI). For children under the age of six, viewing various electronic educational materials is engaging. It increases the involvement of preschoolers in being active, improves the quality of their knowledge development and overall makes the educational process as effective as possible. In this way, the quality of educational services in the field of preschool education is improved and the effectiveness of partnership between the subjects of the educational process is ensured: preschool teachers as providers and children as recipients, ultimately facilitating a harmonious integration of pupils into the information society (Chen & Ding, 2024; García Laborda et al., 2023; Zhang, 2024).

The questions of how e-learning impacts personality development in preschool age and the development of soft skills, which are essential for harmonious integration into society, remain largely overlooked.

The research problem is to find out how e-learning contributes to the development of soft skills in six-year-old children just before they start school, which ensures a full and harmonious life.

The research focus is on determining, from the perspective of preschool teachers, how e-learning is applied in real practice through the organization of the educational process to develop six-year-old children's interaction with the environment as soft skills.

9.2 Research Design

9.2.1 *The Essence of Using e-Learning in the Organization of the Educational Process with Preschool Children*

The research problem raised focuses on the processes of using e-learning to support the achievement of learning goals, creating opportunities to improve children's digital literacy knowledge and helping them "adapt to the learning environment in the digital age" by overcoming time and location constraints and increasing the flexibility of the educational process (Zhang, 2024).

It is important to consider the principle of integrating e-learning resources into traditional methodology of organizing the educational process with preschool children, considering their age and psychological abilities (Behnamnia et al., 2023; García Laborda et al., 2023; Rahiem, 2021). Such resources should be accessible for children to understand, raise their interest and require "minimal effort" from the preschool teachers (García Laborda et al., 2023).

The features of using e-learning as one of the styles in the educational process have been considered by researchers (Behnamnia et al., 2023; García Laborda et al., 2023; Hastürk & Özdemir, 2021; MacDonald et al., 2020; MacDonald et al., 2021; Rahiem, 2021; Raviv & Galili, 2024) in the context of selecting thematic technologies, determining the conditions for their implementation, and identifying specific resources that can be used with preschool children. In particular, research has been conducted on the impact of e-learning on:

- the development of creativity and the search for non-standard approaches to solving tasks (Behnamnia et al., 2020);
- the development of children's thinking and learning skills (Behnamnia et al., 2023);
- the speech development of children and the formation of their vocabulary within the conditions of preschool education (Baltzaki & Chlapana, 2023);
- the cognitive and creative growth of the students' personality in the course of applying the educational program "Preschoolers: Digital Adventures" (Chen & Ding, 2024);
- moral education of senior preschoolers and their moral consciousness development (Lokhvytska et al., 2022);
- social-emotional behavior skills development (Gözen & Cınk, 2017).

Thus, the conducted analysis of the theoretical material gives reason to conclude that e-learning will have value when it provides convenient access to educational materials and contributes to improving the quality of the educational process (Behnamnia et al., 2023; Chen & Ding, 2024; García Laborda et al., 2023). In view of the above, it is important to comply with the requirements for the use of e-learning with preschool children (dosage, considering the educational needs and requests of pupils, alternating types of e-resources, etc.). These positions will form the basis for the empirical study.

9.2.2 *Specifics of Developing Interactions with the Environment in Six-Year-Old Children as Soft Skills*

The quality of education is directly dependent on establishing interaction between the subjects of the educational process—children and preschool teachers. Joint activities are seen as the key to success in children's knowledge acquisition when they exhibit activity and apply their own experience (Boström et al., 2023). The development of interaction skills with others develops the ability to listen to one another, understand the other's position and express one's own position without offending anyone.

In their research G. Gözen and İ. Cırık (2017) found that the use of digital storytelling as a form of e-learning is appropriate for the development of socio-emotional behavior in preschool children. The scientific conclusions of E. Cerezo et al. (2024) are in line with this study, proving the necessity for the search and development of innovative technologies for the development of children's social skills, including interaction with others. Such interaction is based on the development of interpersonal communication skills while performing creative tasks in a team, which in turn contributes to the process of children's socialization. In the course of such joint activities, certain social skills are developed: "effective and productive interpersonal interactions, negotiation, conflict resolution, and persuasion skills" (Cerezo et al., 2024). This is ensured by organizing cooperation between children and social skills development through their involvement in an electronic gaming environment. The process of joint gaming activities requires children to make contact with each other, learn to negotiate and interact with one another (Blanco et al., 2022).

Teaching social skills is an important part of the process of children's socialization and the development of their life competence, which should be based on the social interaction development, that emerges between them through the use of electronic learning resources by preschool teachers (Blanco et al., 2022; Cerezo et al., 2024; Gomez-García et al., 2023). In this regard, attention should be drawn to the correlation between behavioral problems in children aged 3–6, the relationships with others and communication skills development, which form the basis of social competence and interaction with the environment (Masaroğulları & Çerkez, 2014).

Therefore, we consider the following question relevant: in what way, in preschool teachers' opinion, are the opportunities of e-learning applied to develop in 6-year-old children such soft skills as interaction with the environment, which is important for their adaptation to the new social environment of primary school.

9.2.3 *Sample of Research*

Preschool teachers from PEIs in Pereiaslav Nr 7 "Berizka" and Cherkasy Nr 29 "Lastivka" (Ukraine) and the Bilingual Montessori Preschool in Vincennes

(France), who organize the educational process with six-year-old children, participated in the survey. The choice is based on the fact that at the age of six, children in these countries go to other institutions—primary schools, and therefore are equal. It is worth noting that the outlined thematic research occupies a prominent place in the educational activities of these institutions. In these Ukrainian kindergartens, experimental work is being conducted on the topic "Development of the Foundations of Moral Education in Preschool Children", which includes the development of interaction skills with others as one of its components. Similarly, the Bilingual Montessori Preschool in Vincennes (France) pays considerable attention to the development of communication skills and developing cooperation with others, which aligns with students' soft skills development.

The research group consisted of 17 preschool teachers working with six-year-old children in two PEIs in Ukraine and 11 preschool teachers working with children of the same age in a PEI in Vincennes (France). The total number of respondents who participated in the survey was 28.

9.2.4 Research Questions

1. Investigating the use of e-learning in the practical work of PEIs to develop soft skills in children at the age of six through the organization of the educational process by preschool teachers.
2. Conducting an analysis of the feasibility of using e-learning to develop soft skills in children at the age of six for interacting with their environment, based on the generalized results of a survey of teachers in PEIs.
3. Development of conceptual provisions for a model of e-learning implementation in the development of interaction with others as soft skills in 6-year-old children.

9.2.5 Hypothesis

Conducting a pilot study in the form of a survey among preschool teachers on the use of e-learning in the soft skills development in children at the age of six for interacting with their environment will allow us to clarify the current state and outline the tasks for further scientific research. It is important to diagnose the real state from the perspective of self-analysis by preschool teachers who organize the educational process with preschoolers as providers of educational services. Provided that the results obtained from the survey of preschool teachers regarding the use of e-learning are adequate, the survey will be extrapolated to a broader range: for conducting a comparative analysis (involving respondents from different countries), outlining the criteria, indexes and levels of readiness of preschool education professionals to implement e-learning in the educational process with preschool children.

9.2.6 Instrument and Procedures

In the course of the empirical research, the method of analyzing preschool teachers' plans for organizing the educational process and the results of a survey (questionnaire) was used to study the role of e-learning in developing soft skills in 6-year-old children for interacting with their environment. And statistical methods as well: descriptive statistics and the method of average values when calculating average values and percentages during quantitative analysis and performing their qualitative interpretation.

The questionnaire for preschool teachers was based on the scientific positions of A. Raviv and I. Galili (2024) regarding self-assessment of their own attitude towards e-learning for preschool children, particularly the substantive application and use of methodological support, as well as S. Gomez-García et al. (2023) on the feasibility of collecting information among preschool education specialists based on carefully considered questions and the analysis of the obtained responses.

The author's thematic questionnaire contained a total of 7 questions (see Appendix 1): 5 questions with multiple-choice answers (with three answer options each), from which respondents were required to choose only one and 2 open-ended questions to explore preschool teachers' views and their pedagogical experience in using e-learning to develop skills of interaction with the environment in 6-year-old children. Based on the research of S. Gomez-García et al. (2023), the content of the questions was formulated in such a way as to clarify the psychological and pedagogical conditions and didactic means for implementing e-learning in the organization of the educational process with children.

The awareness of the survey among preschool teachers was given through notifications in the professional community's social network—in a Telegram group. Respondents were asked to give online responses to the questions of the survey at any time. At the same time, the purpose of the study and explanations were not provided in advance, which made it possible to eliminate external influence on the results.

9.2.7 Data Analysis

The study was conducted in three stages: the first stage, which lasted from February to April 2024, involved the analysis of the plans for organizing the educational process and the development of the questionnaire for preschool teachers in PEIs; the second stage (May 2024) was focused on collecting empirical data (filling out of the questionnaire by the respondents); and the third stage, which occurred in June and July 2024, involved processing of the obtained research data, conducting quantitative analysis and qualitative interpretation, presenting the results and developing a concept for further research.

9.3 Results of Research

We present the results obtained on the basis of the collecting empirical material: the analysis of preschool teachers' plans for organizing the educational process and processing of questionnaire data. During the study of the content of planning work with children in PEIs, it was found out that Ukrainian preschool teachers systematically record this form of work—64.71%; partially it takes place in 35.29%. Accordingly, somewhat different indexes are observed in the plans of French preschool teachers who work with 6-year-old children: at the same quantitative level, the share of those who plan systematically and those who do it from time to time—45.45% each, and in 9.1% such forms of work were not recorded at all. The obtained results indicate that in the Ukrainian system of preschool education, much more attention is paid to the use of e-learning to implement educational tasks with 6-year-old children.

This is confirmed by the conducted quantitative and qualitative analysis of the survey among preschool teachers. Given the fact that no invalid (damaged) questionnaires were found, all participants' responses were considered. According to the results of the quantitative analysis, which is further presented in Figs. 9.1, 9.2, 9.3, 9.4, 9.5, 9.6, 9.7, and 9.8, a qualitative interpretation of the obtained data was carried out regarding the use of e-learning in the development of interaction with others as soft skills in 6-year-old children on the example of the educational process organization in PEIs in Ukraine and France.

Thus, to the first question of the questionnaire: "Do you use e-learning in the organization of the educational process with 6-year-old children?", the responses were distributed in the same way (see Fig. 9.1), as the results obtained during the analysis of the plans for the organization of the educational process.

Based on the responses to this question, it was found that the vast majority of preschool teachers in Ukrainian PEI (64.71%) implemented e-learning in the

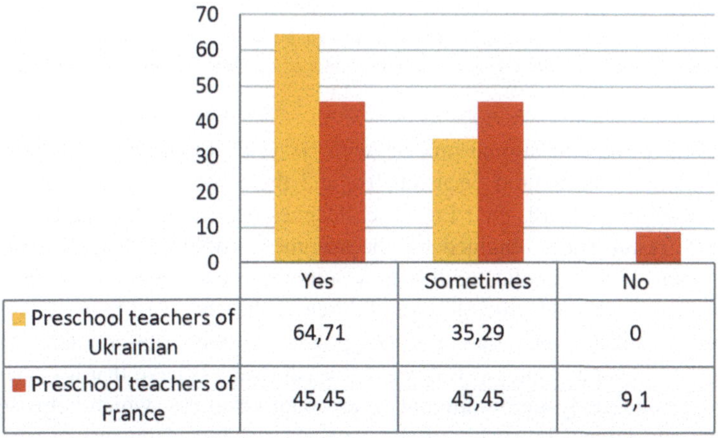

Fig. 9.1 Quantitative indexes of the survey of preschool teachers on the frequency of e-learning in the educational process organisation with 6-year-old children. (Source: Own work)

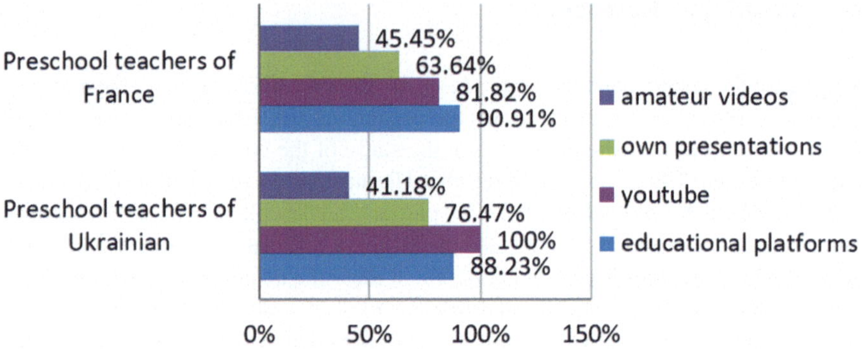

Fig. 9.2 Distribution (according to preschool teachers' responses) of e-learning types in developing interaction skills in six-year-old children. (Source: Own work)

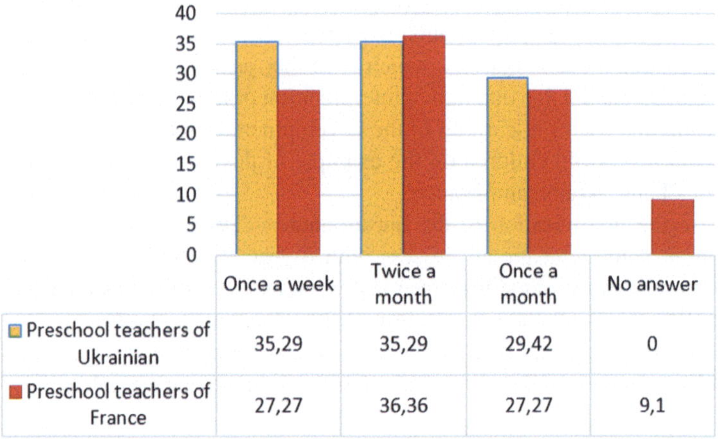

Fig. 9.3 The frequency of preschool teachers' use of e-learning in organizing the educational process with six-year-old children to develop their interaction skills. (Source: Own work)

organization of the educational process with six-year-old children; 35.29% of the respondents use this method occasionally, and there were no negative responses recorded at all. In contrast, their French colleagues responded differently: the same quantitative results were obtained for the statement—45.45% of teachers and the same number (45.45%) indicated that they sometimes use e-learning in the organisation of the educational process with children at the age of 6; however, 9.1% of respondents indicated that they did not practice it with preschoolers. We can interpret the obtained results as follows: most educators believe that it is necessary to use e-learning alongside traditional educational resources, which helps in organizing the educational process with six-year-old children. However, a portion of them (over a third of educators in Ukrainian preschool institutions and almost half in French preschool institutions) occasionally implement this method of working with

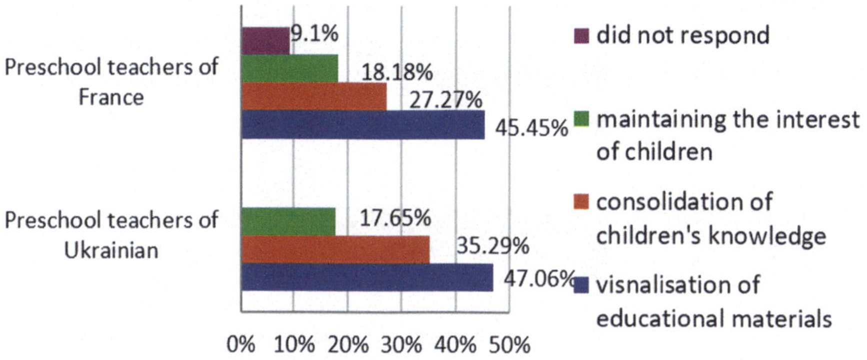

Fig. 9.4 Preschool teachers' determination of the purpose of using e-learning in developing interaction skills in six-year-old children. (Source: Own work)

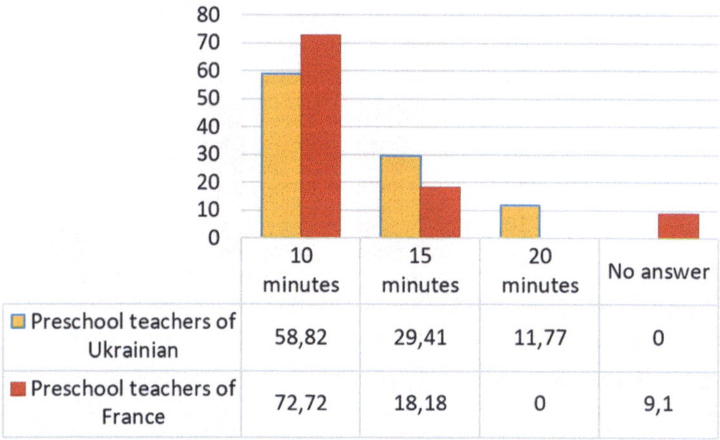

Fig. 9.5 Distribution of the duration of e-learning sessions aimed at developing interaction skills in six-year-old children. (Source: Own work)

preschoolers, and a small number (only 9.1%) of French respondents do not use it at all. Thus, alongside the active use of modern methods, traditional educational trends still persist in the real practice of preschool institutions.

To the second question of the survey, which was open-ended: "What types of e-learning do you use to develop interaction skills in six-year-old children?", preschool teachers provided their own responses (have chosen those types of e-learning, which they use more frequently), which are summarised in a generalised form and classified according to the essence of the statement into the following: "educational platforms", "YouTube", "personal presentations" and "amateur videos". The quantitative results about the most frequent types of e-learning which preschool teachers use in the work, with children (not one, but all those that were

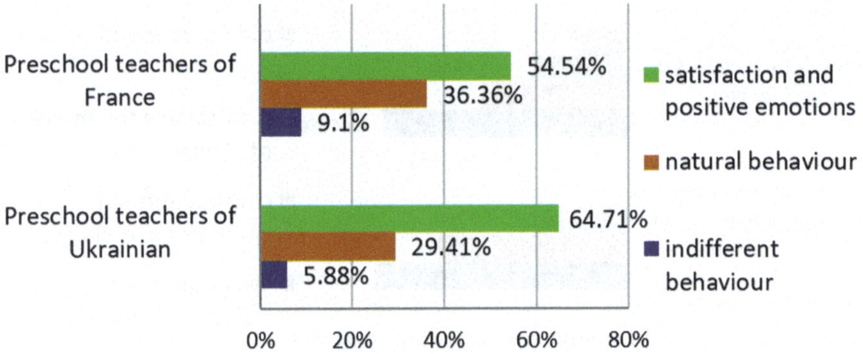

Fig. 9.6 Preschool teachers' determination of children's reactions to the use of e-learning in the educational process. (Source: Own work)

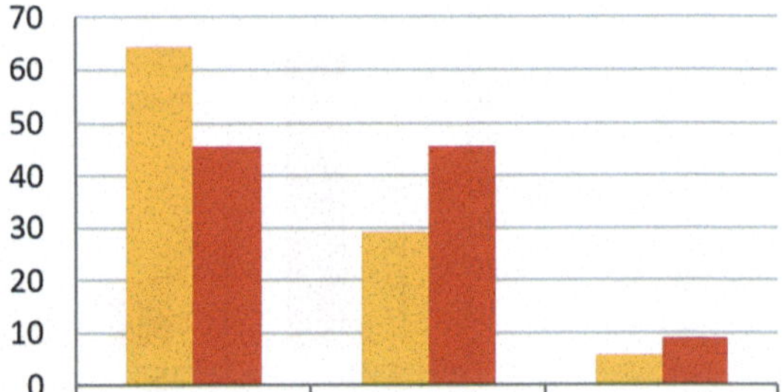

	Yes	Partially	No
Preschool teachers of Ukrainian	64,71	29,41	5,88
Preschool teachers of France	45,45	45,45	9,1

Fig. 9.7 The expediency (according to the teachers' answers) of using e-learning in developing the skills of interaction with environment in children at the age of 6. (Source: Own work)

indicated in the questionnaire were considered), obtained from this synthesis are presented in Fig. 9.2.

The responses to the second question of the survey demonstrated that the vast majority, namely 88.23% of preschool teachers in Ukrainian PEIs and 90.91% in French preschool institutions actively use various educational platforms. Ukrainian preschool teachers mentioned the following: "Korysni pidkazky" (PLUSPLUS TV channel https://1ll.ink/jvr1h), "Tsikavi istorii" (Nadia TV channel https://1ll.ink/tCNNh), "Slukhanky dlia ukrainskoi malechi", an audio project by the "Fair Country" foundation and the National Radio Company of Ukraine (https://1ll.ink/FpZvT), "Lessons from Aunt Owl", a Ukrainian children's educational animated series produced by "Masks" production studio (https://1ll.ink/pzInb). In contrast, their

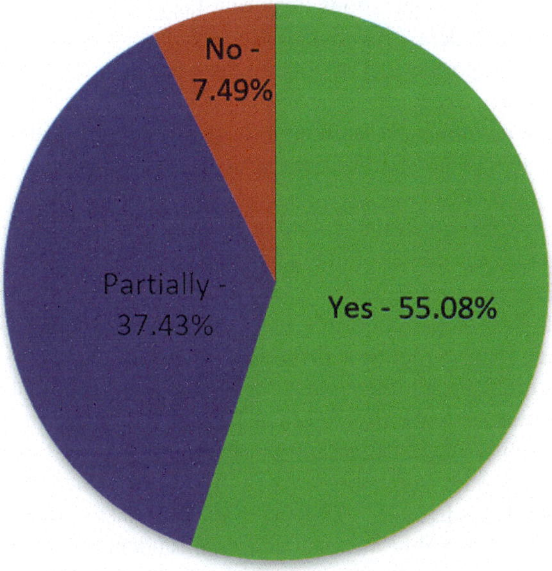

Fig. 9.8 Average value regarding the appropriateness of using e-learning to develop interaction skills in six-year-old children (according to preschool teachers' responses). (Source: Own work)

French colleagues indicated the following: Khan Academy, Duolingo, Kodable, SuzyApp (https://1ll.ink/h860Z).

The obtained data convincingly prove that appropriate educational and developmental materials for preschool children are developed on special electronic media. The second place in the ranking of the participants' responses (100% of preschool teachers in Ukrainian preschool institutions and 81.82% of French preschool teachers) holds YouTube. According to them, YouTube is a convenient and easy-to-use e-learning tool for preschool children. YouTube provides a wide range of videos to watch, in accordance with the topic of developing interaction skills in six-year-old children.

Next in line were presentations independently created by preschool teachers with relevant content. This response was given by 76.47% of respondents from Ukrainian preschool institutions and 63.64% from French ones. Comments indicated that such "visually live" information about developing interactions with others becomes more accessible for children to understand. Bright images, musical or verbal accompaniment on the monitor screen helps to activate the children's knowledge and life experience, moving from an explanatory and illustrative method of informational content to an activity-based one, where the child starts talking about certain situations they have experienced.

According to the generalized data, less than half of the respondents—41.18% of Ukrainian preschool teachers and 45.45% of French preschool teachers—indicated that they use thematic amateur videos created and tested by others and posted on the Internet to organize the educational process. These amateur video files typically depict various situations (featuring fairy tale or literary characters) that could occur in their students' real lives.

An important aspect of the study was obtaining a response to the third question of the survey regarding the preschool teachers' systematic use of e-learning: "How often do you use e-learning in organizing the educational process with six-year-old children to develop their interaction skills?" The quantitative results are presented in Fig. 9.3.

The responses to the third question suggest that e-learning for the development of 6-year olds' skills in interacting with others is conducted on an ongoing basis, but with varying frequency Among the respondents, about a third of the respondents (35.29% of preschool teachers in Ukrainian preschool institutions and 27.27% of French ones) indicated that this educational work takes place once a week, which is also reflected in their plans for organizing the educational process. An equal proportion, 35.29% of Ukrainian preschool teachers and slightly more—36.36% of French teachers—plan and conduct e-learning 2 times a month, i.e. in fact every other week. And approximately less than a third of respondents (29.42% of preschool teachers in Ukrainian preschool institutions and 27.27% of French colleagues) use this method with their students much less frequently—only once a month. It is worth noting that the responses provided by preschool teachers of French PEIs were calculated from the number of 10 respondents, as one respondent (9.1%) indicated in his questionnaire that he did not use e-learning in the organisation of the educational process with children.

The obtained results prove the unanimous respondents' opinion regarding the necessity and significance of using e-learning supplementary didactic tools for developing interaction skills in six-year-old children. With the help of e-learning, there is an alternation in the selection of various types of e-resources, which facilitates the implementation of educational tasks on the part of providers of educational services (preschool teachers) and satisfies the requests and interests of those getting education (children of preschool age). However, the frequency of its use varied: it was almost equally distributed across three gradations: once a week, twice a month and once a month. Planning and the use of e-learning in the development of the skills of interaction with others as soft skills in 6-year-old children in a quantitative dimension depends on the set goals and conditions (if, for instance, children in Ukrainian preschools spend a significant amount of time in a shelter as a safe place during an alert, it is quite appropriate for preschool teachers to engage them by demonstrating thematic videos).

During the survey of preschool teachers, it was important to determine the purpose of using e-learning in developing interaction skills in six-year-old children (the fourth open-ended question of the survey). Participants provided their own responses, which were grouped into three categories based on their content: "to visualize educational material", "to consolidate children's knowledge" and "to maintain children's interest". The obtained quantitative results are presented in Fig. 9.4.

In response to the fourth question, preschool teachers expressed their views on the purpose of using e-learning to develop interaction skills in six-year-old children.

Nearly half, specifically 47.06% of preschool teachers in Ukrainian preschools and 45.45% of preschool teachers in French preschools, see the main purpose as visualizing educational material for the children to master, thereby promoting better understanding and mastering. In fact, this educational and developmental function is seen as the primary purpose of using e-learning as a didactic tool. About a third of respondents (35.29% of Ukrainian preschool teachers and 27.29% of French ones) stated that e-learning should be used to reinforce children's knowledge, which we also consider valid. Significantly fewer respondents (17.65% of Ukrainian educators and 18.18% of French educators) identified the purpose of e-learning as an effective means to maintain children's interest. One respondent among the French preschool teachers (9.1%) has not responded to this question, which indicates the absence of e-learning in the planning and implementation of the educational process with preschool children. Therefore, interpreting the obtained data, we can assert that the vast majority of preschool teachers clearly understand the purpose of using e-learning in developing interaction skills in six-year-old children, specifically for improving and effectively mastering educational materials and consolidating the knowledge acquired by children. With the help of various e-resources, preschool teachers can choose such didactic tools that make a sufficient impact on the social development of preschoolers and stimulate them to think about how to treat others, the development of friendly relationships, etc. In this sense, e-learning as a kind of educational method performs both an educational and teaching function in the development of soft skills in 6-year-old children.

It was important to determine "How long do e-learning sessions aimed at developing interaction skills in six-year-old children last?" (the fifth question of the survey). The quantitative results are presented in Fig. 9.5.

Analyzing the responses to the fifth question, we note that they varied between preschool teachers in Ukrainian PEIs and their French colleagues. Despite the time limits for classes with the use of e-resources and information and communication technologies, determined by sanitary standards, Ukrainian teachers exaggerate them, in particular, 29.41% of respondents indicated a time of 15 min, while among French teachers this figure was only 18.18%. The duration of 20 min was indicated by 11.77% of the surveyed preschool teachers in Ukrainian preschools, whereas none of their French colleagues reported this duration. It is noted that since one of the respondents (9.1%) from French preschools indicated that they do not use e-learning in organizing the educational process, the calculations were made accordingly: only the responses of those who conduct such work, namely 10 people, were considered. A positive fact was that most of the respondents indicated a duration of 10 min for children's involvement in e-learning, which meets sanitary, hygienic and psychological-pedagogical requirements. Such respondents were 58.82% among Ukrainian preschool teachers and 72.72% among their French colleagues. The slight increase in the use of e-learning up to 15 min and even up to 20 min by Ukrainian educators may be due to their motivation not to interrupt the educational process, especially when children are in shelters during air raids, considering the state of war in Ukraine due to Russia's armed aggression.

The next question, related to the previous one, was the sixth question: "How do children react to the use of e-learning in the educational process?" The calculated data are presented in Fig. 9.6.

Analyzing the responses to the sixth question, we note a predominantly positive trend: more than half of the respondents observed that children demonstrate satisfaction and positive emotions—64.71% of preschool teachers in Ukrainian preschools and 54.54% of preschool teachers in French preschools. About a third of the respondents indicated that children behaved naturally during the use of e-learning: 29.41% of preschool teachers in Ukrainian preschools and 36.36% of French ones. However, there were also those who noted that children are indifferent and do not express any particular emotional reactions to e-learning—5.88% of preschool teachers in Ukrainian preschools and 9.1% of teachers in French preschools. These results indicate the ambiguous perception of educational content through e-learning by children, which will require a special study to identify the main factors. We assume that this could be due to psychological determinants, such as children being in stressful situations and lacking interest, as well as information overload. Children may have uncontrolled and unlimited access to e-resources in the context of family upbringing, which is not controlled or restricted by parents.

By formulating the seventh question: "Do you think it is appropriate to use e-learning to develop interaction skills in six-year-old children?" we aimed to determine whether the respondents consider the use of e-learning for educational purposes to be pedagogically normative. The processed quantitative data are presented in Fig. 9.7.

The results of the responses to the seventh question demonstrate that more than half of the preschool teachers in Ukrainian preschool institutions (64.71%) believe that using e-learning to develop interaction skills in six-year-old children is essential. The same opinion was held by 45.45% of their French colleagues. Regarding the partial implementation of e-learning, 29.41% of Ukrainian preschool teachers expressed this view, which is half the number of those who consider it essential, and 45.45% of French preschool teachers, which is the same proportion as those who are convinced of the necessity of e-learning in organizing the educational process. There were also respondents who considered it inappropriate to use e-learning in developing interaction skills in six-year-old children: 5.88% of Ukrainian preschool teachers and 9.1% of French ones. These results coincide with previously established data regarding the responses to the first question: "Do you use e-learning in organizing the educational process with six-year-old children?" Therefore, it is logical that the preschool teachers who consider using e-learning appropriate for developing interaction skills in six-year-old children are the ones who actually apply it in their real pedagogical practice.

Thus, based on the investigation of how e-learning is used in the real practice of preschool institutions to develop interaction skills in six-year-old children as soft skills (the first research question), the next task in conducting experimental work was to determine the summarized quantitative results and analyze the appropriateness of using e-learning in the organization of the educational process by preschool teachers (the second research question).

Based on the calculations of the summarized quantitative results, the average value was determined of the expediencies of using e-learning by preschool teachers in both Ukraine and France (total number of respondents is 28) to develop interaction skills in six-year-old children. The obtained data are presented in Fig. 9.8.

n = 28

Thus, as clearly shown in Fig. 9.8, more than half of the respondents (55.08%) believe that the implementation of e-learning in the educational process with six-year-old children is essential. Over a third of the surveyed educators (37.43%) consider partial application, meaning they do not deny the feasibility of such a method in preschool institutions. It must be noted that among the respondents, there were also those (7.49%) who did not believe that e-learning should be used for preschool children. The obtained results indicate a predominantly positive attitude of preschool teachers towards the use of e-learning in organizing the educational process and, in particular, in developing interaction skills in six-year-old children.

The conducted pilot study revealed the following general trends recorded in the results of the survey of respondents from the two countries: e-learning is used as an additional educational resource, which is mobile in its form and multifaceted, for the formation of the skills of interaction with others in 6-year-old children by content (such varieties as "educational platforms", "youtube", "own presentations", "amateur videos" were defined). This type of education activates preschoolers, makes it possible to expand their awareness of social reality, stimulates the development of their interaction with the immediate environment as soft skills and should be used in compliance with the relevant requirements: sanitary and hygienic, psychological, didactic, technical and others. Teachers of preschool educational institutions in Ukraine and France were unanimous in their belief that e-learning contributes to the visualization of the educational process, deepening and consolidation of children's knowledge, and also supports their desire to learn about the environment. Despite the position expressed by the overwhelming majority of respondents about the advisability of using e-learning in the educational process with children of preschool age, which as a teaching method will contribute to overcoming educational losses, some French teachers believed that preschool educational institutions should have a traditional system of work. Based on these findings, there is a need to develop conceptual provisions for a model of applying e-learning to develop interaction skills in six-year-old children as soft skills *(the third research question).*

Based on the materials of the analyzed studies and the results of the survey conducted among preschool teachers in Ukraine and France, we outline the initial vision of the investigated phenomenon as follows. The author's model is presented in Fig. 9.9.

We believe that for the successful application of e-learning in preschool institutions, a fully functional electronic learning environment must be ensured, as noted in studies by Blanco et al. (2022), García Laborda et al. (2023), and Zhang (2024). Through the use of various electronic informational educational resources, there will be an impact on the development of the child's Self-concept and self-awareness,

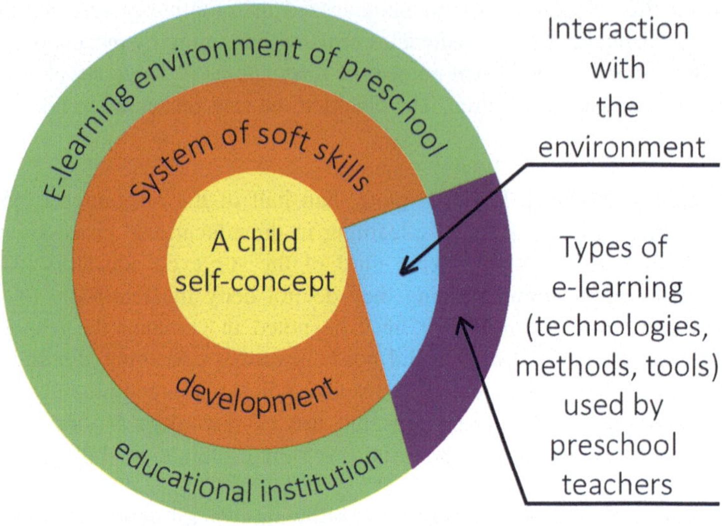

Fig. 9.9 Model for implementing e-learning in preschool institutions to develop interaction skills as soft skills in six-year-old children. (Source: Own work)

which will deepen and expand their knowledge of the surrounding reality and develop a range of life competencies. Among them, we focus on ensuring the development of a soft skills system, the main place among which is interaction with the environment (Boström et al., 2023; Cerezo et al., 2024). For this purpose, it is necessary to develop various types of e-learning that will include innovative technologies, effective methods, and efficient didactic tools. These will be used by preschool teachers in organizing the educational process to develop interaction skills in six-year-old children.

So, at the stage of conducting a pilot study, reliable and adequate results were obtained of a survey of preschool teachers in Ukrainian and French institutions regarding their use of e-learning in the formation of soft skills of 6-year-old children. Given the fact that the obtained results represent only a limited sample of experiment participants, these data cannot be generalized. In order to obtain reliable results, the questionnaire should be extrapolated for the general public of preschool education specialists, in particular, in the aspect of conducting a comparative analysis (with the involvement of respondents from different countries of Europe and the USA), comparing the provisions of the legislative framework of educational policy, social requests, as well as defining criteria and indicators and levels of readiness of preschool teachers to implement e-learning in the educational process with children of preschool age.

9.4 Discussion

Research on e-learning in developing interaction skills as soft skills in six-year-old children provides reasons to assert that this aspect of the pedagogical activity of preschool teachers (as exemplified by Ukraine and France) is being studied from various perspectives. Based on the results of the theoretical study of the literature and empirical research, we assert the undeniable need for using e-learning in organizing the educational process with preschool children (Raviv & Galili, 2024).

However, questions remain as it is debatable how to make e-learning interesting and interactive for children aged three to six (Hashem, 2022). There are also concerns among preschool teachers about the implementation of e-learning in preschool institutions due to "lack of knowledge or time" (García Laborda et al., 2023). Therefore, it is necessary to expand the range of electronic educational resources (Behnamnia et al., 2023) and provide targeted training for preschool education professionals to implement e-learning for children (Chen & Ding, 2024). Important for scientific understanding are also questions of optimizing e-learning, implementing advanced pedagogical experience for its use in the broader context of the preschool education system, and developing such electronic educational content designs that would raise children's interest and desire to learn new things with enthusiasm (Zhang, 2024). This involves ensuring the integration of e-learning tools and resources into preschool education programs (Chen & Ding, 2024). The use of e-learning in the development of children's soft skills, the development of the ability to 'independent learning and teamwork', as well as the search for mechanisms to monitor its success should not be left out of the attention of scientists (Zhang, 2024). This should be supported by the development of regulatory documents on quality standards for implementing e-learning in the educational process with preschool children and the establishment of relevant structural units to ensure e-learning, aimed at eliminating the imbalance in the use of various educational resources (Zhang, 2024).

9.5 Conclusions

Thus, we came to the following conclusions: the raised scientific problem of e-learning in the formation of interaction with others as soft skills in 6-year-old children occupies a prominent place in modern psychological and pedagogical research, which developed the basis for its comprehension and outlining of key provisions. The empirical search focused on three main research issues, all of which were fully realized:

1. It has been established that in the real practice of preschool institutions, e-learning is used by preschool teachers in organizing the educational process, particularly to develop soft skills in six-year-old children. The focus was on developing interaction skills in senior preschoolers: the purpose of using e-learning, its frequency and duration, types, children's responses and its relevance, etc.

2. Based on the survey conducted among preschool teachers regarding the effectiveness of using e-learning to develop interaction skills as soft skills in six-year-old children, its essential significance, effectiveness, and the need for dosed implementation in the educational process have been demonstrated.
3. Based on the synthesis of theoretical review materials and the obtained empirical results, conceptual principles for a model of applying e-learning to develop interaction skills as soft skills in six-year-old children have been developed. The testing of this model will enrich the scientific repository.

Thus, the formulated hypothesis is confirmed: the survey of preschool teachers as providers of educational services (on the example of Ukraine and France) on the use of e-learning in the formation of interaction with others as soft skills in 6-year-old children made it possible to find out the real state of the scientific problem in reality. In turn, this gave ground to identify the following promising research objectives: (1) development of an experimental methodology for integrating e-learning into the educational process of preschool education institutions in partial areas of work; (2) identification of psychological and pedagogical conditions and didactic tools (electronic educational resources) for the effective implementation of e-learning with preschool children; (3) selection of diagnostic tools for monitoring the use of e-learning in the preschool educational system.

Acknowledgements We express our gratitude to the management of preschool institutions in the cities of Pereiaslav and Cherkasy (Ukraine) and Bilingual Montessori Preschool in Vincennes (France) for their assistance in collecting data for the empirical research: allowing us to analyze the plans for organizing the educational process with six-year-old children and conducting a survey among preschool teachers.

Appendix 1

Survey for Preschool Teachers

To investigate e-learning in developing interaction skills as soft skills in six-year-old children

1. Do you use e-learning in organizing the educational process with six-year-old children:

   (a) yes; (b) sometimes; (c) no

2. What types of e-learning do you use to develop interaction skills in six-year-old children (specify):____________________________________
3. How often do you use e-learning in organizing the educational process with six-year-old children to develop their interaction skills:

   (a) once a week; (b) twice a month; (c) once a month

4. What is the purpose of using e-learning in developing interaction skills in six-year-old children (please specify):_______________
5. How long do e-learning sessions aim at developing interaction skills in six-year-old children last:

   (a) 10 min; (b) 15 min; (c) 20 min

6. How do children react to the use of e-learning in the educational process:

   (a) satisfied; (b) perceive it as a usual occurrence; (c) indifferent

7. Do you think it is appropriate to use e-learning to develop interaction skills in six-year-old children

   (a) yes; (b) partially; (c) no

Thank you for participating in the survey!

References

Baltzaki, M., & Chlapana, E. (2023). Fostering receptive vocabulary development of kindergarten children with the use of information and communication technologies (ICT). *Education and Information Technologies, 28*, 14019–14049. https://doi.org/10.1007/s10639-023-11707-7

Behnamnia, N., Kamsin, A., Ismail, M. A. B., & Hayati, A. (2020). The effective components of creativity in digital game-based learning among young children: A case study. *Children and Youth Services Review, 116*, 105227. https://doi.org/10.1016/j.childyouth.2020.105227

Behnamnia, N., Kamsin, A., Ismail, M. A. B., & Hayati, S. A. (2023). A review of using digital game-based learning for preschoolers. *Journal of Computers in Education, 10*, 603–636. https://doi.org/10.1007/s40692-022-00240-0

Blanco, B. B., Triviño-Rodriguez, J. L., Trella-López, M., & Marco-Rubio, J. (2022). How do preschoolers interact with peers? Characterising child and group behaviour in games with tangible interfaces in school. *International Journal of Human-Computer Studies, 165*, 102849. https://doi.org/10.1016/j.ijhcs.2022.102849

Boström, J., Hultén, M., & Gyberg, P. (2023). Who counts? Legitimate solutions in construction activities in preschool. *International Journal of Technology and Design Education, 33*, 1309–1344. https://doi.org/10.1007/s10798-022-09784-9

Bylieva, D., & Nordmann, A. (Eds.). (2022). Technology, innovation and creativity in digital society. In *XXI professional culture of the specialist of the future*. Springer. https://doi.org/10.1007/978-3-030-89708-6. Hardcover ISBN 978-3-030-89707-9 Softcover ISBN 978-3-030-89710-9.

Cerezo, E., González-González, C. S., & Bonillo, C. (2024). Empowering soft skills in children with ADHD through the co-creation of tangible tabletop games. *Universal Access in the Information Society, 23*, 3–21. https://doi.org/10.1007/s10209-023-01041-7

Chen, Y., & Ding, Z. (2024). Effects of digitalization in preschool education on the creative and cognitive development of children. *Education and Information Technologies*. https://doi.org/10.1007/s10639-024-12730-y

García Laborda, J., Madarova, S., & Calle Martínez, C. (2023). Technology for the CLIL preschool. In A. Otto & B. Cortina-Pérez (Eds.), *Handbook of CLIL in pre-primary education* (Springer International Handbooks of Education) (pp. 495–512). Springer. https://doi.org/10.1007/978-3-031-04768-8_31. Print ISBN 978-3-031-04767-1 Online ISBN 978-3-031-04768-8.

Gomez-García, S., Marín-Flores, R., Vallejos-Poma, K., & Iraola-Real, I. (2023). Teaching experience on virtual dramatization in the development of social skills in preschoolers. In M. Botto-Tobar, M. Z. Vizuete, S. M. León, P. Torres-Carrión, & B. Durakovic (Eds.), *Applied technologies. ICAT 2022. Communications in computer and information science* (Vol. 1757, pp. 45–54). Springer. https://doi.org/10.1007/978-3-031-24978-5_5. Print ISBN 978-3-031-24977-8 Online ISBN 978-3-031-24978-5.

Gözen, G., & Cırık, İ. (2017). Impact of digital storytelling on social-emotional behaviours of preschool children. *Elementary Education Online, 16*(4), 1882–1896. https://doi.org/10.17051/ilkonline.2017.358215

Hashem, F. A. R. A. H. (2022). E-learning in early childhood. *International Journal of Childhood and Women's Studies, 2*(2), 27–52. https://doi.org/10.21608/ijcws.2022.246495

Hastürk, G., & Özdemir, O. (2021). Investigation of prospective preschool teachers' attitudes towards science education and learning styles. *Journal of Educational Issues, 7*(1), 260–281. https://doi.org/10.5296/jei.v7i1.18471

Lokhvytska, L. V., Andriushchenko, T. K., Martovytska, N. V., & Joniliene, M. (2022). The importance of ICT in senior preschoolers' moral education. *Information Technologies and Learning Tools, 89*(3), 1–15. https://doi.org/10.33407/itlt.v89i3.4715

MacDonald, A., Huser, C., Sikder, S., & Danaia, L. (2020). Effective early childhood STEM education: Findings from the "little scientists on" evaluation. *Early Childhood Education Journal, 48*(3), 353–363. https://doi.org/10.1007/s10643-019-01004-9

MacDonald, A., Danaia, L., Shukla, S., & Huser, C. (2021). Early childhood educators' beliefs and confidence regarding STEM education. *International Journal of Early Childhood, 53*(3), 241–259. https://doi.org/10.1007/s13158-021-00295-7

Masaroğulları, N., & Çerkez, Y. (2014). The mediating role of mothers' conflict resolution skills in the relationship between mothers' communication skills and behavioral problems of children aged 3–6 years. *Current Psychology, 43*, 21032–21045. https://doi.org/10.1007/s12144-024-05888-8

Olisah Kingsley, S., & Mohamed Ismail, Z. (2015). Web based E-learning system for pre-school kids. *International Journal of Information System and Engineering, 3*(1), 194–207. https://doi.org/10.24924/ijise/2015.11/v3.iss1/219.232

Rahiem, M. D. H. (2021). Storytelling in early childhood education: Time to go digital. *International Journal of Child Care and Education Policy, 15*, Art. 4. https://doi.org/10.1186/s40723-021-00081-x

Raviv, A., & Galili, I. (2024). Preschool teachers' attitudes towards the implementation of science and technology studies in preschool. *Early Childhood Education Journal, 52*, 575–585. https://doi.org/10.1007/s10643-023-01461-3

Zhang, W. (2024). The impact of online learning on children's education and optimisation strategies. *Lecture Notes in Education Psychology and Public Media, 41*(1), 283–289. https://doi.org/10.54254/2753-7048/41/20240818

Chapter 10
The Use of Artificial Intelligence in the Professional Activity of University Teachers in Ukraine

Svitlana Skvortsova, Tetiana Symonenko, Tetiana Britskan, and Yana Haievets

Abstract The modern global landscape is rapidly advancing, driven by the significant potential of information technology, particularly artificial intelligence (AI) systems.

Contemporary AI systems can serve as powerful tools in higher education, contributing to the development of innovative educational technologies and teaching methods, as well as acting as knowledgeable assistants to teachers.

In this article, we present a range of effective AI tools and outline AI's technological capabilities in automating routine tasks for educators. A distinctive feature of this study is the inclusion of data from an empirical experiment aimed at assessing educators' attitudes toward the use of AI in their professional activities and their awareness of AI-based tools.

For the first time, we present data from a survey of 205 educators from Ukrainian universities, summarized by specific indicators and criteria.

Keywords Artificial intelligence · AI tools in higher education · New information technologies

10.1 Introduction

Artificial intelligence (AI) is understood as the capability of a machine or computer program to think, act, and react like a human; to learn, reason, solve problems, understand language, and analyse information. It is important to note that a single,

S. Skvortsova (✉) · Y. Haievets
Faculty of Primary Education, South Ukrainian National Pedagogical University named after K. Ushynsky, Odesa, Ukraine

T. Symonenko
Bohdan Khmelnytsky National University at Cherkasy, Cherkasy, Ukraine

T. Britskan
Faculty of Pedagogy, Izmail State University of Humanities, Izmail, Ukraine

E. Smyrnova-Trybulska et al. (eds.), *E-Learning and Enhancing Soft Skills*,
https://doi.org/10.1007/978-3-031-82243-8_10

universally accepted definition of AI has not yet been established, despite ongoing efforts by leading scientists such as Yoshua Bengio, Luciano Floridi, Demis Hassabis, Geoffrey Hinton, Yann LeCun, Ray Kurzweil, Andrew Ng, Jürgen Schmidhuber, Josh Tenenbaum, and Thomas Wiegand to define the concept and its structure.

Although AI currently performs only limited tasks and has not yet achieved human-level intelligence, researchers are actively exploring its applications across various domains, particularly in education. Recently, educators have increasingly focused on using AI to implement individualized teaching technologies. Researchers (Phillips et al., 2022) discuss natural language processing networks, such as chatbots (for example, ChatGPT (OpenAI), Claude Anthropic, Google Bard, Microsoft Copilot, and Perplexity), which can assist in delivering individualized, personalized education. These programs simulate real dialogues with users, enhancing the learning experience.

10.2 Tools of AI in the Professional Activities of the Teacher

With the help of voice assistants, in particular Chat GPT OpenAI, students can access information, receive tips on various issues, and the chatbot is able to motivate them. At the same time, it should be noted that Chat GPT can function offline, since it has its own database. Claude can perform the same function, but this neural network is able to highlight key concepts, facts, questions and answers to them, and offer students personalized tasks and tests, which are adapted to the level and needs of the student can also be generated by Google Bard and Bing/Copilot. Unlike Chat GPT, these chatbots are connected to the Internet, and therefore have a newer database. To write creative content—essays, poems, stories, teachers can use the Perplexity chat bot, and the Phind chatbot is also useful for recognizing and creating images, uploading them, and generating text descriptions, questions related to pictures and answers.

It is important to take into account the fact that chatbots search the Internet for information and compile responses, which may not always be reliable. Teachers, with their knowledge in their subject areas, are generally able to evaluate the accuracy of information critically. Students, however, often lack this essential knowledge, making it challenging for them to assess the reliability of information received from chat bots. Therefore, it is crucial to approach AI-generated information with caution and verify its accuracy. It should be noted that Google Bard provides links to the original sources of its information, which can help in this verification process.

Therefore, AI systems, in particular neural networks, enable online communication at convenient times for the student and with the amount of educational information corresponding to their intellectual abilities, thus creating conditions for the implementation of personalized education. In addition, AI can provide personalized guidance for a student's learning activities, providing instant feedback, assistance, and correction throughout the learning process. By adapting educational content according to individual learning models, AI can take into account the student's existing level of knowledge (Hwang et al., 2020).

In general, AI is able to analyse a student's knowledge level, skills, learning abilities, style, and pace to create personalized learning plans, recommendations, tasks, and feedback. For example, Quizbot, a tool within Google Forms, allows teachers to create tests by entering a topic, keywords or a link to the source. Quizbot then generates questions and answer options, which teachers can edit and add to their own forms.

Some AI tools, such as CogBooks, enable the creation of dynamic and personalized courses on various subjects. This online service tracks the student's motivation and progress, adapting content and difficulty accordingly.

10.2.1 Analysis of Scientific Research

AI creates opportunities for personalised learning (Nye, 2015), and facilitates the implementation of automated assessment, the introduction of facial recognition systems and predictive analytics (Akgun & Greenhow, 2022). AI is widely used in language learning, in training programs for professionals, including doctors, engineers, lawyers and educators.

Professor Pierre Dillenbourg from École Polytechnique Fédérale de Lausanne is involved in the TinkerLamp project, where he develops and evaluates AI tools for teaching and learning. These tools include robots, desktop interfaces, and virtual reality systems that create immersive collaborative learning scenarios. His research explores cognitive and social aspects of human-machine interaction (Dillenbourg, 2024), examining how learners perceive and interact with AI systems, how these systems provide feedback and guidance, and how they facilitate collaboration and creativity.

Pierre Dillenbourg uses AI for educational innovations, including through the MOOC strategy of EPFL, which expands access to high-quality online education. He also launched the Swiss EdTech Collider, an incubator for learning technology start-ups, and co-founded several companies in the field of AI and education.

Yi-Shan Tsai's research focuses on how AI, along with digital and data technologies, can improve learning, teaching and educational environments (Tsai, 2024). This exploration is part of projects such as Sense-making of data-based feedback and self-regulated learning and Enhancing data-based feedback through storytelling.

Carolyn Penstein Rose designs and evaluates AI systems that facilitate online learning communities, including projects, such as Dynamic Support for Collaboration (Rose, 2024) (1). Through the Rapid Authoring of Intelligent Tutors for Collaborative Learning project, she develops intelligent tutors that provide adaptive feedback and guidance for students and teachers (2). Her research also explores automated analysis of collaborative learning processes, such as how to use text mining and machine learning techniques to measure and model student discourse, cognition, and social interaction.

Rose Luckin leads the EDUCATE project, which explores how AI can improve human learning and collaboration. She also develops and evaluates AI tools to support teachers and students, such as the Ecolab project, a game environment that helps students learn about ecosystems, and the Emotionally Responsive Learning Environment project, a system designed to detect and respond to students' emotions and interests.

Consequently, academic attention is focused on how AI can improve learning, teaching and educational environments through digital and data processing technologies. Notably, researchers are concentrating on the development of AI tools and examining the interactions between the teacher and students during the learning process when using AI tools.

At the same time, scientists are concerned about the ethical issues related to the use of AI. The development of AI in education raises ethical risks, such as personal data and student autonomy, and the academic integrity of both students and teachers. Scientists (Nguyen et al., 2023) analyse the opportunities offered by AI in education and potential ethical issues and propose a set of ethical principles. It is worth noting that UNESCO has developed ethical standards for AI, which were agreed upon and signed by its 193 member states on 25 November 2021 (UNESCO, 2021).

10.2.2 Routine Tasks of a University Teacher

At the same time, although researchers emphasize that AI can assist teachers in performing routine tasks (Popenici & Kerr, 2017), this issue remains underexplored. The routine responsibilities of a university teacher include organizing class schedules, recording student attendance, preparing and delivering lectures, seminars, and consultations, assessing assignments, developing and updating curricula and teaching materials, drafting scientific articles and monographs, organizing and participating in academic conferences and seminars, collaborating with colleagues, administration, and students, and engaging in professional development.

Notably, our previous publications (Skvortsova, 2024a, b; Skvortsova & Britskan, 2024) present research on the application of AI tools in managing these routine tasks in the professional activities of university teachers.

Artificial Neural Networks (ANN) function as analogs of the human brain, mimicking the activity of biological neurons. These networks consist of interconnected nodes, each acting as an artificial neuron, which processes signals and transmits them to other neurons. Neurons are organized into layers, each performing various data transformations. This type of AI is utilized for tasks such as pattern recognition, including object identification in images or videos. For example, ANN can help teachers monitor students' emotions during classes, thereby supporting their cognitive engagement.

ANNs learn from examples and can detect complex patterns in data, making them valuable in data mining. Similar functions are performed by Expert Systems (ES), which use predefined rules and facts to solve specific problems within narrow domains.

Expert Systems (Mesbah et al., 2019) emulate human reasoning in problem-solving. In higher education, these systems can assist teachers in designing and evaluating educational activities such as games, essays, and simulations. AI tools can adapt learning materials, methods, and pacing to the individual needs of students, taking into account prior knowledge and knowledge gaps. These tools can generate personalized tasks, automatically assess student work, and provide tailored assistance. For instance, the ALEKS system in mathematics creates individualized learning plans based on student assessments, while Math Expert assists in solving and verifying mathematical problems. Geometer's Sketchpad allows manipulation and exploration of geometric shapes and properties.

In language instruction, Expert Systems analyse and correct speech, identify student errors, provide hints and explanations, generate and check texts, simulate various speech situations, and create realistic communication scenarios. Tools like Rosetta Stone support listening, speaking, reading, and writing practice in English. Chatbots, such as "Replika," offer basic dialogue creation for practicing speech scenarios on various topics, while Grammarly checks spelling, contextualizes messages, and suggests stylistic improvements.

Psychometric Expert software enables psychologists to administer tests, analyse results, and create custom assessments based on existing templates, allowing the development of new testing rules via the Expert Systems service.

Consequently, AI can enhance learning in several ways: it can respond to student queries, serve as a simulator for skill practice, and assess student work while providing feedback.

In addition to ANNs, natural language processing (NLP) technology is capable of understanding and processing human speech. NLP focuses on language comprehension and manipulation by computers, involving tasks such as translation, speech recognition, semantic analysis, question answering, and text generation. NLP applications include automatic translation, text analysis, and voice search.

Examples of NLP tools already used by university professors include ChatGPT (OpenAI), Claude (Anthropic), Google Bard, Microsoft Copilot, Perplexity, and Phind. These tools can answer questions, generate texts, and provide information, though it is important to remember that they may not always provide reliable answers as they search for information online.

Computer Vision (Guo et al., 2022), the ability of a computer to recognise, interpret, and understand visual information, is another type of AI that can be applied in education, such as reading graphs, charts, and recognising handwritten text.

Moreover, certain AI tools, such as Gradescope, facilitate peer assessment among students, offering structured feedback.

AI systems also streamline the analysis of class attendance (Roll & Winne, 2015; Fong et al., 2019), as seen in Ukrainian universities that conduct online distance learning through platforms like Teams, Moodle, Google Meet, and Zoom, which integrate AI tools for attendance tracking.

For instance, Teams generates attendance reports in Excel format for each class and allows for features like face recognition and emotion detection during lessons through Computer Vision tools. Although not all platform features rely on AI, enterprise platforms are increasingly incorporating AI to optimize routine teaching tasks. However, in the context of looking at a university teacher solving routine tasks, it is worth noting that the scheduled classes on this platform are recorded in the calendar, which essentially represents the teacher's class schedule. Also, during classes in MS Teams, you can turn on transcription and receive a text file with a summary of the class; you can record a video and it is placed in the group for a certain time. On this platform, when creating a group, e-mails of students are entered, they are automatically invited to it, and in this way an electronic magazine is generated. On the platform, the teacher posts tasks and notes the deadlines and indicates the maximum number of points; students upload completed assignments to the platform. After the teacher checks the tasks, the number of points scored by the student for each task is automatically displayed in the electronic journal. Moreover, the teacher always has the opportunity to review the student's work again. So far, on this platform, the teacher personally checks the student's work. Awarding a certain number of points, but it has the ability to provide the student with a comment with instructions in text, voice or video form. These functions are usually implemented through queries to the data warehouse and interaction with the web interface, rather than through AI technologies. However, the capabilities of enterprise platforms are constantly expanding through the use of AI tools, and in the future, university professors will be able to further optimise the process of solving routine tasks in their professional activities.

Thus, as evidenced, AI already plays a significant role in organizing the professional activities of university teachers and holds potential for further assisting in routine tasks.

Research Questions

1. What is the attitude of university teachers towards the use of AI in professional activities, considering both its benefits and limitations?
2. Does teachers' awareness of AI tools contribute to the efficiency of solving routine tasks?

The hypothesis of this research posits that the use of AI tools in the professional activities of higher education teachers can enhance the efficiency of performing routine tasks.

Research Methodology

In the first half of 2024, we conducted a survey involving 205 Ukrainian university teachers. The purpose of this survey was to investigate the perception and use of AI tools among Ukrainian university educators within the educational process.

The purpose of this questionnaire is to investigate the problem of perception and use of AI tools by teachers of Ukrainian universities in the educational process.

The research objectives were to clarify the current state of pedagogical practice in Ukrainian higher education institutions, focusing on: (1) university teachers' attitudes towards AI in professional activities, considering both its advantages and limitations; and (2) awareness and use of AI tools in addressing routine tasks.

The survey was developed based on an analysis of the professional activities of Ukrainian university teachers, identifying routine tasks (Skvortsova et al., 2022; Skvortsova, 2024a, b; Skvortsova & Britskan, 2024) that could be optimized through AI. The survey design also incorporated an analysis of the types of AI and services that could aid in these professional tasks. The questionnaire, created with assistance from Bing/Copilot and ChatGPT, consisted of 11 closed-ended questions with optional open-ended responses. It was distributed via Google Forms (https://docs.google.com/forms/d/1nIixlaPT5wD0q41x_mRu1bDjXw9fnaq_H4Hgv8JpHnY/edit?ts=65bfeb23) and shared on social networks such as Facebook, Viber, Telegram, Messenger, and the website of the International Public Association "Ukrainian Researchers in Austria."

The questionnaire included questions related to: (1) the respondent's teaching discipline; (2) tenure at the university; (3) frequency of AI usage in professional activities; (4) AI tools currently used or desired by the respondent; (5) self-assessment of AI skills; (6) perceived benefits of AI use; (7) risks and challenges associated with AI; (8) attitudes towards AI; (9) sources of AI-related information; (10) AI training needs; and (11) the purpose of AI use in professional activities.

10.3 The Results of the Study

10.3.1 Analysis of the Results of the Questionnaire of Ukrainian University Teachers

A total of 205 university teachers from Ukraine participated in the survey, with 71.1% having more than 10 years of experience at the university, and only 5.4% having between 1 and 3 years of experience. The distribution of respondents by specialty is as follows (Fig. 10.1): 12.3% representatives of humanities, 22.5%—teachers of pedagogy, 10.8%—mathematical sciences, 22.5%—teaching methods, 13.7%—psychology, 6.4%—social sciences, 4.9%—natural sciences, 2.5%—technical sciences. The survey participants, university teachers in Ukraine, possess considerable experience of working at the university before AI was integrated into the educational field. Moreover, the majority of teachers do not represent exact sciences—mathematics, natural sciences and technology.

The survey results indicate that only 30.4% of teachers do not use AI tools, while 5.4% use them daily, 15.2% weekly, 18.6% monthly, and 30.4% several times a month (Fig. 10.2). It is important to note that the responses of 30.4% of respondents may be questionable, as they might not even be aware that they are using AI. In Ukraine, due to the full-scale aggression of the Russian Federation, many students

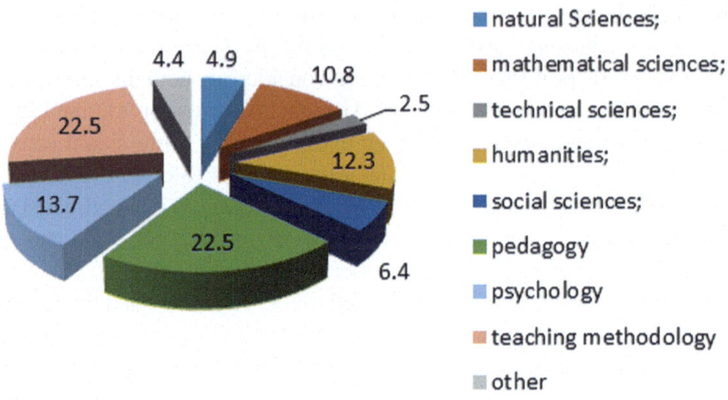

Fig. 10.1 Distribution of respondents by academic disciplines. (Source: Own research)

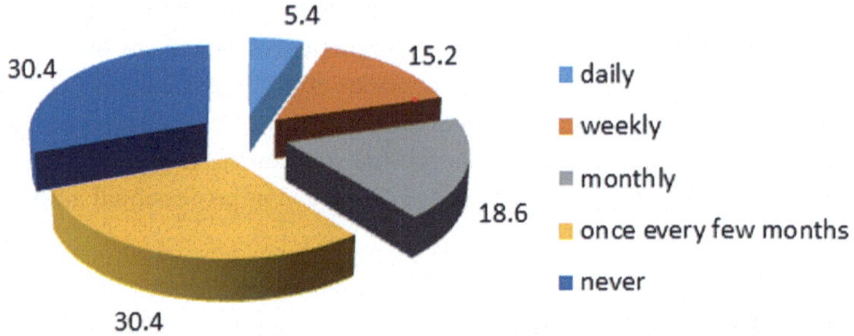

Fig. 10.2 Frequency of use of AI in professional activities. (Source: Own research)

are studying remotely, with teachers conducting real-time online classes on educational platforms. Universities operate on platforms like Teams, Moodle, Google Meet, and Zoom, which already incorporate built-in AI systems, particularly NLP tools that enable features such as speech-to-text transcription, text translation, and more. For example, the Teams platform includes several AI-driven functionalities that assist teachers in managing routine tasks. These include the automatic generation of class schedules that allow teachers to start lectures or practical lessons, and the automatic creation of journals for students' assignments. Additionally, the platform generates tables that contain information on submission deadlines, the number of points students have earned, and it calculates the percentage of successfully completed assignments by group. Moreover, teachers can review, evaluate, comment on student work, and either accept or request revisions at any time. AI tools built into the platform allow teachers to provide feedback in text, audio, or video format, significantly reducing the teacher's workload while offering students detailed expert assessments of their work. Teachers and students can also use these tools to transcribe speech into text, which only requires minor edits to create a lecture summary. While students still need to work with the finished text—highlighting key points, creating an outline, etc.—the process is streamlined by AI assistance. The platform also allows for video recording of meetings, enabling students to review lesson content at their convenience.

Google Meet offers similar features, including "live captions" for real-time subtitles during video calls. Google Meet also supports automatic translation from English to languages like German, French, Spanish, and Portuguese, allowing users to read the translation through "live captions."

The Zoom platform also includes built-in AI tools that enable conference organizers to create video meeting notes and share them via team chat and email without recording the conversation. Similar to Google Meet, Zoom allows users to work with subtitles, and speech translation is possible if the audio channels for simultaneous translation are pre-configured.

On the Moodle platform, AI systems are used to automate student assessments. AI helps determine the number of grades received per semester, the attendance record, the total number of classes, the total points earned by a student, and the percentage of points obtained relative to the maximum possible. Based on these results, teachers can generate a rating of student performance.

For example, more than a third of Ukrainian university teachers are unaware of the AI capabilities built into the platforms they use almost daily in their professional activities. It is therefore plausible that 30.4% of educators do not utilize AI features such as subtitling, simultaneous translation, voice-to-text conversion, and other functions, and may not employ AI in their professional activities.

Moreover, 71.1% use or wish to use AI for content generation (text, images, audio, video, etc.); 32.3%—for data analytics (statistics, visualization, forecasting, etc.); 42.3%—for process automation (evaluation, verification, feedback, etc.); 24.4%—for adaptive training (individualization, personalization, recommendations, etc.). Interestingly, 18.4% of teachers use or want to use voice and visual technologies (recognition of speech, faces, gestures, etc.) (Fig. 10.3). Additionally, among our own answer options, we included the use of AI for text translation.

Evaluating the answers of university teachers, we can state that the most common application of AI is in generating text, video, and images. AI tools are widely used by teachers to search for information, to visualize it through relevant images or videos based on specific descriptions or requirements. Teachers also use speech-to-text transcription tools to generate their own text, significantly speeding up the writing process. In addition, AI tools, such as Claude Anthropic, Microsoft Copilot, offer features for text editing and translation into other languages. However, voice and visual technologies for recognizing speech, faces, and gestures are used to a lesser extent. We believe this limitation is largely due to the technical capabilities of the platforms used by Ukrainian universities. However, research on recognizing students' emotions during class, such as in the Emotionally Responsive Learning Environment project by Rose Luckin, suggests that these tools may soon be integrated into educational platforms.

As evidenced by the results of the survey, teachers widely use AI to automate assessment, verification and feedback processes. This is made possible by integrating appropriate AI tools into educational platforms. As a result, respondents use data analytics, in particular to monitor student attendance. The Teams platform eliminates

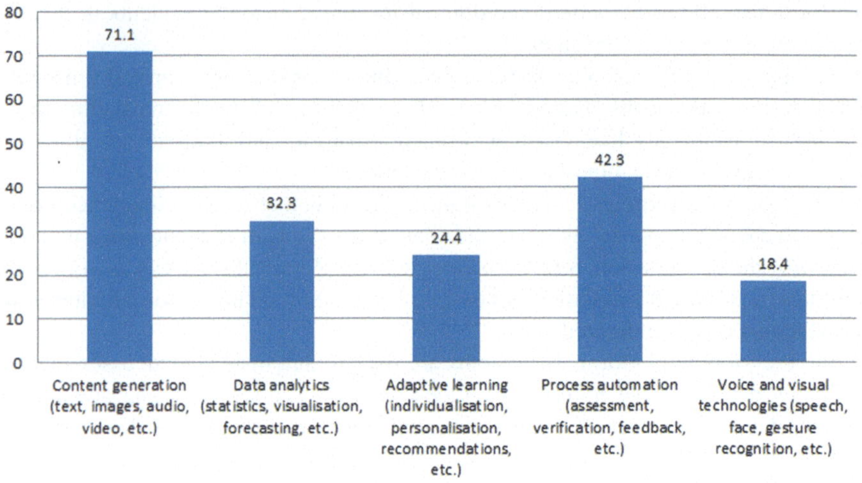

Fig. 10.3 Types of AI tools used by teachers. (Source: Own research)

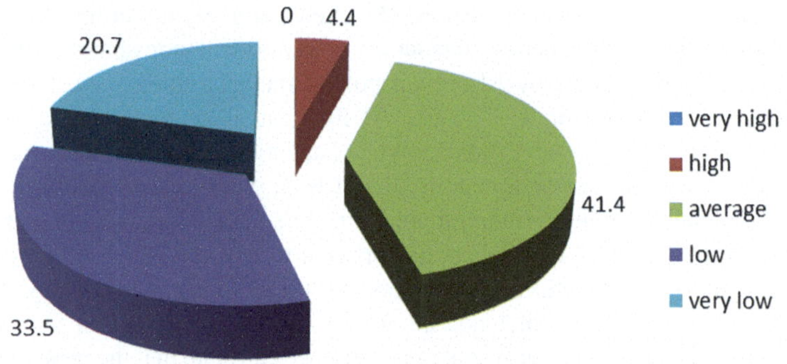

Fig. 10.4 Teachers' self-perception of their level of AI knowledge and skills. (Source: Own research)

the need for traditional roll call, providing minute-by-minute attendance information. Similar possibilities are available on the Moodle platform.

In recent years, the technology of personalized learning paths has been introduced in Ukrainian universities. More than a fifth of respondents use or want to use AI tools for adaptive learning, and personalization.

This analysis indicates that Ukrainian teachers are determined to use AI tools in their professional activities.

One of the questions in the questionnaire involved the assessment by university teachers of their own level of knowledge and skills in the field of AI (Fig. 10.4). The fact that the majority of teachers have either an average (41.4%), low (33.5%), or very low (20.7%) level is evident. At the same time, 4.4% of respondents noted that they have a high level of knowledge and skills in the field of AI. Therefore,

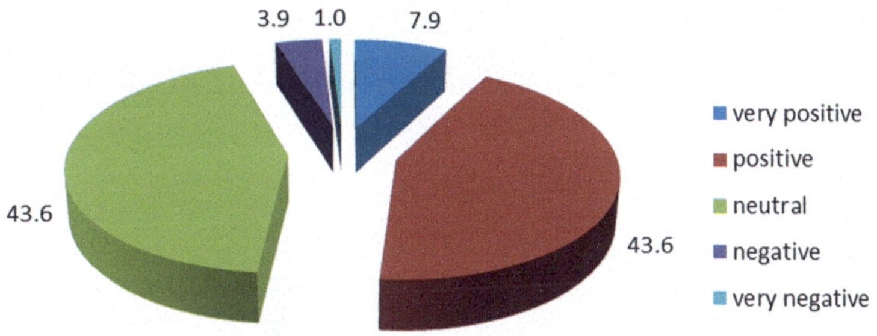

Fig. 10.5 Teachers' attitudes towards the use of AI in education in general. (Source: Own research)

university teachers are just beginning to master the tools of AI to improve the efficiency of their own professional activities.

Among the advantages of using AI, teachers see: Expanding learning opportunities and accessibility (73.9%), increasing efficiency and productivity (50.7%), stimulating creativity and innovation (38.4%), improving the quality of education (25.6%). Among their own answers, the teachers noted the development of critical thinking and the reduction of routine work. At the same time, teachers see the greatest risk of AI in the reduction of critical thinking and interaction (75.1%); in violation of ethics and human rights, the collection of personal data—40.8% of respondents; in loss of control and autonomy—29.9%. To a lesser extent, teachers are concerned about violation of academic integrity (0.5%), increasing inequality and discrimination (8.5%). In general, 2% of teachers do not see any risks. Among their own answers, a decrease in independence in thinking was noted. In general, 7.9% of respondents are very positive about the use of AI in education, 43.6% are positive, 43.6% are neutral, and 3.9% are negative. At the same time, it should be noted that 1% of the teachers marked the answer as "very negative" (Fig. 10.5).

As noted above, when considering the risks of using AI in education, 40.8% of university professors are concerned about ethics and human rights violations due to the collection of personal data. In particular, the ethics and legality of the use of text generated by AI, especially in relation to its copyright attribution. In addition, 29.9% of teachers cite loss of control and autonomy among the risks of using AI. Of course, the use of chatbots affects the creativity of both students and teachers, especially when content creation is reduced to simply formulating a query to an AI program. Easy access to information without critical reflection reduces the motivation to learn new things, e.g. to learn and improve a foreign language, because of the possibility of quick automatic translation. Quick access to theoretical issues reduces the motivation to memorise formulas, laws, etc.

Responding the question "What sources of information about AI do you use or would like to use?", the largest percentage of respondents indicated educational courses and webinars (69.2%). As many as 48.8% of respondents want to learn about AI independently through professional sites and blogs, 44.3%—through scientific journals and articles, 31.3%—through social networks and forums.

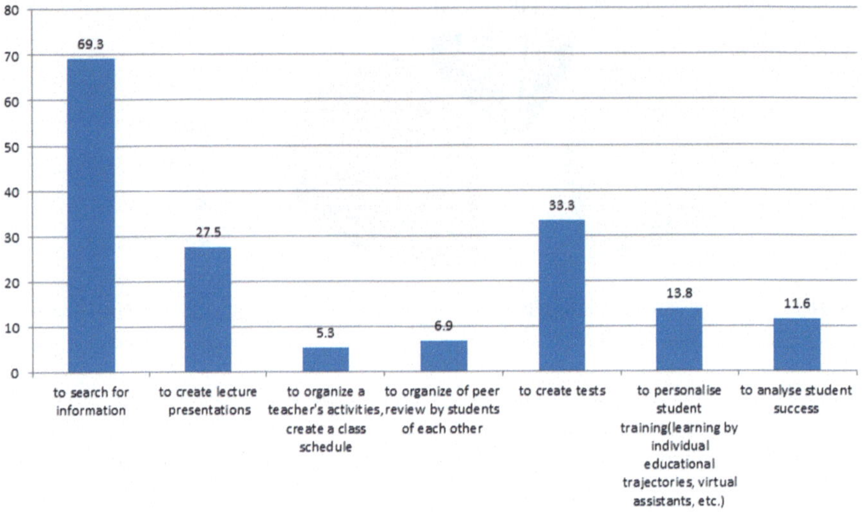

Fig. 10.6 The purpose of using AI in the professional activity of a university teacher. (Source: Own research)

In mastering AI teachers require the following: AI-based tools that allow the teacher to automate the performance of routine tasks (76.4%), pedagogical strategies and approaches to learning using AI (54.7%), knowledge about the basics of AI and its application (general information about AI and its possibilities to facilitate the performance of professional functions of a university teacher) (40.9%), consideration of ethical and social aspects of the use of AI for training students (33.5%). Additionally, respondents emphasised the need for the development of pedagogically oriented AI systems.

The survey results indicate that almost all teachers at Ukrainian universities use AI to perform professional functions (Fig. 10.6). The specific uses of AI include: searching for information—69.3%, creating tests—33.3%, developing lecture presentations—27.5%, personalizing student training through individual educational trajectories, virtual assistants, etc.—13.8%, analyzing student success—11.6%, organizing peer review among students—6.9%, and managing the teacher's activities, such as creating class schedules—5.3%. Additionally, respondents mentioned using AI for translation, designing online courses, and creating training programs. Only 20 respondents indicated that they do not use AI in their professional activities.

Certain responses to question 11 correlate with those in question 4, highlighting interesting trends. For example, 71.1% of respondents use or want to use AI for content generation, which includes information searching used by 69.3%. However, it should also be noted that there are differences in the results, in particular, 24.4% of respondents use or want to use AI tools for organizing adaptive learning (question 4), while only 13.8% use AI for personalizing learning, a function closely related to adaptive learning (question 11). Similarly, while 32.3% of teachers use or want to use AI for data analytics, statistics, visualization and forecasting (question 4), only

11.6% of teachers use AI for performance analysis (question 11). These discrepancies can be explained by the fact that question 4 includes not only teachers who use AI tools, but also those who express a desire to use them for addressing routine tasks.

10.4 Conclusions

The survey results reveal that Ukrainian university teachers show a clear interest in using AI, reflecting a broader trend among educators worldwide who increasingly focused on integrating AI into the educational process. This interest has spurred educators in Ukraine and abroad to develop AI tools that support both university teachers and students. However, the survey also indicates that many Ukrainian university teachers are unaware that AI tools are already integrated into the educational platforms they use daily. For instance, by setting up an e-classroom group, uploading educational materials and assignments, or by creating an e-journal linked to student work, teachers often use AI tools without realizing it. These tools enable teachers to provide feedback in various formats (text, audio, video), generate performance statistics, and personalize learning. Currently, these platforms can track student attendance in online classes by analyzing system data and presenting it in tabular form. In the future, AI tools could potentially track students' emotions to support their cognitive engagement during lessons. AI tools also facilitate speech-to-text transcription, language translation during online lectures, and video recordings, making note-taking easier for both students and teachers.

Each year, these platforms are enhanced with new AI tools, thanks in part to research projects like Ecolab, Emotionally Responsive Learning Environment, CoWriter, TinkerLamp, Dynamic Support for Collaboration, and Rapid Authoring of Intelligent Tutors for Collaborative Learning. However, while many Ukrainian teachers see the potential in advanced AI applications such as emotion tracking during lessons, peer evaluation, collaborative learning, and intelligent tutoring systems, they view these as future possibilities rather than current realities. For now, most teachers use AI primarily for information retrieval, critical analysis, generating educational and scientific content (such as transcribing speech to text, editing, and translating), and to create multimedia presentations, tests, and personalized learning experiences.

The survey results confirm the initial hypothesis and validate the study's objectives, leading to the formulation of several key conclusions.

The use of AI tools by university teachers holds significant potential for improving educational processes, adapting materials to student needs, and enhancing research efficiency. Ukrainian teachers recognize that AI, along with a certain level of knowledge and skills in using AI tools, can streamline routine tasks.

AI greatly enhances the efficiency of education, enriches the educational process, automates administrative tasks, and contributes to scientific advancement. However, the use of AI also presents challenges, such as academic integrity issues when AI-generated texts are misrepresented as original work, inaccurate information from chatbots, and decreased motivation to learn due to AI's easy access to information. There is also a growing need for legal regulations on the use of AI in education and research, especially concerning its impact on cognitive motivation in the information society. Additionally, there are concerns about personal data collection, an issue that worries educators worldwide, including in Ukraine (Holmes et al., 2022).

Our research findings align with those of Qin and Wang (2022), who highlighted the benefits of AI for personalized learning, targeted teaching, and decision support for educational administrators. However, they also identified four significant challenges: incomplete algorithm design, insufficiently labeled data, excessive reliance on technology, and inadequate security guarantees. AI should ultimately aim to support students and teachers by providing tools that facilitate effective learning and equitable access to education.

Given AI's advantages and potential drawbacks, teachers need tools that automate routine tasks. Effective AI use in education requires teachers to acquire new skills and understand AI's basic principles. Teachers must be proficient in AI tools, such as learning management systems and educational applications, while recognizing that these tools are meant to complement, not replace, traditional pedagogical theories, strategies, and methods. Developers of AI tools must prioritize data privacy and security, maintain human oversight, and enable intervention if AI deviates from established rules and guidelines.

Universities should provide access to professional development resources and programs to support faculty in this transition. Teachers should embrace AI-based educational innovations that can automate many routine tasks and offer personalized learning opportunities for students, while acknowledging that AI cannot replace the human element essential for effective teaching and learning. The integration of AI into education, particularly higher education, should be approached thoughtfully, with careful consideration of its potential limitations and ethical implications.

Future research should focus on developing a comprehensive set of AI tools to assist teachers with routine tasks such as creating presentations, designing tests, organizing peer evaluations, and reviewing assignments.

References

Akgun, S., & Greenhow, C. (2022). Artificial intelligence in education: Addressing ethical challenges in K-12 settings. *AI and Ethics, 2*, 431–440. https://doi.org/10.1007/s43681-021-00096-7

Dillenbourg, P. (2024). *AI for Good*, blog post. Retrieved from https://aiforgood.itu.int/speaker/pierre-dillenbourg/. Accessed 23 June 2024.

Fong, K. Y., Yang, J., Sips, R.-J., Valle Torre, M., Lofi, C., Bozzon, A., & Houben, G.-J. (2019). Phonon heat transfer across a vacuum through quantum fluctuations. *Nature, 576*(7787), 243–247. https://doi.org/10.1038/s41586-019-1800-4

Guo, M. H., Xu, T. X., Liu, J. J., et al. (2022). Attention mechanisms in computer vision: A survey. *Computational Visual Media, 8*, 331–368. https://doi.org/10.1007/s41095-022-0271-y

Holmes, W., Porayska-Pomsta, K., Holstein, K., et al. (2022). Ethics of AI in education: Towards a community-wide framework. *International Journal Artificial Intelligence in Education 32*, 504–526. https://doi.org/10.1007/s40593-021-00239-1

Hwang, Y. I. J., Arnold, S., Srasuebkul, P., & Trollor, J. (2020). Understanding anxiety in adults on the autism spectrum: An investigation of its relationship with intolerance of uncertainty, sensory sensitivities and repetitive behaviours. *Autism, 24*(2), 411–422.

Mesbah, S., Yang, J., Sips, R.-J., Valle Torre, M., Lofi, C., Bozzon, A., & Houben, G.-J. (2019). Training data augmentation for detecting adverse drug reactions in user-generated content. In *Proceedings of the 2019 conference on empirical methods in natural language processing and the 9th international joint conference on natural language processing (EMNLP-IJCNLP)*. Hong Kong, Association for Computational Linguistics. https://doi.org/10.18653/v1/d19-1239

Nguyen, A., Ngo, H. N., Hong, Y., et al. (2023). Ethical principles for artificial intelligence in education. *Education and Information Technologies, 28*, 4221–4241. https://doi.org/10.1007/s10639-022-11316-w

Nye, B. D. (2015). Intelligent tutoring systems by and for the developing world: A review of trends and approaches for educational technology in a global context. *International Journal of Artificial Intelligence in Education, 25*(2), 177–203. https://doi.org/10.1007/s40593-014-0028-6

Phillips, N., Hughes, L., Vindrola-Padros, C., Petrella, A., Fern, L. A., Panel-Coates, F., & Taylor, R. M. (2022). Impact of leadership on the nursing workforce during the COVID-19 pandemic. *BMJ Leader, 7*(1), 21–27. https://doi.org/10.1136/leader-2021-000588

Popenici, S. A. D., & Kerr, S. (2017). Exploring the impact of artificial intelligence on teaching and learning in higher education. *Research and Practice in Technology Enhanced Learning, 12*(1). https://doi.org/10.1186/s41039-017-0062-8

Qin, H., & Wang, G. (2022). Benefits, challenges and solutions of artificial intelligence applied in education. In *Proceedings of the 11th International Conference on Educational and Information Technology (ICEIT)*. Paphos: Chengdu, Institute of Electrical and Electronics Engineers. https://doi.org/10.1109/ICEIT54416.2022.9690739

Roll, I., & Winne, P. H. (2015). Understanding, evaluating, and supporting self-regulated learning using learning. *Analytics Journal of Learning Analytics, 2*(1), 7–12. https://doi.org/10.18608/jla.2015.21.2

Rose, C. P. (2024). *CMU School of Computer Science*, blog post. Retrieved from https://www.cs.cmu.edu/~cprose/. Accessed 23 June 2024.

Skvortsova, S. (2024a). Artificial intelligence as a means of implementing the professional functions of a teacher. In *Education and science through today's challenges*. Zaporizhzhia Regional Institute of Postgraduate Pedagogical Education. https://drive.google.com/file/d/1LrbISw_plulGG5NZJ74b-iXBpBuNh6Ia/view

Skvortsova, S. (2024b). Artificial intelligence in the professional activity of a university lecturer. In *Modern education in a globalised world: Realities, challenges, prospects* (pp. 137–139). T.H. Shevchenko National University «Chernihiv Colehium». https://drive.google.com/file/d/1ciHMN1HNVkliA9-o4dZ3bwTpSIE0ZaN0/view

Skvortsova, S., & Britskan, T. (2024). The model of training future primary school teachers for the use of ICT in teaching subjects (integrated courses) to primary schoolchildren. In *Artificial intelligence in science and education* (pp. 257–261). Ukrainian Institute of Scientific and Technical Expertise and Information. https://doi.org/10.35668/978-966-479-141-7

Skvortsova, S., Symonenko, T., & Britskan, T. (2022). Methodology for the use of digital services in the organisation of online and offline education of primary school children. In *E-learning in the transformation of education in digital society* (pp. 229–241). STUDIO NOA. https://doi.org/10.34916/el.2022.14.17

Tsai, Y.-S. (2024). *Monash University*, blog post. Retrieved from https://research.monash.edu/en/persons/yi-shan-tsai. Accessed 23 June 2024.

Report of the Social and Human Sciences Commission (SHS) (programme and meeting document UNESCO. General Conference 41 st, C/73). (2021). Retrieved from https://unesdoc.unesco.org/ark:/48223/pf0000379920.page=14. Accessed 23 June 2024.

Chapter 11
New Approach to Learning Cyber-Security in Secondary School

Kornel Chromiński, Małgorzata Przybyła-Kasperek, and Rafał Skinderowicz

Abstract Ensuring an adequate level of cyber-security is becoming one of the key aspects these days, requiring a solid awareness and understanding of the risks as digital technologies evolve. Cyber-security education in high schools is essential to prepare students to use the Internet safely. This study proposes an example-based teaching method that allows students to see the dangers of the Internet in a hands-on way, as well as learn best practices for safe Internet use. The method includes workshops designed to demonstrate the risks of not taking sufficient care of online safety. In practice, workshops conducted in this way have proven more effective than traditional teaching methods. The actual demonstration of how easy it is to become a victim of a cyber-attack improved the memorization of information and showed how easy it is to fall victim to a cyber-attack.

Keywords Cyber-security · Online security · Interactive learning · Teaching methodology

11.1 Introduction

In today's digital age, cyber-security is of crucial importance because we are all vulnerable to cyber-attacks, not excluding children and teenagers. As digital technologies become integral to our personal and professional lives, ensuring the security of our systems and data has never been more critical. In this article, we propose a method for working with schoolchildren and teaching by example that raises awareness of the dangers of the digital world and presents good practice and introduces habits for security.

Cyber-security involves measures and protocols to safeguard information technology (IT) systems from cyber-threats such as hacking, data breaches, and malware (Thakur, 2024). The significance of cyber-security has grown with the promotions of

K. Chromiński (✉) · M. Przybyła-Kasperek · R. Skinderowicz
Institute of Computer Science, University of Silesia in Katowice, Sosnowiec, Poland
e-mail: kornel.chrominski@us.edu.pl; malgorzata.przybyla-kasperek@us.cdu.pl; rafal.skinderowicz@us.edu.pl

E. Smyrnova-Trybulska et al. (eds.), *E-Learning and Enhancing Soft Skills*,
https://doi.org/10.1007/978-3-031-82243-8_11

digital data and the increasing sophistication of cyber-attacks. With the development and increased activity of cyber-attacks, cyber-security techniques are also developing (Safitra et al., 2023).

The latest techniques include applications of Artificial Intelligence (AI) in cyber-security (Kaur et al., 2023). AI enhances cyber-security by automating the detection and response to threats, predicting potential security breaches, and providing insights into security practices (Sarker, 2023). It helps organizations stay ahead of cyber-criminals by continuously analyzing vast amounts of data for unusual patterns that may indicate a security breach (Ozkan-Ozay et al., 2024). AI systems can detect threats by analyzing network traffic, user behavior, and system logs to identify anomalies and potential security incidents (Aggarwal et al., 2023). The initial response to detected threats, such as isolating affected systems, notifying security personnel, and initiating predefined countermeasures, can also be automatized by using AI. AI can scan systems for known vulnerabilities, prioritize them based on risk levels, and suggest or implement fixes to prevent exploitation. By monitoring user behavior, AI can detect deviations that might indicate compromised accounts or insider threats (Rangaraju, 2023). Files and network traffic can be scanned with AI to identify malware, even if it has not been previously encountered (zero-day threats).

As organizations migrate to cloud environments, ensuring the security of these platforms has become crucial (Jimmy, 2024). Unauthorized access to sensitive data stored in the cloud can lead to data breaches, potentially exposing personal information, intellectual property, and other confidential data. Cyber-criminals can access cloud accounts through phishing, credential theft, or exploiting vulnerabilities, leading to unauthorized access and control over cloud resources. Latest trends also consider the zero trust architecture (Phiayura & Teerakanok, 2023), which assumes that threats could be internal or external and verifies each request as though it originates from an open network.

Educational institutions hold vast amounts of sensitive data, including personal information of students and staff, research data, and intellectual property. The education sector has become a target for cyber-attacks due to the value of this information and often less robust security measures compared to other sectors (Haque et al., 2023). However, this is not the particular focus of this article. The problem is how to teach school-age children and teenagers to be aware of the dangers of cyber-attacks. In the modern digital world, cyber-security education is crucial as cyber-attacks become increasingly common and sophisticated. Young students, being active users of digital technologies, are particularly vulnerable to cyber-threats. Teaching cyber-security in schools equips students with essential skills and knowledge to protect themselves and understand the importance of digital safety (Ayyash et al., 2024; Triplett, 2023). Introducing cyber-security concepts early helps students develop a fundamental understanding of online safety and responsible digital behavior. Equipping children and students with cyber-security skills prepares them for future careers in the growing field of cyber-security.

Teaching cyber-security in schools is essential for preparing students to navigate and protect themselves in the digital world. By integrating cyber-security education into school curricula, providing hands-on learning opportunities, and addressing challenges such as resource constraints and the need for trained educators, schools can effectively equip students with the knowledge and skills needed for a secure digital future.

Nowadays, cyber-security is taught in high schools as part of IT subjects, devoting several hours to the issue, which usually allows only basic information to be discussed, most often in theoretical terms. After familiarizing students with online dangers and how to secure themselves online, most often students have the conviction that online threats do not apply to them. Conducting a discussion with students and surveys, it can be deduced that, yes, students are not the least familiar with cyber-security issues, but know only the theoretical side, while they mostly do not apply the rules related to the safe use of the Internet. Therefore, the authors decided that they should try a different approach to learning about computer network risks. This article presents a new approach to learning cyber-security. The developed approach was verified in a group of dozens of high school students. The following part of the article describes the methodology of the workshop itself, as well as the results of the feedback in the form of surveys conducted.

11.2 Introduction to the Problem

Integrating cyber-security topics into existing subjects such as computer science, social studies, and mathematics is critically important (Crabb et al., 2024). Encouraging students to undertake projects that involve solving real-world cyber-security problems, fostering practical understanding and skills. However, recent studies have highlighted the global inconsistencies in cyber-security education for children. Sağlam et al. (2023) found that the cyber-security skills taught worldwide vary greatly. They suggest a curriculum based on six broad categories of cyber-security awareness, developed with input from both experts and children. Innovative teaching methods like gamification can enhance traditional learning, though their effectiveness needs more evaluation.

Many projects in school education have been very successful. One example is the CyberPatriot program (Wagner, 2023), initiated by the Air Force Association, is a national youth cyber-education program that includes competitions, camps, and educational materials designed to inspire K-12 students toward careers in cyber-security. This program has been successful in increasing student interest and competency in cyber-security. The National Security Agency (NSA) and the National Science Foundation (NSF) sponsor GenCyber (Bovee & Read, 2024) camps across the United States to provide cyber-security education to students and teachers. These camps focus on hands-on learning experiences and have been effective in raising cyber-security awareness and skills among participants.

Alnajim et al. (2023) reviewed the use of virtual reality (VR) and augmented reality (AR) in cyber-security education. These technologies can make learning more engaging and practical. The study emphasizes the benefits and challenges of using VR and AR, noting their potential to significantly improve traditional teaching methods by offering hands-on experiences. Cyber-competitions also play a key role in cyber-security education. Balon and Baggili (2023) highlighted the effectiveness of these competitions in making learning fun and practical. They recommend improving competition platforms' interoperability and reducing participation barriers to make them more inclusive. These findings are also supported by Triplett (2023) who noticed their success in increasing student interest and engagement in cyber-security careers.

Another set of ideas come from the AI field that brings a lot of potential in improving the effectiveness of cyber-security education. Grover et al. (2023) explored integrating AI and machine learning (ML) into high school cyber-security curricula with their 'AI & Cyber-security for Teens' (ACT) initiative. This program helps students understand AI and its ethical implications while learning about cyber-security through programming activities and games.

It is worth noting that studies on the effectiveness of cyber-security education strategies show varying results. Ondrušková and Pospíšil (2023) found that one-off training sessions had limited long-term effects on children's online behavior. They recommend continuous integration of cyber-security education into the curriculum.

Educators' perspectives are also crucial for successful cyber-security education. Ayanwale et al. (2024) found that teachers' personal relevance and self-efficacy in cyber-security influence their willingness to learn and teach the subject. Teacher training programs should emphasize the personal and societal importance of cyber-security to encourage proactive teaching attitudes.

In conclusion, integrating cyber-security education into school curricula requires innovative teaching methods, continuous curriculum development, and active involvement of educators. By using diverse educational strategies and assessing their effectiveness, schools can better prepare students for the digital world and contribute to a safer cyber-environment.

11.3 Workshop Methodology

The cyber-security workshop was developed for students in the final classes of high school. The workshops conducted were implemented outside the standard high school curriculum. Students participated in them voluntarily. The standard way of learning about the field of cyber-security is implemented based on a theoretical discussion of the dangers of computer networks, and a discussion from the theoretical side on how to prevent attacks and protect their data. Unfortunately, this approach results in knowledge of the dangers themselves but has little impact on the practical aspects of online protection. Students know what they should do, but treat online threats as something that does not concern them and therefore do not see the need to follow the recommendations. As a result, it turns out that this type of cyber-security instruction does not improve students' safety at all, and does not encourage them to pay more attention to security aspects. The main problem is that a significant number of school students believe that just carrying out attacks or data creation is very difficult to carry out, requires extensive knowledge and no one is likely to spend so much time trying to hack into their devices connected to the Internet network. The authors decided, as a starting point for the workshop, to place the greatest emphasis on showing the ease of carrying out different types of attacks. At the same time, they decided to completely abandon the presentation of theoretical aspects (which are quite familiar to students). The way the workshop was conducted was divided into several parts. The main division was based on the different types of threats, of course, due to the limited time of the workshop, it was necessary to choose the most relevant types of attacks, in fact, those to which we are most exposed daily. The workshop selected the following types of threats:

- What happens to our data when it is stolen;
- The creation and complexity of passwords;
- The phishing problem;
- The problem of sniffing;
- Dangerous security applications and software.

Each of these topics was divided into several parts. Within the first part, it was decided to replace the standard theoretical introduction with a discussion of the given risks. The approach was that the workshop leader was only the moderator of the whole discussion. The purpose of this approach was to encourage students to discuss. It is much easier for students to discuss with people who are on the same level of dependency, i.e. their peers than with teachers. In the latter case, there is a teacher-student relationship, which is a form of subordination. The teacher's role is only to moderate the discussion in such a way as to direct it to certain issues, while the teacher should participate in the discussion as little as possible. Of course, in this approach, the teacher must sometimes correct the conclusions reached in the discussion, but this should not have the overtones of teaching, only of guiding.

The second part was to demonstrate how easy it is to carry out a given type of attack. The idea was to show students that with the tools available today, carrying out attacks does not require much IT knowledge or time. The first practical part was to make students aware of what happens to their stolen data. At the same time, information about the dark net was introduced. As part of the demonstration of what happens to the data, a darknet site was presented that offered the opportunity to purchase data from eBay accounts, credit cards, etc. (see Fig. 11.1).

The issue of password creation and complexity was another major topic discussed. As with all of the topics discussed, a discussion of what our passwords should look like and why they should be complex is planned at the beginning. The final element of the discussion is a presentation of the time needed to crack passwords of different complexity. In the second part, a practical presentation shows how easy it is to crack passwords using a free program downloaded from the Internet. The students were made aware of the need to pay attention to the complexity of their passwords by cracking a password of low complexity in a few seconds.

The third issue discussed was the threat of phishing. This is an issue related to sending appropriately crafted messages that pretend to come from a trusted

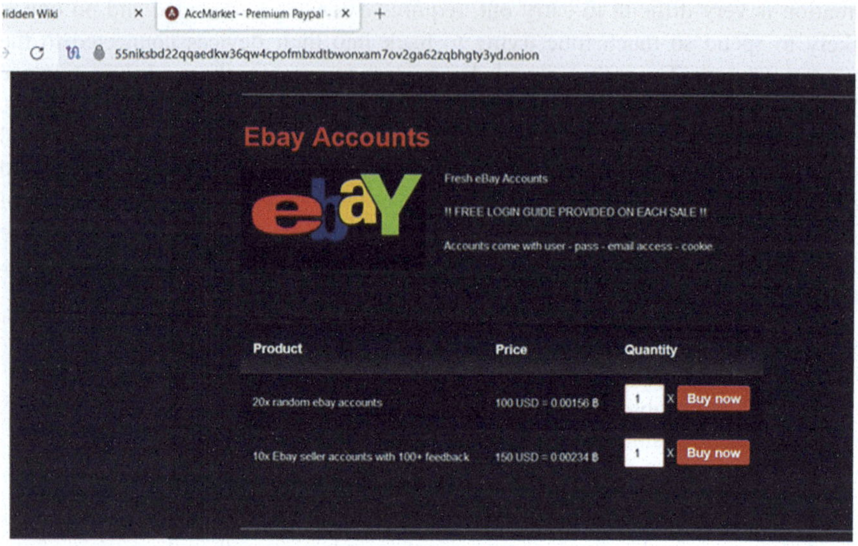

Fig. 11.1 Stolen data for sale on Darknet site

organization (such as a bank, social network, etc.). Such a message usually contains a spoofed link that redirects the unsuspecting user to a correspondingly spoofed page. Most often, such attacks are carried out to steal credentials to various portals and sites (e.g. e-banking sites, email, and social networks). An interesting conclusion from the discussion was that practically every student has encountered a phishing attack, either via email or SMS. As part of the practical part, a well-prepared phishing attack was shown. The ease of execution was also discussed. First, students are shown a fake message purportedly sent from one of the social networking sites. An important element here is to show the sender of the message, which has also been crafted so that the message does not arouse suspicion. An example of the message with the sender visible is shown in Fig. 11.2.

In the next step, the teacher shows the students the page that will open when they click on the link. For the implementation of the lesson, the crafted page was opened from a local disk, so the page address was hidden in the browser. After displaying the crafted social network page (see Fig. 11.3), the instructor asked the students if they saw anything suspicious in the displayed page. An interesting feedback from some of the students was that they do not know what the page should look like, because they only use the application on mobile devices. Therefore, there were statements that the real page will not be so simple. In connection with such comments, the presenter showed the real site for comparison, only then the conclusions were that there were no differences. During the presentation, the teacher also explained the purpose of such attacks.

The presenter explained how this type of attack is prepared and how much time is needed to prepare this type of attack. Most of the students, to the initial question of

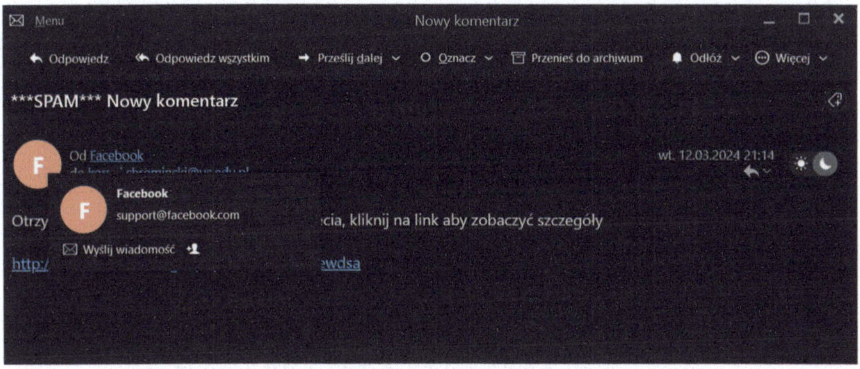

Fig. 11.2 Message prepared for the presentation of a phishing attack

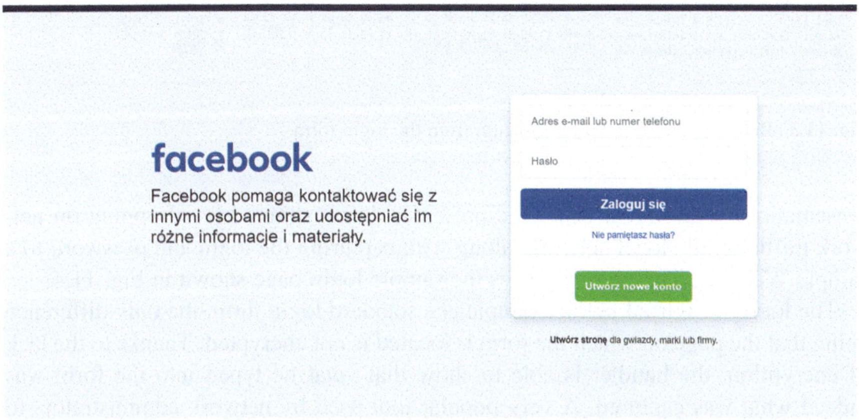

Fig. 11.3 A copy of a popular social networking site for the workshop

how much time we need to prepare a phishing attack like the one presented in class, that a minimum of several to several hours for the creation of the site itself. What came as a surprise was the information that the page itself can be copied in about 2 min with the help of appropriate software. This software does not require any additional special knowledge since it has a graphical interface and its use is limited to entering the address of the page we want to copy and pressing the copy button.

Another threat discussed was the possibility of eavesdropping on network traffic. This threat is especially dangerous when connecting to untrusted networks. This is very popular these days. We like to connect to networks in hotels, restaurants, and many other public places. Often, users of these unsecured networks are unaware of how easily the information they send can be intercepted. As with previous cyber-security topics, and in this case, the discussion of the topic begins with a discussion and some questions from the moderator. One of the most important questions is whether the students have been using public networks. As part of a hands-on

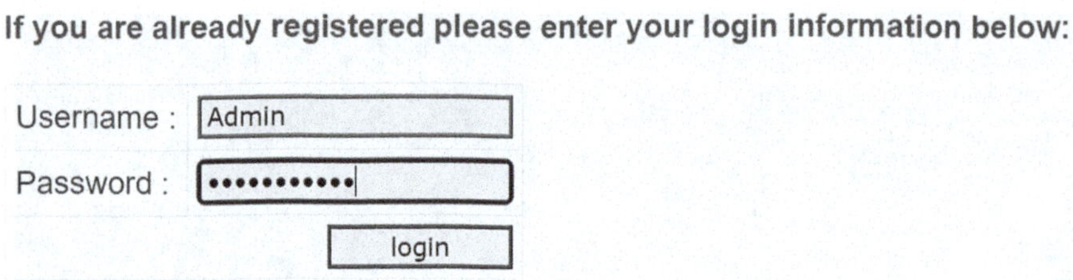

Fig. 11.4 Login form used in the workshop

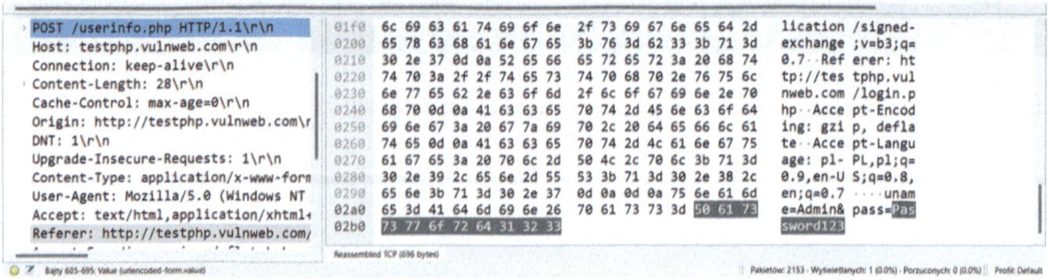

Fig. 11.5 Reading the captured information from the login form

presentation of the threat itself, the presenter demonstrates eavesdropping on network traffic on the local network, along with capturing the login and password to a sample Web site. The presenter uses the sample login page shown in Fig. 11.4.

The login form used in the example is a standard login form, the only difference being that the page on which the form is located is not encrypted. Thanks to the lack of encryption, the handler is able to show that what he typed into the form was indeed what was captured. A very popular tool used by network administrators to monitor computer networks is used to capture network traffic in the example. The result of capturing information from a login form is shown in Fig. 11.5.

It is possible to read the address of the site where the attempt to log in was made, as well as the user name and password, based on Fig. 11.5. Of course, the example shown by the presenter requires additional commentary, because in this way it is possible to read only the information sent from pages without encryption, in case of encryption of the page it would still be necessary to break the encryption.

The last topic discussed was security for devices using the Internet. Only a moderated discussion was held on this topic. An interesting conclusion that could be drawn was that practically all the students had some kind of security on their computers (some kind of antivirus program, firewall), while practically no one had such programs on their mobile devices, which are more vulnerable these days, and in addition, they very often contain all our information, our personal data, credit card numbers, electronic banking access, etc.

The workshop ends with an opportunity for students to ask questions and an open discussion. Closing questions range from clarification of some of the issues discussed during the workshop to how to protect oneself from the dangers of the

Internet. For many students, this illustrative form of threat forced them to analyze their behavior when using the Internet and to review the security features of their own devices.

After the workshop, students were asked to complete a questionnaire about learning cyber-security. The surveys could be administered in this case because the workshop was aimed at upper secondary students, and therefore adults. Completion of the questionnaire was voluntary, and the results of the survey are presented in the following section.

11.4 Results

The survey was conducted to verify whether this form of conducting workshops would be a more interesting option for introducing students to cyber-security issues. The questionnaires were completed voluntarily in a group of students who participated in the workshops conducted by the authors. The students who participated in the survey were students in their final year of secondary school, so they were soon graduating. In addition, they were adults, which meant that they already belonged to the group of people who use electronic banking, and electronic ID cards, which undoubtedly increases the risk of data theft.

Cyber-security issues are discussed only within the compulsory computer science course for secondary school students in Poland. The scope of the subject depends largely on the teacher in charge, and as a result, the subject is often discussed only superficially or not at all. The problem of not teaching cyber-security is particularly common in humanities-based courses that require a minimum number of hours of computer science. The workshop conducted by the authors took place in fifteen high schools in the Silesian region of Poland, surveys were conducted on a group of 50 students. These were additional classes outside the compulsory curriculum. As part of the survey, students answered several questions about their previous cyber-security education. Because all students were in the same age group and no gender differences were observed in the students' responses, it was decided to omit these two parameters in the following section. The first question was whether the students had taken any cyber-security courses. The percentage of responses is shown in Fig. 11.6.

From Fig. 11.6, it can be seen that more than 60% of the students have not taken part in any kind of course on network security so far. This result clearly shows how cyber-security is being neglected in schools, even though we are all vulnerable to attacks by cyber-criminals, and the theft of data and money through various types of fraud has become an everyday occurrence. Students report their knowledge of cyber-security to be quite high, even though the majority of students have not taken cyber-security classes (see Fig. 11.7).

Students' self-assessment of their knowledge of cyber-security is shown in Fig. 11.7. A response of 0 indicates a complete lack of knowledge, while a response of 5 indicates a very high level of knowledge. The majority of students will rate their

Fig. 11.6 Percentage of students who have participated in cyber-security classes

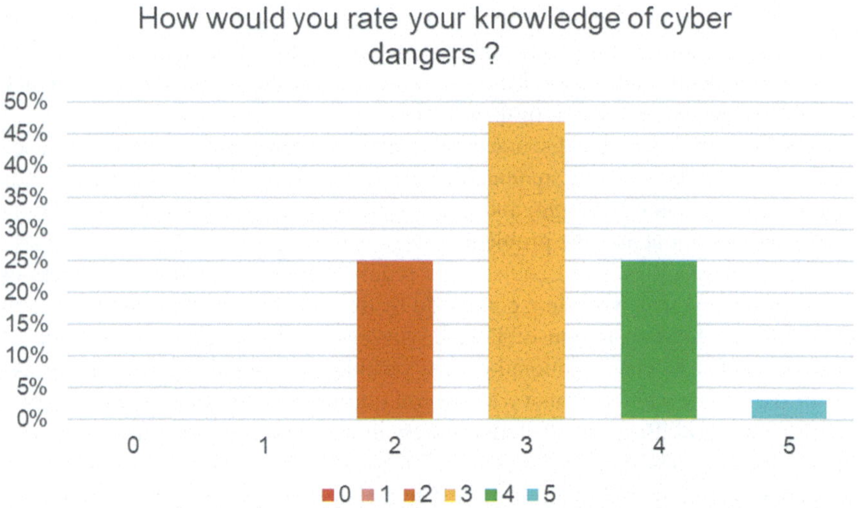

Fig. 11.7 Students' self-assessment of cyber-security knowledge

knowledge at an average level, while there were no responses with a complete lack of knowledge or very little knowledge. This is probably related to the fact that the topic of Internet threats often appears in the mass media or in the form of warnings on the Internet itself. The next question was designed to detail the types of cyber-security activities in which students had participated (see Fig. 11.8).

Unfortunately, of the group of students who took some kind of cyber-security courses, most used online courses followed by lectures. The authors, knowing the core curriculum for teaching computer science in high schools, expected such a result, hence the motivation to develop a more practical form of teaching about

Fig. 11.8 Students' participation in cyber-security classes

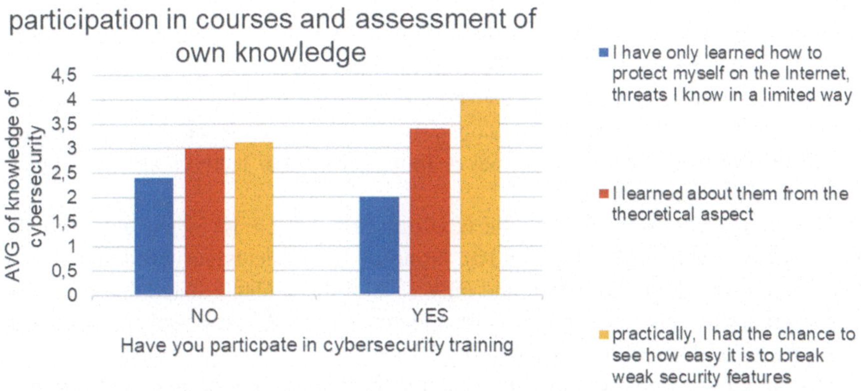

Fig. 11.9 Attendance at cyber-security courses and assessment of students' cyber-security skills

cyber-security. During the workshop, the authors had no problems getting theoretical information about Internet security through discussions, but the situation was much worse when questions were asked about what such a threat is, how easily we can fall victim to it, or how to protect ourselves effectively. Students did not pay much attention to the implementation of the principles of safe use of the Internet, mainly because they thought that it was very difficult to carry out an attack and steal data, and no one would try to do it to them. Showing how easy it is to carry out an attack was a kind of shock for most of the students. Based on Fig. 11.9, it can be concluded that students who were either exposed to the practice and demonstration of how easy it is to steal data, hack into a system as part of their courses or self-study

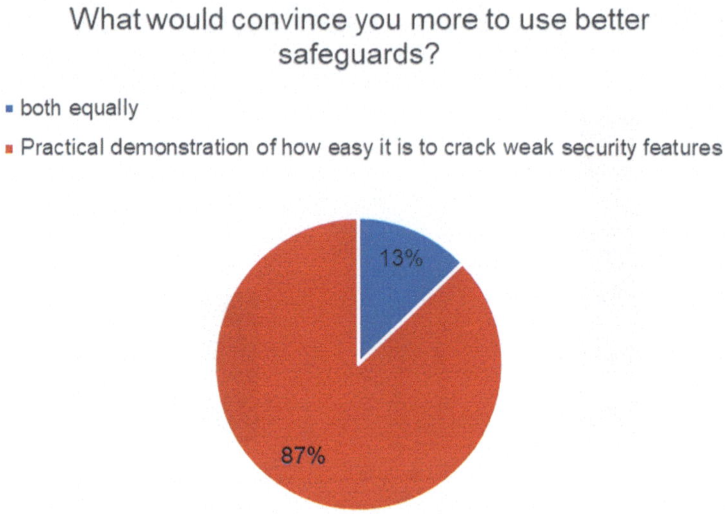

Fig. 11.10 Students' responses to the questionnaire question "Was the theoretical or practical cyber-security course better for you?"

rated their knowledge of cyber-security higher than their counterparts who only learned about Internet security issues from the theoretical side.

The results presented in Fig. 11.9 show that the proposed approach to teaching cyber-security based on practical demonstrations of how easy it is to launch an attack on an unsuspecting Internet user makes more sense than lectures and theoretical discussions of threats alone.

This thesis was confirmed by the next question, which asked students directly which method was better (Fig. 11.10). None of the students surveyed stated that theoretical instruction was better, and only 13% of the students stated that they saw no particular difference. The majority of students felt that the proposed approach was a better way of discussing cyber-security issues.

Most of the students surveyed also said that they would be willing to participate in more extensive hands-on workshops conducted in the manner presented in the article, especially if it allowed them to improve their safety when using the Internet (see Fig. 11.11).

As part of the analysis of the survey results, the authors decided to perform an exploratory data analysis. Figure 11.12 shows the k-means classification tree. The tree confirms the previous conclusions that students who have taken practical courses in cyber-security rate their knowledge on the topic higher than students with only theoretical knowledge.

Students who were also exposed to a hands-on presentation of Internet security issues were more likely to participate in a more in-depth, but hands-on, workshop. These students also paid more attention to their security, such as using stronger passwords and being more aware of how they use the Internet.

The results of the surveys showed that a hands-on approach to learning about cyber-security works better than traditional methods based on theory.

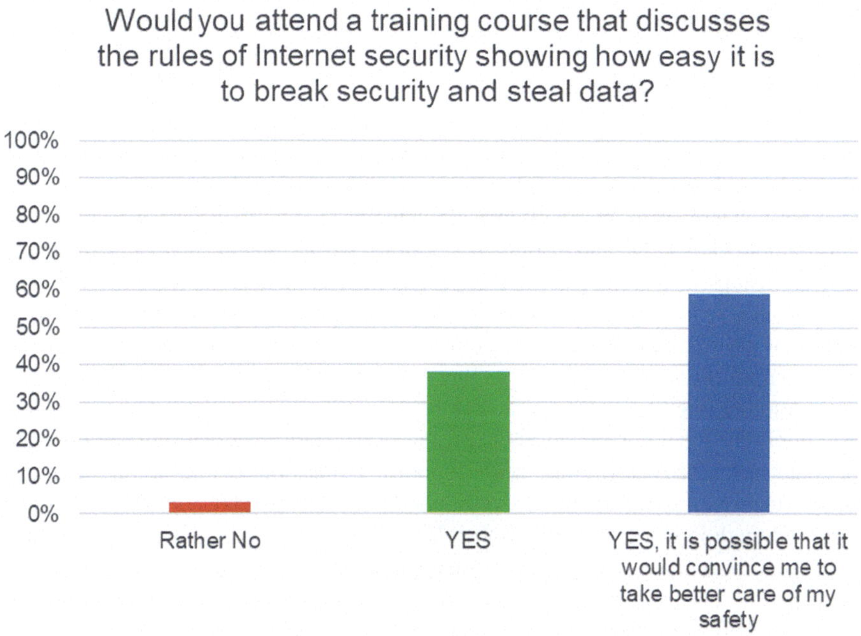

Fig. 11.11 Students' responses to the questionnaire question "Would you take a practical class on cyber-security?"

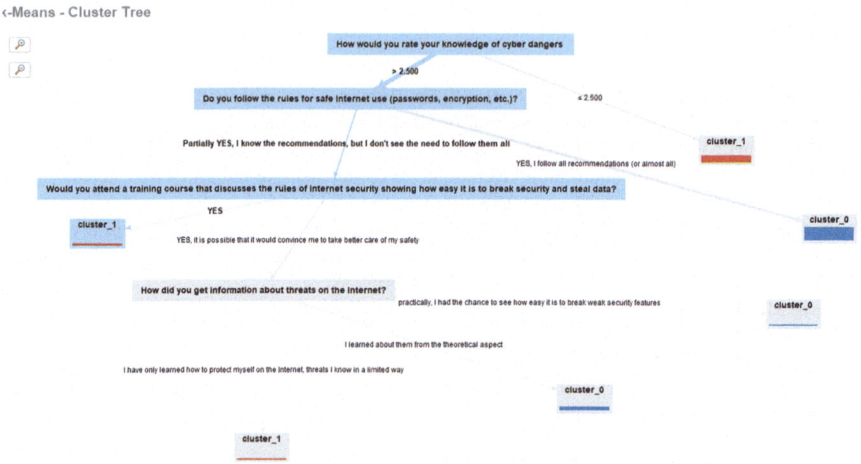

Fig. 11.12 A classification tree generated for data with survey results

11.5 Summary

The article presents a new approach to teaching a safe use of the Internet. In the classical approach, which is the one most often used, students are introduced to the dangers and how to deal with them from the theoretical side. As a result of this approach, students are aware of the dangers, while treating them as something that does not concern them, realizing that any attempt to steal their data would require considerable knowledge and resources, yet they have nothing of such value to devote so many resources to. The proposed approach includes a practical demonstration of how easy it is to launch an attack on an unsuspecting user, as well as what happens to the stolen information. The authors completely abandoned the lecture form; instead of presenting the knowledge in a lecture form, it was decided to use a discussion form. Another important aspect was that the discussion was conducted in such a way as to minimize the involvement of the presenter. The presenter was only supposed to moderate the discussion, guide with supporting questions, or in some situations correct incorrect information. Surveys conducted at the end confirmed that the proposed method works better than the classical approach. An additional advantage is that it is more interesting for the students than lectures alone.

As part of further work, it is planned to expand the workshop to include additional topics, as well as practical presentations on how to easily protect oneself from the dangers of the Internet.

References

Aggarwal, D., Sharma, D., & Saxena, A. B. (2023). Role of AI in cyber security through anomaly detection and predictive analysis. *Journal of Informatics Education and Research, 3*(2).

Alnajim, A. M., Habib, S., Islam, M., AlRawashdeh, H. S., & Wasim, M. (2023). Exploring cybersecurity education and training techniques: a comprehensive review of traditional, virtual reality, and augmented reality approaches. *Symmetry, 15*(12), 2175.

Ayanwale, M. A., Sanusi, I. T., Molefi, R. R., & Otunla, A. O. (2024). A structural equation approach and modelling of pre-service teachers' perspectives of cybersecurity education. *Education and Information Technologies, 29*(3), 3699–3727.

Ayyash, M., Alsboui, T., Alshaikh, O., Inuwa-Dute, I., Khan, S., & Parkinson, S. (2024). *Cybersecurity education and awareness among parents and teachers: A survey of Bahrain* (Vol. 12, pp. 86596–86617). IEEE Access.

Balon, T., & Baggili, I. (2023). Cybercompetitions: A survey of competitions, tools, and systems to support cybersecurity education. *Education and Information Technologies, 28*(9), 11759–11791.

Bovee, M., & Read, H. (2024, June). Remotely the same? Going virtual with a Cybercamp in a pandemic. In *European conference on cyber warfare and security* (Vol. 23, No. 1, pp. 86–95).

Crabb, J., Hundhausen, C., & Gebremedhin, A. (2024, March). A critical review of cybersecurity education in the United States. In *Proceedings of the 55th ACM technical symposium on computer science education V. 1* (pp. 241–247).

Grover, S., Broll, B., & Babb, D. (2023, March). Cybersecurity education in the age of AI: Integrating AI learning into cybersecurity high school curricula. In *Proceedings of the 54th ACM technical symposium on computer science education v. 1* (pp. 980–986).

Haque, M. A., Ahmad, S., John, A., Mishra, K., Mishra, B. K., Kumar, K., & Nazeer, J. (2023). Cybersecurity in universities: An evaluation model. *SN Computer Science, 4*(5), 569.

Jimmy, F. N. U. (2024). Cyber security vulnerabilities and remediation Through cloud security tools. *Journal of Artificial Intelligence General science (JAIGS), 2*(1), 129–171, ISSN: 3006-4023.

Kaur, R., Gabrijelčič, D., & Klobučar, T. (2023). Artificial Intelligence for cybersecurity: Literature review and future research directions. *Information Fusion, 97*, 101804.

Ondrušková, D., & Pospíšil, R. (2023). The good practices for implementation of cyber security education for school children. *Contemporary Educational Technology, 15*(3), ep435.

Ozkan-Ozay, M., Akin, E., Aslan, Ö., Kosunalp, S., Iliev, T., Stoyanov, I., & Beloev, I. (2024). A comprehensive survey: Evaluating the efficiency of Artificial Intelligence and machine learning techniques on cyber security solutions. *IEEE Access, 12*, 12229–12256.

Phiayura, P., & Teerakanok, S. (2023). A comprehensive framework for migrating to zero trust architecture. *IEEE Access, 11*, 19487–19511.

Rangaraju, S. (2023). Secure by intelligence: Enhancing products with AI-driven security measures. *EPH-International Journal of Science and Engineering, 9*(3), 36–41.

Safitra, M. F., Lubis, M., & Fakhrurroja, H. (2023). Counterattacking cyber threats: A framework for the future of cybersecurity. *Sustainability, 15*(18), 13369.

Sağlam, R. B., Miller, V., & Franqueira, V. N. (2023). A systematic literature review on cyber security education for children. *IEEE Transactions on Education, 66*(3), 274–286.

Sarker, I. H. (2023). Machine learning for intelligent data analysis and automation in cybersecurity: Current and future prospects. *Annals of Data Science, 10*(6), 1473–1498.

Thakur, M. (2024). Cyber security threats and countermeasures in digital age. *Journal of Applied Science and Education (JASE), 4*(1), 1–20.

Triplett, W. J. (2023). Addressing cybersecurity challenges in education. *International Journal of STEM Education for Sustainability, 3*(1), 47–67.

Wagner, P. (2023). CyberEducation-by-design. *Cybersecurity Pedagogy & Practice Journal, 2832*, 1006.

Chapter 12
Solution Time and Confidence in the Assessment of Algorithmic and Logical Thinking of First-Year Computer Science Students

Štefan Gubo and Ladislav Végh

Abstract Computer programming belongs to the important profile subjects in the curriculum of first-degree study programs of both Applied Informatics and Teaching of Informatics at the J. Selye University. In order to acquire computer programming competencies, students need to gain well-developed algorithmic and logical thinking skills. In this article, we describe an assessment related to algorithmic and logical thinking which was conducted in the academic year 2023/2024, and involved 75 first-year computer science students. Participants completed the measurement instrument online using a specifically designed framework. This instrument contained 16 tasks related to algorithmization, problem-solving and logical thinking, which can be grouped into four types: (1) tasks related to the execution of a given algorithm, (2) tasks associated with the analysis of an algorithm, (3) tasks associated with the development of an optimal algorithm, and (4) tasks related to problem-solving and logical thinking. During the evaluation of the survey, we focus on the comparison of students' responses to each task type, by considering the solution times. We examine solution time differences between correct and incorrect answers, and determine the relation between solution time and expressed confidence. Finally, we draw our conclusions and mention our future research plans.

Keywords Algorithmic thinking · Logical thinking · Programming · Solution time · Confidence

Š. Gubo (✉) · L. Végh
Faculty of Economics and Informatics, Department of Informatics, J. Selye University, Komárno, Slovak Republic
e-mail: gubos@ujs.sk; veghl@ujs.sk

E. Smyrnova-Trybulska et al. (eds.), *E-Learning and Enhancing Soft Skills*,
https://doi.org/10.1007/978-3-031-82243-8_12

12.1 Introduction

In recent years, an increasing focus has been placed on algorithmic thinking in computer science school education. Algorithmic thinking is a specific type of thinking that, along with problem solving and logical thinking, forms one of the fundamental components of computational thinking (Wing, 2006; Korkmaz et al., 2017; Park & Jun, 2023). Lockwood et al. (2016) define algorithmic thinking as "a logical, organized way of thinking used to break down a complicated goal into a series of (ordered) steps using available tools," emphasizing the usage of procedural steps to solve a given goal.

As a fundamental skill, algorithmic thinking should be incorporated into the curriculum not only at primary and secondary levels but also at higher education level. In order to acquire computer programming competencies, introductory computer science students need to gain well-developed algorithmic and logical thinking skills. Cachero et al. (2020) emphasize the need for training at the university level in aspects related to algorithmic thinking, as it is not developed naturally. Bacelo and Gómez-Chacón (2023) characterized algorithmic thinking in a university context by unplugged activities, where students of mathematics had to solve tasks in which they had to create an algorithm. Lafuente Martínez et al. (2022) developed and validated a tool to assess algorithmic thinking of adults with and without a background in STEM. According to Futchek (2006), tasks suitable for the development of algorithmic thinking should not be too simple, but the task statements should be clear and easily comprehensible.

Logical thinking is a process during which one can draw conclusions based on preexisting, incomplete knowledge. According to Wu (2011), logical thinking is the active reflection of objective reality as a reasonable cognitive process using concepts, judgments, reasoning, and other forms of thinking. Tan (2020) defined logical thinking as a thought process that derives a possible solution or solutions using clear reasoning or inference. To gain appropriate logical thinking skills is important for pupils of every age. As stated by Szarka et al. (2021), in the absence of such skills, students have problems with general information processing, especially with comprehension and interpretation of texts.

Item solution (or response) times are a natural kind of data to investigate cognitive processes underlying cognitive test performance (De Boeck & Jeon, 2019), which can be used, among other things, to detect test-taker engagement and identify rapid guessing behavior. Michaelides and Ivanova (2022) in their study aiming to describe test-taking effort in the Mathematics and Reading PISA 2015 assessments used a fixed threshold of 5 s and a liberal 15% Normative Threshold (NT15) in determining rapid guessers. Lasry et al. (2013) measured the time taken by students to respond to physics conceptual questions and examined response time differences between correct and incorrect answers, and also determined the relation between response time and expressed confidence. We carried out a similar investigation for tasks requiring algorithmic and logical thinking, and in this paper we introduce a computer-based assessment and show data suggesting that solution times, which are usually not available in traditional paper-based surveys, can provide additional insight into students' algorithmic and logical thinking. We obtained these times using a self-developed framework that allowed the measurement of the time a student needed to answer an individual item from task onset to providing a response. Thus, solution time was calculated as the difference in time stamps between the displays of consecutive items.

In our study, we have formulated the following two research questions:

RQ1: Are there any solution time differences between correct and incorrect answers on tasks belonging to different categories?

RQ2: Are there any solution time differences between correct and incorrect answers at different levels of student confidence?

12.2 Methods

The online survey was conducted in September and October 2023 at the Faculty of Economics and Informatics of the J. Selye University (Komárno, Slovakia) and involved 66 first-year BSc students of applied informatics and 9 first-year BSc students of teaching informatics. All respondents participated on a voluntary basis, without additional motivation.

Data acquisition was performed on computers in a computer lab under the personal supervision of a teacher. Participants completed the measurement instrument online using a specifically designed framework that allowed test takers to securely log in with a unique pre-generated username and password. This framework (Fig. 12.1) displayed only one test task at a time, and after selecting an answer and indicating the level of confidence in the answer, the next task was displayed. The students were not allowed to return to a previously solved task.

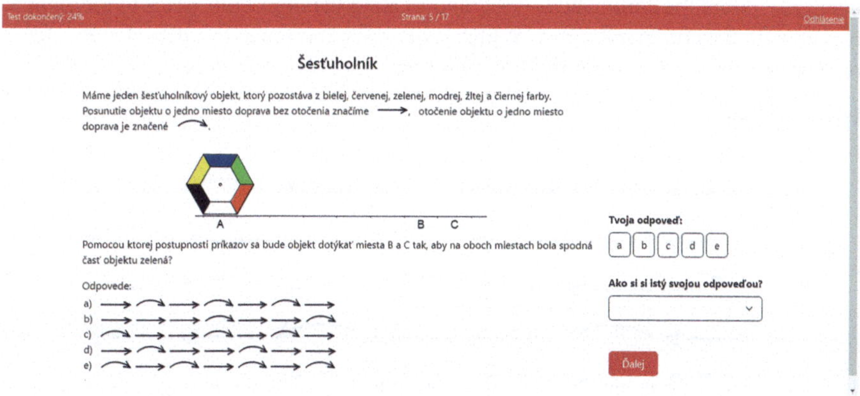

Fig. 12.1 The framework used for data acquisition in the survey displays one test task (in Slovak)

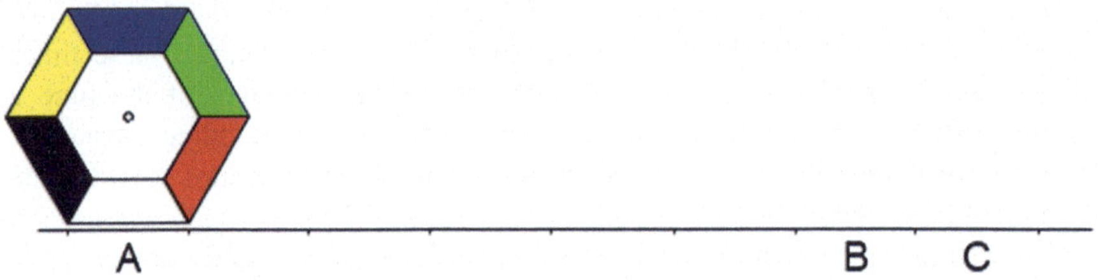

Fig. 12.2 A task ("*Hexagon*") related to the execution of a given algorithm

The measurement instrument, the algorithmic and logical thinking test (ALTT), contained altogether 16 tasks; fourteen of these tasks were multiple-choice tasks with 5 options, one was a multiple-choice task with 8 options, and one was a closed task with only one possible correct answer. The tasks could be solved in a few minutes without any prior knowledge on computer programming and computer science.

The tasks of the ALTT were classified into four different categories: **(1)** tasks related to the execution of a given algorithm (*Three robots*, *Hexagon*, *Drawing crosses*, *Ladybug robot*, *Airplane*, *Colorful walls*, and *Maze*), **(2)** tasks associated with the analysis of an algorithm (*Queueing*, *Colorful blankets*, and *Water lilies*), **(3)** tasks associated with the development of an optimal algorithm (*River crossing*, *Weighing*, and *Prisoners*), and **(4)** tasks related to logical thinking and problem solving (*Social network*, *Knights and scoundrels*, and *Decoding*). In the following section, we briefly describe these four task categories and give an example for each of them.

1. **Tasks related to the execution of a given algorithm**

This category of tasks contained probably the easiest assignments. Students had to follow some simple instructions and imagine how an object (or objects) would move on a board. As an example, we show the task "*Hexagon*" in the following paragraphs.

Hexagon In Fig. 12.2, we have a hexagonal object consisting of white, red, green, blue, yellow and black colors. Moving the object one place to the right without rotating it is marked with the command ⟶, and rotating the object one place to the right is marked with the command ↷. Which sequence of commands will make the object touch ranges B and C so that the bottom of the object is green at both ranges?

Answer Choices

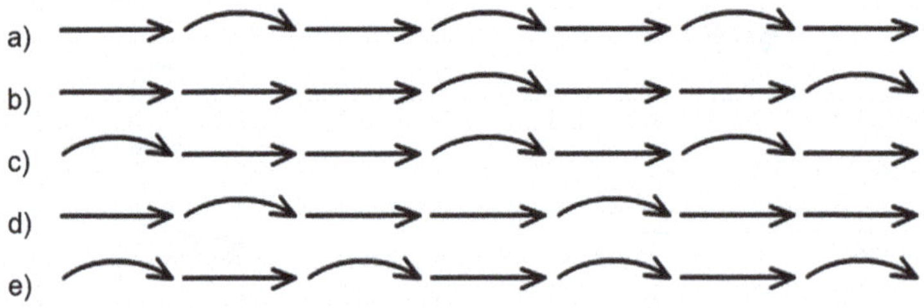

Solution: Note that the hexagonal object must make exactly two rotations for the green colored part to be at the bottom. Since the bottom of the object must be green in both ranges, the last command cannot be a rotation. Thus, the correct answer is **(d)**.

Fig. 12.3 Examples of correct and incorrect patterns in the task "*Colorful blankets*"

In the ALTT, there were six other tasks related to the execution of given algorithms. In the task "*Three robots*," three robots simultaneously had to execute a given sequence of instructions and pick up different fruits in an array. In the task "*Drawing crosses*," students had to decide which pattern of crosses will be drawn by executing a given algorithm. In the task "*Ladybug robot*," students have to choose the correct sequence of instructions that moves the ladybug robot to a given position. In the task "*Airplane*," sequences of instructions were given to students, and they had to choose the correct route of the airplane after executing the instructions. In "*Colorful walls*," a ball moved in a labyrinth based on the colors of the walls inside the labyrinth, and students had to decide where the ball would exit the labyrinth. And finally, in the task "*Maze*," students had to choose how many times they were needed to repeat a sequence of given instructions to exit the maze.

2. **Tasks associated with the analysis of an algorithm**

To solve this category of tasks, students need not only to execute some algorithms but also to analyze and understand them. As an example, we introduce the task "*Colorful blankets*" in the following paragraphs.

Colorful Blankets The fairytale IT spider weaves triangular colored blankets made up of red, blue, and green hexagonal patches. In a blanket pattern, each hexagon and the two beneath it must be the same color or three different colors (e.g. all three are red) or all three are different colors (one is red, the other is blue, and the third is green). In Fig. 12.3, we can see examples of correct and incorrect patterns.

With these rules in mind, decide how many blue colored hexagonal patches are there in the following blanket if you only know the colors of the bottom row.

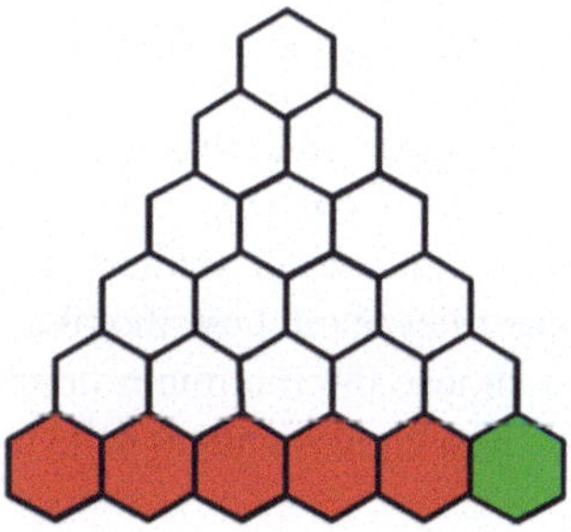

Answer Choices (a) 1 (b) 3 (c) 5 (d) 8 (e) 9

Solution: We have to color all the patches of the blanket. Let us start with the first hexagon of the penultimate row. Since beneath this hexagon there are two red hexagonal patches, it must be the same color (red) as them. In a similar way, we can see that the 2nd, 3rd and 4th hexagonal patches of the penultimate row are also red. However, the last hexagon of the penultimate row must be blue, because beneath it there are red and green hexagonal patches. So the colors of the hexagonal patches in the penultimate row of the blanket are RRRRB. Based on similar considerations, we get that the colors of the hexagonal patches in the next rows of the blanket are RRRG, RRB, RG, and B, respectively. Thus, the correct answer is **(b)**.

There were two other tasks associated with the analysis of algorithms. In the task "*Queuing*," students had to decide the minimum number of adjacent swaps needed to reverse the order of animals standing in a row. In the task "*Water lilies*," a growing water lily doubles in size every day, and students had to determine on what day of its life the lake was half covered.

3. Tasks associated with the development of an optimal algorithm

This category of tasks requires students not only to provide a working algorithm for a given problem, but to find that algorithm which delivers the best solution. As an example, we show the task "*Prisoners*" in the following paragraphs.

Prisoners Four prisoners (A, B, C, D) decide to escape from prison. To do this, they must go through a dark tunnel. The prisoners can cross the tunnel at different times: prisoner A in 1 min, prisoner B in 2 min, prisoner C in 5 min, and prisoner D in 8 min. The prisoners have only one flashlight, and by its light, only two people can cross the tunnel at the same time (when two people cross the tunnel, they both go at the slower speed). What is the least time, in minutes, required to get all prisoners through the tunnel?

Answer Choices (a) 14 (b) 15 (c) 16 (d) 17 (e) 18

Solution: The key insight here is that when prisoner D crosses the tunnel, it will take 8 min no matter whoever goes with him. So the best person to go with him will be the second slowest, prisoner C. But after they have gone across, someone else will have to take the flashlight back to prison. This should be the faster prisoner. So prisoner A and prisoner B will go through the tunnel, prisoner A (or prisoner B) will go back with a flashlight, prisoner C and prisoner D go across, and prisoner B (or prisoner A) will take the flashlight back. We now have the slowest two escaped and the fastest two in prison. As a last step, prisoner A and prisoner B cross the tunnel together. Thus, the correct answer is **(b)**.

In the ALTT, there were two more tasks that focused on the development of an optimal algorithm. In the task "*River crossing*," students had to find the minimum number of river crossing steps to move all soldiers to the other side of the river, taking into consideration some rules about how the river can be crossed. In the task "*Weighing*," students had to choose the minimum number of weighings that were needed to find a heavier ball from eight identical balls using a two-armed scale.

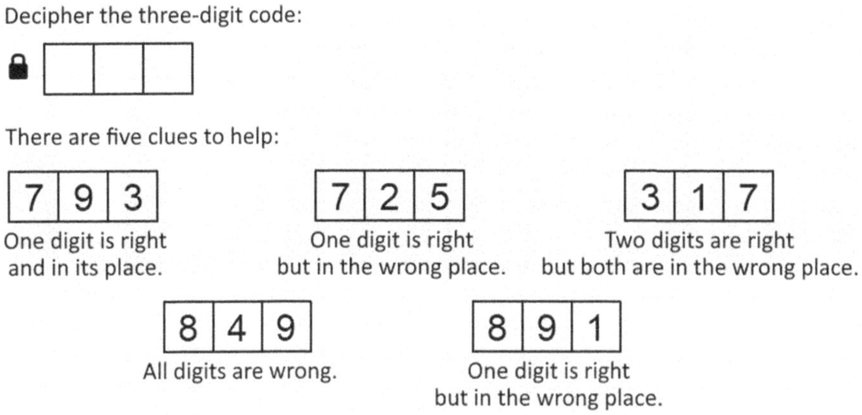

Fig. 12.4 A task ("*Decoding*") related to logical thinking and problem solving

4. **Tasks related to logical thinking and problem solving**

To solve this category of tasks, logical thinking and problem-solving skills are required. As an example, we can see the task "*Decoding*" in Fig. 12.4.

Solution: In this task, five sets of clues are given to decode the correct lock code. Let us start with the first two sets of clues. Since the first one says that the only one correct digit is correctly placed and the second says that the only one correct digit is wrongly placed, digit 7 can be ruled out from the lock code. According to the third clue, two digits are correct but wrongly placed. Since digit 7 is already eliminated, the remaining digits 1 and 3 must be in the code. The fourth set tells us that digits 4, 8 and 9 are not in the code, so according to the first clue, digit 3 is in the third position of the code. Using the result that digit 1 is in the code together with third and fifth sets of clues, we find that digit 1 is in the first position of the code. Now we have to find the digit in the second position of the code. Since it is stated in the second clue that one digit is correct but wrongly placed, we can eliminate digit 2, as it cannot be in the second position of the code. The only digit that remains is 5, and it goes to the second position of the code. The three-digit lock code is thus **153**.

In the ALTT, there were two more tasks related to problem-solving and logical thinking. In the task "*Social network*," students had to find persons to share content with using specific rules and a graph displaying acquaintances in a social network. In the task "*Knights and scoundrels*," students were to figure out who is a knight and who is a scoundrel, if knights always say the truth and scoundrels always lie.

12.3 Results and Discussion

The average of the overall results of the students participating in the survey was 63.8%. Figure 12.5 illustrates the percentage of students who answered each task correctly, and Table 12.1 contains the percentage distribution of correct and incorrect answers given by students to all fifteen multiple-choice tasks of the ALTT.

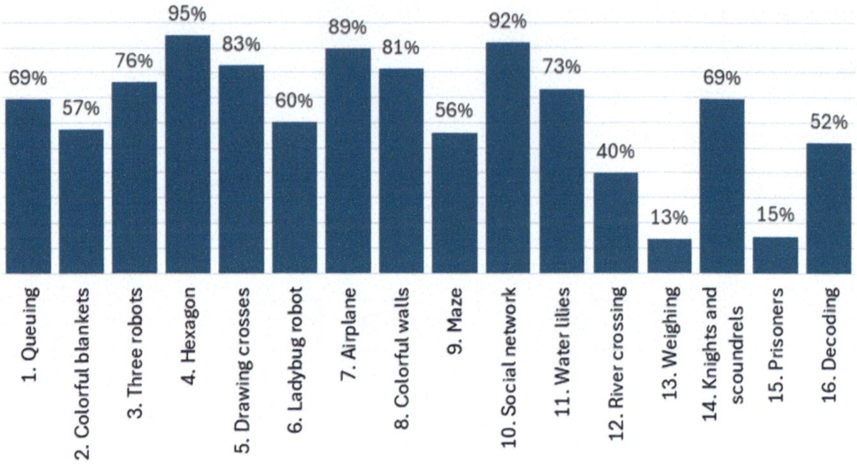

Fig. 12.5 Percentage of correct solvers on tasks of the ALTT. (Source: Own work)

Table 12.1 Percentage distribution of correct (in bold) and incorrect answers to multiple-choice tasks

		Answer choices							
Task	n	a	b	c	d	e	f	g	h
Queuing	75	15%	7%	**69%**	5%	4%			
Colorful blankets	75	7%	**57%**	21%	7%	8%			
Three robots	75	7%	**76%**	13%	4%	0%			
Hexagon	75	1%	3%	1%	**95%**	0%			
Drawing crosses	75	0%	0%	**83%**	17%	0%			
Ladybug robot	75	3%	**60%**	16%	9%	12%			
Airplane	75	0%	3%	**89%**	1%	7%			
Colorful walls	75	0%	1%	0%	**81%**	8%	8%	0%	1%
Maze	75	12%	7%	7%	**56%**	19%			
Social network	75	4%	1%	3%	0%	**92%**			
Water lilies	75	3%	9%	11%	**73%**	4%			
River crossing	75	27%	24%	**40%**	9%	0%			
Weighing	75	15%	**13%**	28%	37%	7%			
Knights and scoundrels	75	4%	8%	**69%**	12%	7%			
Prisoners	75	9%	**15%**	16%	47%	13%			

Source: Own work

Table 12.1 clearly shows how difficult the tasks related to the development of an optimal algorithm (*River crossing*, *Weighing*, and *Prisoners*) proved to be for the participating students. This suggests that at the beginning of their university studies, students do not have deeper experience in programming and algorithmization. The "*Prisoners*" task turned out to be the second most difficult for the students. Almost half of them marked answer **(d)**, probably assuming that the optimal solution would be obtained if the fastest prisoner escorted the other prisoners through the tunnel one by one. Remarkable is the relatively high number of students who marked answer **(c)**, since no crossing sequence can be found for the time value specified here. However, we consider it important to emphasize that if we accepted not only the optimal solution but any correct solution, i.e., answers **(d)** and **(e)**, the percentage distribution of accepted answers would be 75% instead of 15%.

For the non-multiple choice task "*Decoding*", the analysis of the results showed that more than half of the students (52%) gave the correct solution. Notable is that just a bit less than a quarter of the students (23%) got the code 173, which apparently does not correspond to the third clue. The same digits were given by 4% of the students, but in a different order (code 713). Another 4% of the students gave as solution 123, which is inconsistent with the second clue, and 3% of students provided the same digits in a different order (code 213), which does not fit with the third clue. Finally, 3% of the students entered the code 791, but this could easily have been excluded as it does not match the fourth clue. Other solutions (137, 163, 213, 283, 723, 731, 741, 793, and 973) were also created; however, they were each provided just by a single student.

Students solved the entire set of tasks in an average of 39.4 min. Table 12.2 summarizes the basic statistical indicators of solution time on each task individually, and Table 12.3 presents the mean solution times in seconds measured for multiple-choice tasks broken down to individual answer choices. Upon analyzing the data presented in these tables, it can be stated for all four tasks that test takers have spent

Table 12.2 Basic statistical indicators of solution time by tasks

	Statistical indicator				
Task	n	Mean	SD	Rel. SD	Median
Queuing	75	165.41	92.619	0.560	148
Colorful blankets	75	168.51	79.087	0.469	154
Three robots	75	169.11	95.922	0.567	146
Hexagon	75	138.65	73.091	0.527	124
Drawing crosses	75	85.09	38.776	0.456	77
Ladybug robot	75	210.43	105.013	0.499	189
Airplane	75	86.12	45.166	0.524	71
Colorful walls	75	114.68	40.177	0.350	102
Maze	75	174.53	80.788	0.463	165
Social network	75	144.99	50.009	0.345	134
Water lilies	75	97.17	61.417	0.632	76
River crossing	75	187.39	83.116	0.444	177
Weighing	75	107.43	72.024	0.670	94
Knights and scoundrels	75	140.87	72.691	0.516	132
Prisoners	75	123.41	57.410	0.465	119
Decoding	75	253.13	183.421	0.725	177

Source: Own work

Table 12.3 Mean solution time taken (in seconds) for correct (in bold) and incorrect answers in multiple-choice tasks

		Answer choices							
Task	n	a	b	c	d	e	f	g	h
Queuing	75	197	191	**158**	161	143			
Colorful blankets	75	145	**193**	132	149	126			
Three robots	75	144	**156**	200	350	–			
Hexagon	75	217	141	194	**137**	–			
Drawing crosses	75	–	–	**83**	97	–			
Ladybug robot	75	100	**207**	213	186	270			
Airplane	75	–	92	**81**	167	136			
Colorful walls	75	–	63	–	**115**	129	96	–	149
Maze	75	140	211	164	**185**	156			
Social site	75	207	130	178	–	**142**			
Water lilies	75	81	106	91	**98**	87			
River crossing	75	162	155	**215**	225	–			
Weighing	75	91	**119**	129	98	85			
Knights and scoundrels	75	179	121	**135**	127	223			
Prisoners	75	105	**102**	109	142	111			

Source: Own work

enough time reading and understanding the task text. For a given task, a very small amount of solution time could indicate a rapid guess, i.e., the student simply marked the answer without reading the task description and question. In our study, for all tasks of the ALTT, the number of rapid guessing test-takers under the fixed 5-s threshold used by Michaelides and Ivanova (2022) was zero, as each respondent spent at least 14 s on each task. Furthermore, it can be said for each test task that more than 70 students spent at least 34 s on the solution.

As presented in Table 12.2, in all tasks, the median value is lower than the mean value. High standard deviation (SD) values indicate that there are notable personal differences between students when comparing the solution times (Tóth et al., 2021).

As shown in Table 12.2, the task "*Decoding*" has the highest mean solution time of 253.13 s, which can be explained by the fact that this task is a non-multiple choice task. Here, the students have to read carefully and pay close attention to the clues, and then use them to determine the lock code. The high relative standard deviation value of this task also implies that the text was more challenging for students to get the correct solution.

Unexpectedly, tasks "*Weighing*" and "*Prisoners*," which proved to be tasks with the two lowest average scores, had low mean solution times; there were only 3 tasks (*Drawing crosses, Airplane* and *Water lilies*) with lower mean solution times. In the task "*Prisoners*," as displayed in Table 12.3, the mean solution time was the shortest in the group of students who gave the correct answer and the longest in the group of students who marked the incorrect answer **(d)**. It seems that students from the latter group quickly recalled and then subsequently used the incorrect crossing strategy, where the fastest prisoner leads the others one at a time through the tunnel.

In order to answer RQ1, we analyzed for each task of the ALTT the difference in the solution time values between the groups of students who did and did not solve the given task correctly. For normality testing, we used the Shapiro-Wilk test, which is also appropriate for small sample size ($n < 50$) as it has more power to detect the non-normality (Mishra et al., 2019). Then we used the nonparametric Mann-Whitney U-test for the comparison of medians to evaluate the statistical significance of solution time differences. The results are summarized in Table 12.4. The data in this table show that for 12 out of 16 tasks (including the tasks "*Hexagon*" and "*Prisoners*"), the median solution time is greater for incorrect answers than for correct answers, but the differences are not statistically significant. Contrarily, for the tasks "*Colorful blankets*," "*River crossing*," "*Colorful walls*," and "*Decoding*," we found that the median solution time is greater for correct answers than for incorrect answers, and for the former two, the difference is statistically significant ($p < 0.05$).

Summarizing the obtained results for each task category, it can be said that in the category of tasks related to the execution of a given algorithm and also in the category of tasks related to logical thinking and problem solving there are no significant solution time differences between successful and unsuccessful students. For the remaining two task categories (tasks associated with the analysis of an algorithm and tasks associated with the development of an optimal algorithm), it

Table 12.4 Comparison of the solution times between successful and unsuccessful students

Task	Median solution time		Mann-Whitney U	Z	p
	incorrect answers	correct answers			
Queuing	148 (n = 23)	142 (n = 52)	554.500	−0.500	0.617
Colorful blankets	130 (n = 32)	193 (n = 43)	385.500	−3.241	**0.001**
Three robots	172 (n = 18)	145 (n = 57)	416.500	−1.197	0.231
Hexagon	190 (n = 4)	121 (n = 71)	81.500	−1.427	0.154
Drawing crosses	84 (n = 13)	73 (n = 62)	307.000	−1.344	0.179
Ladybug robot	195 (n = 30)	185 (n = 45)	640.500	−0.373	0.709
Airplane	115 (n = 8)	68 (n = 67)	162.500	−1.811	0.070
Colorful walls	98 (n = 14)	103 (n = 61)	401.000	−0.354	0.724
Maze	172 (n = 33)	156 (n = 42)	629.500	−0.678	0.498
Social network	161 (n = 6)	134 (n = 69)	134.500	−1.416	0.157
Water lilies	90.5 (n = 20)	71 (n = 55)	458.500	−1.096	0.273
River crossing	162 (n = 45)	210 (n = 30)	443.500	−2.504	**0.012**
Weighing	97 (n = 65)	77.5 (n = 10)	300.500	−0.382	0.703
Knights and scoundrels	140 (n = 23)	114.5 (n = 52)	537.000	−0.701	0.483
Prisoners	119.5 (n = 64)	119 (n = 11)	284.500	−1.011	0.312
Decoding	173 (n = 36)	193 (n = 39)	564.000	−1.463	0.143

Source: Own work

holds the same; a significant difference can be observed in only one of the three examined tasks.

Based on these results, it can be concluded for tasks of the ALTT that the occurrence of significant solution time differences between correct and incorrect answers can only be observed occasionally. We examined the relationship between the students' total solution times and the total scores achieved on the ALTT and found no statistically significant correlation.

The average of the confidence levels indicated by the students (0—I was just guessing, 1—I'm not quite sure, and 2—I'm absolutely sure) on the entire test, expressed as a percentage, was 77.1%. Figure 12.6 displays the same values broken down by individual test tasks. In the case of a task, this value was obtained by averaging the level of confidence the students indicated and then converting it to a percentage. If, for example, everyone had marked "I was just guessing" for a task, then the value calculated for that task would be 0%. And if everyone had indicated "I'm absolutely sure," then the calculated value would be 100%.

The data in Fig. 12.6 and also in Table 12.5 show that only a small part of students claimed that they were only guessing; the proportion of such students is below 20% for all test tasks. It is also visible in this figure that the students solved the tasks related to the execution of a given algorithm and also tasks requiring logical thinking and problem solving with high confidence (in average of 83% and 80%, respectively); they were slightly less confident in solving tasks associated with the analysis of an algorithm (in average of 74%) and were the least confident in solving the tasks requiring the creation of an optimal algorithm (in average of 64%).

In the following paragraphs, for reasons of space, we do not deal with all tasks of the ALTT but rather give a brief evaluation of the four tasks mentioned in Sect. 12.2. Figures 12.7, 12.8, 12.9, and 12.10 display the number of correct and incorrect answers on the tasks "*Hexagon*," "*Colorful blankets*," "*Prisoners*," and "*Decoding*," respectively, for each confidence level (0—I was just guessing; 1—I'm not quite

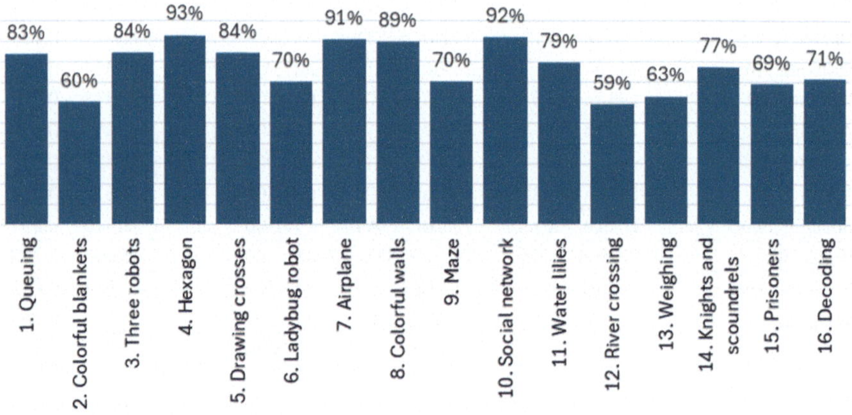

Fig. 12.6 Average confidence levels marked by the students in the individual tasks of the ALTT. (Source: Own work)

Table 12.5 Comparison of the solution times between successful and unsuccessful students for answers with low confidence level (0—I was just guessing)

Task	Median solution time		Mann-Whitney U	Z	p
	incorrect answers	correct answers			
Queuing	– (n = 0)	– (n = 0)	–	–	–
Colorful blankets	180 (n = 11)	103 (n = 2)	5.000	−1.186	0.236
Three robots	338 (n = 4)	268 (n = 3)	2.000	−1.414	0.157
Hexagon	– (n = 0)	275 (n = 1)	–	–	–
Drawing crosses	75 (n = 1)	107.5 (n = 2)	1.000	0.000	1.000
Ladybug robot	194 (n = 11)	207 (n = 3)	12.000	−0.701	0.484
Airplane	71 (n = 1)	124 (n = 2)	0.000	−1.225	0.221
Colorful walls	165 (n = 1)	153.5 (n = 2)	1.000	0.000	1.000
Maze	182.5 (n = 10)	121.5 (n = 2)	8.000	−0.430	0.667
Social network	161 (n = 2)	– (n = 0)	–	–	–
Water lilies	81.5 (n = 6)	62 (n = 1)	2.000	−0.500	0.617
River crossing	205 (n = 11)	185 (n = 2)	7.000	−0.790	0.430
Weighing	98 (n = 13)	79 (n = 1)	2.000	−1.116	0.264
Knights and scoundrels	154.5 (n = 4)	– (n = 0)	–	–	–
Prisoners	77.5 (n = 8)	119 (n = 1)	2.000	−0.775	0.439
Decoding	166.5 (n = 4)	– (n = 0)	–	–	–

Source: Own work

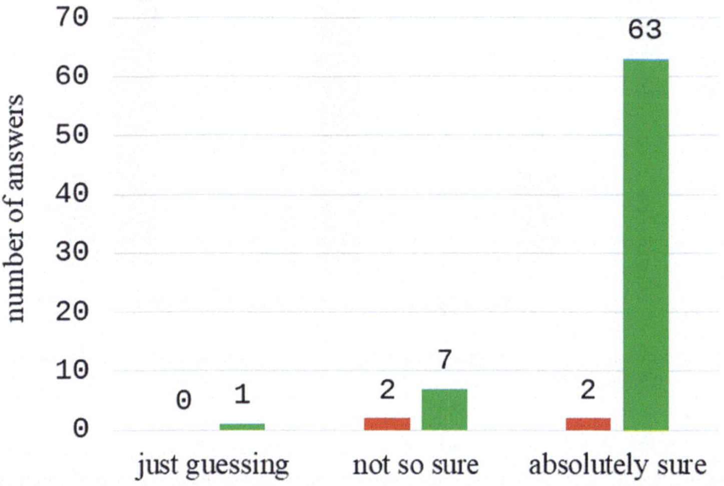

Fig. 12.7 The number of correct (green) and incorrect (red) answers on the task "*Hexagon*" for each level of confidence. (Source: Own work)

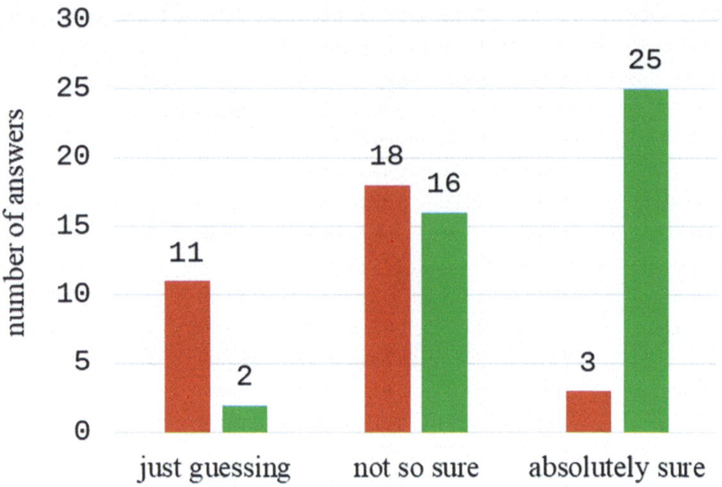

Fig. 12.8 The number of correct (green) and incorrect (red) answers on the task "*Colorful blankets*" for each level of confidence. (Source: Own work)

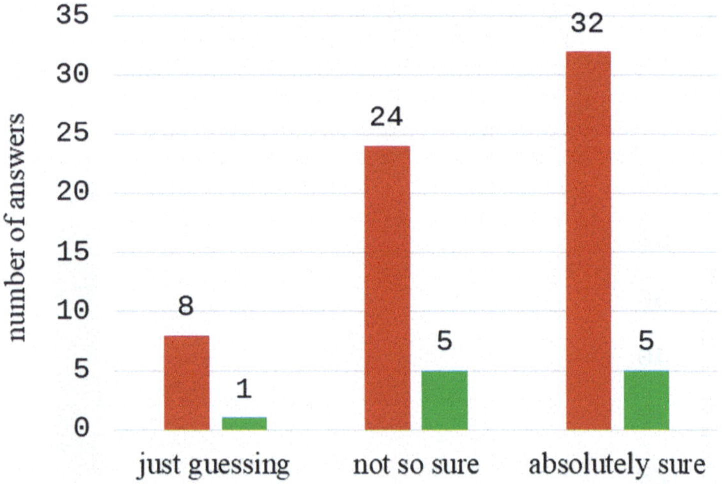

Fig. 12.9 The number of correct (green) and incorrect (red) answers on the task "*Prisoners*" for each level of confidence. (Source: Own work)

sure; 2—I'm absolutely sure). The data in these figures show that, for all four tasks, that the number of correctly guessing participants is very low. Figure 12.7 shows that in the task "*Hexagon*," almost all students answered correctly and with high confidence. In the task "*Colorful blankets*," as Fig. 12.8 displays, the number of students who rated their answers with the highest confidence greatly decreased, but the vast majority of them could give the correct solution. From this figure, it can also be seen

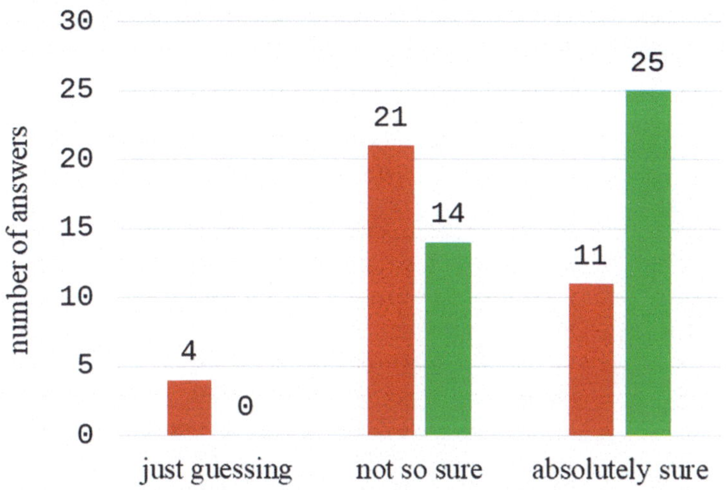

Fig. 12.10 The number of correct (green) and incorrect (red) answers on the task "*Decoding*" for each level of confidence. (Source: Own work)

that the number of students who gave the correct answer and were not quite sure in their answers is roughly equal to the number of students who marked any incorrect answer option with the same confidence. We obtained a similar result in the task "*Decoding*," where, as Fig. 12.10 shows, just the number of highly confident students who answered incorrectly was somewhat higher, and the number of incorrectly guessing students was about the same lower number. Finally, the data in Fig. 12.9 show for the task "*Prisoners*" that only a very small proportion of students who were pretty sure or not so sure in the correctness of their answers were able to give the correct solution. Here we would like to emphasize again that this result can primarily be explained by the fact that only the optimal solution provided by the students was accepted as correct.

In order to answer RQ2, we analyzed for each task of the ALTT the difference in the solution time values between the groups of correct and incorrect solvers at different levels of confidence. The results for low (0—I was just guessing), medium (1—I'm not quite sure), and high (2—I'm absolutely sure) confidence levels are summarized in Tables 12.5, 12.6, and 12.7, respectively. We consider it an interesting result that, as data in Table 12.5 show, even the students who guessed spent a relatively large amount of time answering, from which it can be concluded that they at least read the entire text of the tasks properly.

For students who rated their answers with the lowest confidence (0—I was just guessing), the results are limited by the small number of such students. According to Table 12.5, for each test task, there were no more than 3 students in the group of students who guessed correctly. Thus, it is not surprising that no statistically significant solution time differences were found between successful and unsuccessful students for answers at this level of student confidence.

Table 12.6 Comparison of the solution times between successful and unsuccessful students for answers with medium confidence level (1—I'm not quite sure)

Task	Median solution time		Mann-Whitney U	Z	p
	incorrect answers	correct answers			
Queuing	148 (n = 13)	160.5 (n = 12)	72.500	−0.299	0.765
Colorful blankets	122 (n = 18)	183 (n = 16)	57.500	−2.985	**0.003**
Three robots	111.5 (n = 6)	215 (n = 4)	6.000	−1.279	0.201
Hexagon	205.5 (n = 2)	87 (n = 7)	0.500	−1.911	0.056
Drawing crosses	133 (n = 6)	99 (n = 12)	15.000	−1.968	**0.049**
Ladybug robot	202.5 (n = 14)	223 (n = 3)	20.000	−0.126	0.900
Airplane	147 (n = 5)	110 (n = 3)	5.000	−0.745	0.456
Colorful walls	100 (n = 7)	139 (n = 3)	5.000	−1.254	0.210
Maze	167 (n = 13)	127 (n = 8)	38.500	−0.978	0.328
Social network	232.5 (n = 2)	149 (n = 6)	4.000	−0.667	0.505
Water lilies	106.5 (n = 10)	142 (n = 7)	20.000	−1.464	0.143
River crossing	134 (n = 22)	216 (n = 14)	80.000	−2.401	**0.016**
Weighing	96 (n = 26)	285 (n = 2)	9.000	−1.517	0.129
Knights and scoundrels	163 (n = 11)	100 (n = 15)	53.500	−1.506	0.132
Prisoners	115 (n = 24)	136 (n = 5)	58.000	−0.115	0.908
Decoding	171 (n = 21)	365 (n = 14)	65.500	−2.744	**0.006**

Source: Own work

The data in Table 12.6 show that in the group of students who were not quite sure of the correctness of their answer, the median solution time for 10 out of 16 tasks is greater for correct answers than for incorrect answers, and the difference is statistically significant ($p < 0.05$) for the tasks "*Colorful blankets*," "*River crossing*," and "*Decoding*." Among the remaining six tasks where incorrect answers took longer than correct answers, the only statistically significant difference was observed in the task "*Drawing crosses*." We note that the difference is also quite large in the task "*Hexagon*" but does not reach the significance level ($p = 0.056$).

As Table 12.7 indicates, in the group of students who expressed the highest confidence level (2—I'm absolutely sure), we found only one statistically significant difference, and that is in task "*Prisoners*." Here, the students who answered correctly took shorter to answer than those who answered incorrectly. Based on the low median correct answer solution time for this task, it cannot be ruled out that these students had encountered this task before and already knew what to pay attention to during the solution.

Summarizing the results obtained for the individual task categories, in the category of tasks related to the execution of a given algorithm, in the category of tasks associated with the analysis of an algorithm, and also in the category of tasks requiring logical thinking and problem solving, only one significant solution time difference can be observed between correct and incorrect answers, namely in the

Table 12.7 Comparison of the solution times between successful and unsuccessful students for answers with high confidence level (2—I'm absolutely sure)

Task	Median solution time		Mann-Whitney U	Z	p
	incorrect answers	correct answers			
Queuing	161.5 (n = 10)	132.5 (n = 40)	158.000	−1.019	0.308
Colorful blankets	124 (n = 3)	203 (n = 25)	17.500	−1.486	0.137
Three robots	192 (n = 8)	136 (n = 50)	122.500	−1.748	0.081
Hexagon	140.5 (n = 2)	121 (n = 63)	57.000	−0.228	0.820
Drawing crosses	73.5 (n = 6)	69.5 (n = 48)	140.500	−0.096	0.923
Ladybug robot	185 (n = 5)	166 (n = 39)	94.000	−0.129	0.897
Airplane	76.5 (n = 2)	66 (n = 62)	57.500	−0.174	0.862
Colorful walls	93 (n = 6)	101 (n = 56)	136.000	−0.762	0.446
Maze	162 (n = 10)	171.5 (n = 32)	131.500	−0.842	0.400
Social network	159 (n = 2)	132 (n = 63)	52.500	−0.399	0.690
Water lilies	70 (n = 4)	67 (n = 47)	90.500	−0.123	0.902
River crossing	150 (n = 12)	208 (n = 14)	50.000	−1.750	0.080
Weighing	91 (n = 26)	61 (n = 7)	65.000	−1.145	0.252
Knights and scoundrels	93 (n = 8)	115 (n = 37)	96.000	−1.544	0.123
Prisoners	125 (n = 32)	65 (n = 5)	34.500	−2.022	**0.043**
Decoding	186 (n = 11)	163 (n = 25)	129.000	−0.292	0.770

Source: Own work

group of students who expressed the medium confidence level (1—I'm not quite sure). In the remaining category of tasks requiring the development of an optimal algorithm, two significant solution time differences were found, one in the group of students who expressed the medium confidence level (1—I'm not quite sure) and one in the group of students who expressed the high confidence level (2—I'm absolutely sure).

Based on these results, it can be concluded for tasks of the ALTT that at low and high levels of student confidence there are no significant solution time differences between correct and incorrect answers. At the medium level of student confidence, significant differences can be observed in some tasks, and in most of them, students who answered correctly took longer to answer than those who answered incorrectly. We examined the relationship between the students' total solution times on the ALTT and the sums of confidence levels indicated by students (0—I was just guessing, 1—I'm not quite sure, and 2—I'm absolutely sure) on each test task, expressed as a percentage, and found no statistically significant correlation.

However, a strongly significant correlation was observed between the total scores achieved on the ALTT and the sums of confidence levels ($n = 75$; Pearson correlation $= 0.713$; $p < 0.001$). From this, it can be concluded that the students were mostly able to judge correctly how confident they were in the answers they gave to the tasks of the ALTT.

12.4 Conclusions

The aim of this study was to show that solution time and expressed confidence can provide some additional information about first-year university computer science students' algorithmic and logical thinking. As at this time, we are not aware of any publications that investigate the solution time a respondent takes to respond to individual items in a test measuring algorithmic and logical thinking, we cannot compare the obtained results with the results of studies by other authors.

For our sample, it can be stated that respondents were sufficiently motivated to complete the assessment and spent enough time reading and understanding the content of the test tasks. Incorrect answers took enough time for all multiple-choice tasks of the test, showing that the vast majority of distractors do not trigger quick and automatic answers from students.

The results of the solution time analysis show that the majority of students had no difficulties solving a task requiring the execution of a specific algorithm in a short time. However, when faced with a task that requires the analysis of an algorithm or a task involving logical thinking and problem solving, the number of correct solvers decreases, and they spend more time answering. And finally, for a task that requires the creation of an optimal algorithm, the number of successful solvers drops notably, and they, contrary to expectations, provide the correct answer in a relatively short amount of time.

Concerning research question RQ1, upon examining solution time differences between correct and incorrect answers to the tasks, no statistically significant differences were found in tasks related to the execution of a given algorithm and in tasks related to logical thinking and problem solving. In the category of tasks associated with the analysis of an algorithm, a statistically strongly significant difference ($p < 0.001$) was found in the task "*Colorful blankets*," where correct answers took more time than incorrect answers. And finally, the second and last statistically significant difference ($p < 0.05$) was discovered in the task "*River crossing*," which belongs to the category of tasks related to the creation of an optimal algorithm, where correct answers required more time than incorrect answers. Based on the obtained results, it can be concluded that in tasks requiring algorithmic and logical thinking, the occurrence of significant solution time differences between correct and incorrect answers is only sporadically observed. The weak correlation between the total solution times and the total test scores suggests that there is no relationship between the time spent on the solution and effectiveness, i.e., longer solution time on a task does not mean that it will be solved correctly.

The findings of the analysis of the solution time at different levels of student confidence show that the great majority of students were not only able to successfully and quickly solve tasks requiring the execution of an algorithm and also tasks requiring logical thinking and problem solving, but were also highly confident in the correctness of the answer. When it comes to a task where an analysis of a given algorithm is needed, the confidence decreases moderately. And finally, in tasks that require determining the steps of an optimal algorithm, the confidence in the correctness of the solution decreases even more, suggesting that the majority of newcomer university students lack in-depth knowledge of programming and algorithmization.

Concerning research question RQ2, significant solution time differences between correct and incorrect answers were observed at the medium level of student confidence in four tasks, and in three of them, the students who answered correctly spent more time on the solution. The task "*Prisoners*," where incorrect answers took longer than correct answers, was the only task with a significant difference in the group of students who expressed the highest confidence level. These results suggest that there is no relationship between total solution times and expressed student confidence levels, which was finally confirmed by correlation analysis. As an additional result, a strongly significant correlation was confirmed between the total test scores and expressed student confidence levels.

However, we are aware that mainly the small number of participating students and, to some degree, the inclusion of the time taken to choose a confidence level within reported solution times make it difficult to draw robust conclusions from the results obtained within this study. Therefore, in our future research, we would like first of all to expand the study sample, even by including university students from other Central European countries in the measurement. Currently, the ALTT test is available in three languages: Slovak, Czech, and Hungarian. Thus, another important research goal would be to investigate whether there are notable differences between the solution times attained by students from different countries in the individual ALTT test tasks. Another interesting research goal would be to examine for multiple-choice tasks of the ALTT the two novel response-based indicators (Unsuccessful Time Management and Successful Time Management) of test-taking behaviors introduced by Papanastasiou and Michaelides (2024).

Acknowledgments The research was supported by grant BGA/145/2024 "Support for the J. Selye University's research activities – social science research on Hungarians in Slovakia" provided by the Government of Hungary.

References

Bacelo, A., & Gómez-Chacón, I. M. (2023). Characterising algorithmic thinking: A university study of unplugged activities. *Thinking Skills and Creativity, 48*, article no. 101284. https://doi.org/10.1016/j.tsc.2023.101284

Cachero, C., Barra, P., Meliá, S., & López, O. (2020). Impact of programming exposure on the development of computational thinking capabilities: An empirical study. *IEEE Access, 8*, 72316–72325. https://doi.org/10.1109/ACCESS.2020.2987254

De Boeck, P., & Jeon, M. (2019). An overview of models for response times and processes in cognitive tests. *Frontiers in Psychology, 10*, article no. 102. https://doi.org/10.3389/fpsyg.2019.00102

Futchek, G. (2006). Algorithmic thinking: The key for understanding computer science. In R. T. Mittermeir (Ed.), *Informatics education—The bridge between using and understanding computers* (pp. 159–168). Springer.

Korkmaz, Ö., Çakir, R., & Özden, M. Y. (2017). A validity and reliability study of the computational thinking scales (CTS). *Computers in Human Behavior, 72*, 558–569. https://doi.org/10.1016/j.chb.2017.01.005

Lafuente Martínez, M., Lévêque, O., Benítez, I., Hardebolle, C., & Dehler Zufferey, J. (2022). Assessing computational thinking: Development and validation of the algorithmic thinking test for adults. *Journal of Educational Computing Research, 60*(6), 1436–1463. https://doi.org/10.1177/07356331211057819

Lasry, N., Watkins, J., Mazur, E., & Ibrahim, A. (2013). Response times to conceptual questions. *American Journal of Physics, 81*(9), 703–706. https://doi.org/10.1119/1.4812583

Lockwood, E., Delarnette, A. F., Asay, A., & Thomas, M. (2016). Algorithmic thinking: An initial characterization of computational thinking in mathematics. In *Proceedings of the 38th annual meeting of the North American chapter of the international group for the psychology of mathematics education* (pp. 1588–1595). The University of Arizona.

Michaelides, M., & Ivanova, M. (2022). Response time as an indicator of test-taking effort in PISA: Country and item-type differences. *Psychological Test and Assessment Modeling, 64*(3), 304–338.

Mishra, P., Pandey, C. M., Singh, U., Gupta, A., Sahu, C., & Keshri, A. (2019). Descriptive statistics and normality tests for statistical data. *Annals of Cardiac Anaesthesia, 22*(1), 67–72. https://doi.org/10.4103/aca.ACA_157_18

Papanastasiou, E. C., & Michaelides, M. P. (2024). Examining successful and unsuccessful time management through process data: Two novel indicators of test-taking behaviors. *Large-scale Assessments in Education, 12*(3), 1–14. https://doi.org/10.1186/s40536-024-00193-z

Park, H., & Jun, W. (2023). A study of development of algorithm thinking evaluation standards. *International Journal of Applied Engineering and Technology, 5*(2), 53–58.

Szarka, K., Fehér, Z., Jaruska, L., & Tóthová Tarová, E. (2021). Investigation of the students' solutions in tasks focused on selected elements of propositional logic in a scientific and mathematical context in teacher education. In *ICERI2021 proceedings* (pp. 822–831). https://doi.org/10.21125/iceri.2021.0256

Tan, E. X. (2020). Improving logical thinking skills through Jigsaw-based cooperative learning approach. In *Preparing 21st century teachers for teach less, learn more (TLLM) pedagogies* (pp. 162–181). https://doi.org/10.4018/978-1-7998-1435-1.ch010

Tóth, P., Horváth, K., & Kéri, K. (2021). Development level of engineering students' inductive thinking. *Acta Polytechnica Hungarica, 18*(5), 107–129. https://doi.org/10.12700/aph.18.5.2021.5.8

Wing, J. M. (2006). Computational thinking. *Communications of the ACM, 49*(3), 33–35.

Wu, B. (2011). *Consumption and management – New discovery and applications*. Woodhead Publishing Limited. 510 p. ISBN: 978-1-907-56807-7.

Chapter 13
An Approach to Automatic Test Generation from Ontologies in the Virtual Education Platform

Todorka Glushkova, Asya Stoyanova-Doycheva, Vanya Ivanova, Irina Krasteva, and Mariya Grancharova-Hristova

Abstract Digitization and semantic processing of knowledge with general meaning is one of the most important tasks related to the development and use of modern educational platforms. One of the appropriate approaches to the semantic modelling of this knowledge is the use of ontologies. Through proper structuring, the ontology can provide automatic generation of test questions. The article presents an approach to building an ontology for structuring knowledge related to Bulgarian history and literature—BulgarianLiteratureAndHistoryOntology (BuLHO), which would enable automatic generation of test questions in correct, literary Bulgarian language.

Keywords Semantic modelling · Ontologies · Artificial intelligence in education (AIEd)

13.1 Introduction

The development and use of modern educational platforms based on cyber-physical and social systems (CPSS) (Yilma et al., 2021) require significant changes in the organization and management of the educational process. In essence, these platforms combine the dynamic changes of the physical environment with the capabilities of virtual and social spaces, thus providing opportunities for personalized and contextual learning.

The creation of a CPSS-based educational platform is a basic task for the team from the DeLC Laboratory of Plovdiv University. A reference software architecture ViPS has been created, on the basis of which a prototype of a CPSS educational

T. Glushkova (✉)
Department of Computer Technologies, University of Plovdiv, Plovdiv, Bulgaria
e-mail: glushkova@uni-plovdiv.bg

A. Stoyanova-Doycheva · V. Ivanova · I. Krasteva · M. Grancharova-Hristova
Department of Computer Systems, University of Plovdiv, Plovdiv, Bulgaria
e-mail: astoyanova@uni-plovdiv.bg; vantod@uni-plovdiv.bg; irina_krasteva@uni-plovdiv.bg; m.grancharova@ghp-plovdiv.org

E. Smyrnova-Trybulska et al. (eds.), *E-Learning and Enhancing Soft Skills*,
https://doi.org/10.1007/978-3-031-82243-8_13

platform called VES (Virtual Education Space) is being developed (Gramatova et al., 2018).

VES is essentially a multi-agent system that processes information about environmental changes, systematizes knowledge in different learning areas, and provides individual learning processes and adapted learning resources to students. The basis for the implementation of these tasks is the need for semantic modelling of knowledge, which in VES is implemented by building a hierarchical structure of ontologies, which are used both for personalized knowledge provision and for feedback—through automatic test generation and questions.

The use of AI in education (AIEd) aims to improve the interaction between learners and intelligent educational systems (Ayeni et al., 2024). CPSS integrates physical and virtual environments, enabling more immersive and adaptive learning. Through the use of ontologies, educational content can be structured and categorized systematically, facilitating better personalization and context-sensitive learning paths.

The article presents an approach to creating an ontology with knowledge related to Bulgarian literature and history, enabling the automatic generation of interrogative and declarative sentences as well as various types of test questions. A basic learning scenario is modeled in the VES educational platform, where the presented approach is implemented.

The structure of this article is as follows: the second section presents the motivation for the development and the related research on the topic. The third section discusses an approach for the automatic generation of test questions from the BulgarianLiteratureAndHistoryOntology (BuLHO), as well as the modelling of the process. The conclusion summarizes the obtained results and outlines future tasks on the topic.

13.2 Motivation and Related Works

13.2.1 Semantic Modelling and Ontologies in Education

The current trend in the development of the Internet is a transition from "documents read by computers" to "documents understood by computers". To realize this goal, it is necessary that computers "understand" the semantic (meaningful) essence of the information transmitted to them by people or other computers. Such "understanding" requires that information sources be semantically structured. The Semantic Web enables data to be represented as knowledge and processed automatically. The Semantic Web is the basis for the development of modern cyber-physical platforms through the formalization of concepts such as (Dameron, 2023)

- Semantics—a core component of information that explores meaning and finds meaning in data.

- Metadata—description and formalization of data for the purpose of machine processing (Garoufallou & Ovalle-Perandones, 2019);
- Ontology—knowledge to process data (Liu & Özsu, 2018) by describing it as a taxonomy with multiple inference rules.
- Logic and inference—for discovering unexpected connections and contradictions, performing automatic logical deduction (reasoning) using reasoning machines (reasoners).
- Software agents that in the global semantic network can receive and perform tasks from the user, search for appropriate resources through communication with other agents, compare and select information according to user preferences and requirements, etc.

Semantic modelling and ontologies have diverse applications in all spheres of modern life where meaningful automatic knowledge processing is required. Education ontologies are formal knowledge models that describe concepts and relationships between concepts in a subject area. They help structure and organize information, as well as provide a semantic interpretation of the data. The application of ontologies in education can be used to solve tasks such as:

- Organization of courses and training programs: Ontologies can be used to create structures and describe the content of training courses and training programs. They can help define learning objectives, competency levels, and relationships between topics.
- Development of educational standards: Ontologies can help in the development and formalization of educational standards. They can define the key competences, the content of the learning subjects, and the expected learning outcomes.
- Information Search and Retrieval: Ontologies can be used to improve the search and information retrieval of educational resources. They can structure knowledge and the relationship between concepts, allowing more accurate finding and analysis of relevant materials;
- Adaptive learning: Ontology can help create adaptive learning systems that can analyze each student's knowledge and needs and suggest personalized learning materials and methods (Stancin et al., 2020);
- Choice systems: The ontology can be used to implement choice systems that can offer students and educators relevant resources, courses, and tools based on their preferences and needs.
- Knowledge testing and assessment: Ontologies can be used to automatically generate test questions.

Semantic technologies used in ELI (Educational Learning Initiative) (Oblinger, 2005) aim to provide greater clarity in data analysis and increase the quality of the educational process. Ontologies are used as a knowledge base, which stores information about the learning process, curricula, classes, students, and teachers. The semantic links created between the concepts improve the possibilities of searching for and discovering hidden educational resources.

Adaptive Learning Systems (ALS) (Ennouamani & Mahani, 2017) are computer platforms that use artificial intelligence to personalize the learning process based on the individual needs and skills of each student. These systems use different methods and techniques based on building a portfolio of each student and determining his profile characteristics. This allows systems to build a specific learning plan for each individual student (Katsaris & Vidakis, 2021). The educational content is specified, based on the initially constructed plan, the needs and interests of the student, as well as the feedback dynamic relationship with the system in the course of the training itself. Separately, the systems actively monitor the progress of each student so that they can enrich and improve the educational strategies for him, expanding the resources provided in the topics with which he struggles (Cholakov, 2021). Such an ALS is presented in (Halim et al., 2024), where adaptive information retrieval and representation is based on domain models that are implemented through ontologies.

Checking the acquired knowledge, in the form of tests, is a service provided by LMS and ALS—learning environments with access to a test system. The automatic generation of relevant questions plays a key role in the educational process. Therefore, many researchers are focusing their efforts on the automatic generation of questions using semantic networks and ontologies. In most cases, the ontology models and question templates used to support the question generation process remain domain-specific. In (Kusuma et al., 2022), the authors focus on forming an ontology generation model and a template model for domain-specific question generation by combining two types of ontologies. It uses template queries to extract information and then translates the results of the query template into natural language questions.

Therefore, the creation and use of a network (or hierarchy) of ontologies can significantly improve the proposed mechanism (Stoyanova-Doycheva et al., 2022), and the presentation of the questions generated by the ontology in the correct literary native language of the learner will inevitably increase the success rate of the educational process. On the other hand, the use of this mechanism in multi-agent platforms makes it possible to increase the adaptability of the learning process (Cholakov, 2020).

In the next section, we will present the work of the DeLC Laboratory team of Plovdiv University in the realization of these tasks.

13.2.2 Virtual Education Space (VES)

Cyber-physical-social space is a fusion of physical, virtual, and social spaces. The evolution of CPSS is a long process that involves solving challenges and problems of a different nature. Today, these systems are developing at an extremely high speed as the transition from "digital twins" to parallel intelligence in CPSS and their connection with block-chain intelligence, smart contracts, and metaverses provide a basis for this development.

The Virtual-Physical Space (ViPS) is being developed by a team of the DeLC laboratory at Plovdiv University as a reference IoT-based architecture that can be adapted to different application areas (Stoyanov et al., 2018). Due to the fact that in this architecture individual software components and agents interact depending on changes in the environment (physical and virtual) providing appropriate services in the interest and at the request of users, ViPS can be considered as a cyber-physical and social space. When building a specific ViPS application, only individual components of the architecture are adapted, adding and integrating modules specific to the subject area under consideration. The ViPS architecture contains two sets of components. The first group is the general-purpose components that deliver common functionality to any CPSS application, regardless of the specific application domain. The second group includes components specific to the application in the respective subject area—agriculture, education, tourism, etc. The ViPS-architecture provides virtualization of real objects that are relevant to the considered domain. In this aspect, the architecture should reflect and present in the virtual world an essentially identical model of the real physical world, in which processes, users, objects and the interaction between them are realized dynamically in a personalized and context-aware manner. This virtualization is supported by the ViPS middleware, comprising several subspaces and components, the core of which is the Digital Libraries Subspace for the representation of domain-specific knowledge.

In the field of education, the team created two adaptations of the ViPS architecture—VES in the field of higher education (Stoyanov et al., 2021) and BLISS (Todorov et al., 2019)—for school education. Despite some differences related to the specifics of educational activities and services, these two educational spaces have a similar software architecture. The need to check the knowledge of the students necessitates the work on creating a prototype of a Test System that uses the information stored in the digital libraries of the space in the form of ontologies and databases In (Stoyanova-Doycheva et al., 2015) the architecture of the Virtual Education Space is considered, as a detailed the Digital Library (Digital Library Subspace—DiLibS) for storing electronic educational resources is presented as:

- Ontology Network (OntoNet)—created ontologies can be organized into hierarchies of ontologies, which allows them to be semantically related.
- Domain Libraries (DoL) are various resources that are used by operational assistants.

The Test System uses DoL and has two functions—automatic generation of test questions from ontologies and their evaluation. The architecture of the automatic question generation environment is detailed in (Stancheva et al., 2017). Test generation is supported by specialized ontologies that are served by two intelligent agents—Questioner Operative (QO) and Assessment Operative (AO). The QO is responsible for generating the questions from the ontologies, and the AO verifies the answers. A formal model is defined that describes rules and algorithms for automatically generating test questions in correct English.

Based on these studies, we set ourselves the task of developing a model and prototype of a test system that uses ontologies in the automatic generation of different types of questions and their presentation in the correct Bulgarian language.

13.3 An Approach to Automatic Generation of Test Questions from the Bulgarian Literature and History Ontology

13.3.1 An Approach to the Development of a Bulgarian Literature and History Ontology (BuLHO)

In the development of an ontology, it is necessary to formulate a strategy to achieve the desired result. When creating an ontology in the field of humanities, it is necessary to present and structure knowledge—about individuals and their professions, the periods during which they lived and worked, the settlements they inhabited, and the results of their work, so that information can easily be processed and conclusions drawn by verifying the truth of a statement. Concepts are defined as a hierarchy of classes that are related to each other through various properties: object, annotation, and data properties. These relations form axioms that shape propositions, structured as subject-predicate-object triplets, enabling inferences about the truth of propositions in the field.

To describe this knowledge, several important tasks are followed:

- Analyzing the domain and deciding which knowledge needs to be described in the ontology. Redundant and unnecessary knowledge should be excluded so as not to burden the processed information;
- Defining the main concepts in the field and creating the corresponding classes. Building a hierarchy of these classes;
- Determining the need for specific properties: object, data, and annotation properties;
- Setting property restrictions;
- Introduction of instances;
- Checking the ontology for consistency and ambiguities.

Once the domain is defined, it is necessary to compile the class organization that describes the main objects in the ontology. Classes in any ontology represent collections of objects with common characteristics. They have a set hierarchical structure and relationship. In BuLHO, classes and Object property hierarchy are presented in Fig. 13.1 (the ontology is created in Bulgarian and the screenshots are also in Bulgarian).

Object properties in an ontology (Fig. 13.1) allow a relationship to be created between the individuals of two classes, forming a Resource Description Framework (RDF) triple: subject-predicate-object, where the subject is the subject of the

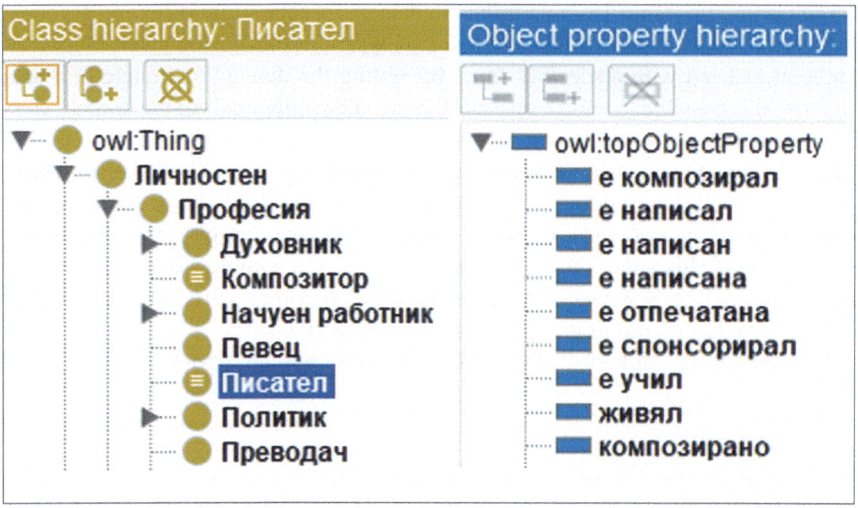

Fig. 13.1 Classes and object property hierarchy. (Source: Own work)

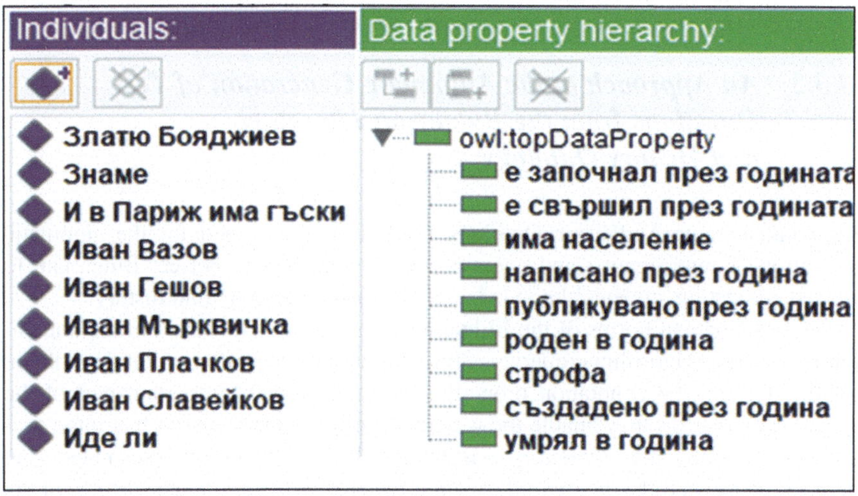

Fig. 13.2 Individuals and data property hierarchy. (Source: Own work)

sentence, and the object plays the role of the complement. By setting the domain and range of the object property, the possibility of building a connection between individuals of the domain class and individuals of the range class through the given property is shown.

Individuals represent concrete objects such as people, works, cities, etc., and correspond to a certain descriptive logic. They belong to the main components that make up the ontology (Fig. 13.2).

Data properties (Fig. 13.2) describe the data of an entity with different Resource Description Framework Schema RDF(S) data types. In this way, a given individual can be associated with a specific value. By setting the domain and range of a data property, it is possible to build a relationship between individuals of the domain class and a value belonging to the type set as the range of the property.

Meta-knowledge can be added to any ontology. It represents knowledge about knowledge itself, i.e., it provides additional information about classes, entities, data properties, object properties, or about itself. In this way, an object is bound to a specific value. This value can be a character string or a complex object (a complex annotation property). We speak of a complex annotation property when an annotation property has one or more sub-annotation properties.

The developed ontology BuLHO contains the following metrics:

- Axioms—5094
- Classes—90
- Object properties—22
- Data properties—9
- Individualities—330
- Annotation properties—12

13.3.2 An Approach to the Automatic Generation of Test Questions from the Bulgarian Literature and History Ontology

BulgarianLiteratureAndHistoryOntology (BuLHO) is designed for the automatic generation of interrogative and communicative sentences in correct Bulgarian, to be used for testing students' knowledge of Bulgarian literature and history.

The grammatical norms of the Bulgarian language are embedded in the annotation properties. Algorithmic queries extract the underlying information from them. For this reason, the annotation properties are an extremely important part of the created ontology, as they enable the generation of grammatically correct interrogative and communicative sentences in Bulgarian. If there are no mistakes in entering the information and the entered information adheres to the grammatical rules of the Bulgarian language, then the generated sentences will be correctly composed.

The information embedded in the annotation properties of classes, individuals, data properties, object properties, and the annotation properties themselves is used to create interrogative or declarative sentences. The syntax of the sentences requires that the information entered in the ontology obey different grammatical rules. In the created annotation properties, these rules of the Bulgarian language are laid down, so that it is possible to create properly structured sentences.

Annotation properties are used with classes, data properties, and object properties in the ontology to represent the grammatical possible forms of nouns and verbs in the Bulgarian language. The grammatical correctness of question generation in

Bulgarian from the ontology depends on the representations in the annotation properties. We use annotation properties to represent the different genders of nouns, their definite articles in full and short forms for different genders, and their agreement. For example, if we take the profession of writer, in Bulgarian, the word is masculine, which means the verb following it must also be in the masculine form. To achieve this, we apply annotation properties for the respective genders and the corresponding verb conjugations to each class and object property.

In Fig. 13.3, you can see the class " Писател – Writer", which is in the masculine gender, but it is also used in the feminine gender as "Писателка – Writer (female)" and in the plural as "Писатели –Writers". For each gender of the word, there are corresponding definite articles—short, full, and neutral forms. An annotation property is also added for the agreement of the class with the object properties (a noun with a verb). The noun and verb must agree according to the gender of the noun (the class from the ontology) that participates in the sentence. The example in Fig. 13.3 illustrates the annotation properties of a class, with similar properties added to each data property, object property, and individual.

The generation of the sentences themselves is done with the help of developed algorithms, in which the rules for sentence generation are specified depending on

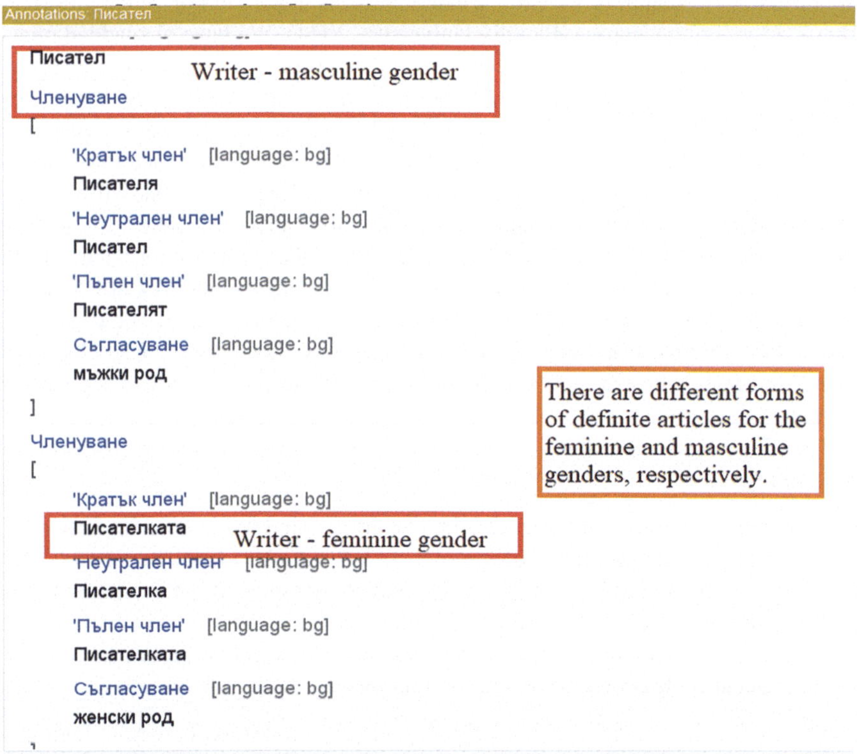

Fig. 13.3 Annotation properties for class "Писател – Writer". (Source: Own work)

startQuestionLabel	answerType	questionSentenceLabelA	join	questionSentenceLabelB	fullMemberLabel
"В кой"@bg	Град	"е родена"@bg	"и"	"е живяла"@bg	"Жената"@bg
"В кой"@bg	Град	"е родена"@bg	"и"	"е живяла"@bg	"Жената"@bg
"В кой"@bg	Град	"е родена"@bg	"и"	"е живяла"@bg	"Жената"@bg
"В кой"@bg	Град	"е родена"@bg	"и"	"е живяла"@bg	"Жената"@bg
"В кой"@bg	Град	"е родена"@bg	"и"	"е живяла"@bg	"Жената"@bg

individualLabel	label	correctAnswer...	wrongAnswerA...	wrongAnswerB...	wrongAnswerCL...
"Леда Милева"@bg	"Отговор"	"София"@bg	"Габрово"@bg	"Пловдив"@bg	"Йорусалим"@bg
"Леда Милева"@bg	"Отговор"	"София"@bg	"Габрово"@bg	"Пловдив"@bg	"Калофер"@bg
"Леда Милева"@bg	"Отговор"	"София"@bg	"Габрово"@bg	"Пловдив"@bg	"Одеса"@bg
"Леда Милева"@bg	"Отговор"	"София"@bg	"Габрово"@bg	"Пловдив"@bg	"Ресен"@bg
"Леда Милева"@bg	"Отговор"	"София"@bg	"Габрово"@bg	"Пловдив"@bg	"Пещера"@bg

Fig. 13.4 Generated questions. (Source: Own work)

which part of the sentence the question is directed to—subject, predicate, object, or circumstance (Research on creating semantic models in the field of humanities, Phd Thesis, Marya Grancharova, Plovdiv University, 2024). We have also added question words to each class, object property, and data property in the form of interrogative pronouns. These are used to start a question directed at the respective part. The various algorithms are developed using the SPARQL language (https://www.w3.org/TR/rdf-sparql-query/). When an algorithm is executed, the returned information is structured in a table containing rows and columns (Fig. 13.4). Each row of the table represents a unique sentence and the possible answers offered for it. The columns of the table are the separate components that, when connected, create lexically and logically coherent interrogative or declarative sentences in Bulgarian and their possible answers. The algorithms can be executed a single time, and the information returned from them can be saved in JSON (JavaScript Object Notation) format or another format, and subsequently used by testing systems designed for training.

The developed algorithms for creating interrogative sentences provide an opportunity to:

- generation of a question sentence with one correct and three incorrect answers;
- generation a question sentence with two correct and two incorrect answers;
- generation of an interrogative sentence in which two subjects are in relation to the object, by means of a common object property;
- generation of an interrogative sentence in which two object properties are used for the connection between the subject and the object.

In order to be able to create a meaningful sentence, a detection between "Interrogative Pronoun" and "Interrogative Verb" based on the sub-annotation property "Agreement" is used. When there is no such detection between the two complex annotation properties, it is not possible to generate a question that meets the grammatical and syntactic norms in the Bulgarian language for matching the predicate with the subject in the sentence.

The complex annotation property "Verb in Communicative Sentence" describes all possible concordances and is used by the algorithms that generate:

- communication sentence and the answer is required to be added;
- declarative sentence that makes a statement and the answer must be affirmed or denied.

Restrictions are set for some classes from the ontology. They represent a relationship to other classes or entities through properties. In this way, each class is defined semantically. Each subclass inherits all the restrictions of its superclasses. Constraints form anonymous (unnamed) classes in the ontology.

The constraints created in the ontology belong to one of the following types: existential, universal, or cardinal. Existential constraints ("some" or some values from) represent a relationship through a given property to at least one class. The class to which the constraint applies is the base class, and the classes to which it is linked via the property are restrictive classes. Universal restrictions ("only", "all values from") restrict the relationship through a given property to be only with a given class, and Cardinal restrictions represent the number of relationships. In the ontology, data and object properties are used to set such criteria. As a result of the described approach, different types of test questions can be generated from the created ontology in different humanities academic disciplines—history, literature, philosophy, etc.

13.3.3 Modelling of the Test Generation Process

The process of automatically generating test questions from the developed ontology in the educational space VES can be modeled in advance under dynamic consideration of various environmental changes by using the mathematical notation of Calculus of Context-aware Ambients (CCA) (Siewe et al., 2010). The advantages of this modelling is the possibility of simulating different scenarios and multiple preliminary testing and verification of the processes. Modeled scenarios are programmed and executed using the ccaPL language interpreter in console, visual or animated mode.

The modelling process in CCA involves individual identities—Ambients that can communicate with each other (taking into account the current context), dynamically form ambient hierarchies, and change their location and interaction with other Ambients into the current relationship: sibling, child, or parent.

Figure 13.5 presents the general VES infrastructure needed for the process of automatic test question generation. The main identities (Ambients) are:

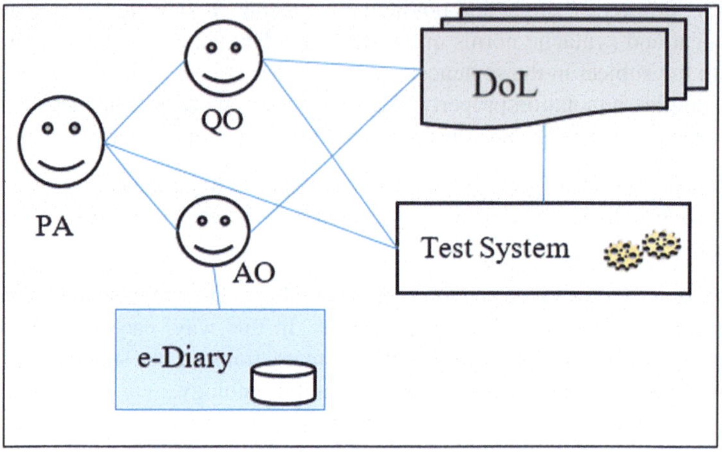

Fig. 13.5 VES test generation infrastructure. (Source: Own work)

- PA: The student's personal assistant that represents them in the system.
- DoL: The component in the DiLib subspace of VES that supports the specialized ontologies.
- QO (Questioner Operative Agent): This agent generates questions and sentences from the ontology when interacting with DoL.
- Test System (TSys): A component in VES that selects and creates a personalized test from a set of questions on a certain topic.
- E-Diary (eDiary): An electronic diary storing information on the individual progress and improvement of each student.
- AO (Assessment Operational Agent): This agent, after receiving the student's answer (via their PA), checks its correctness when interacting with specialized ontology and sends the student's result to the system's e-Diary

In the modelled process, the student, through his PA, sends a request to the QO to generate a test question to check his knowledge of the topic. After interacting with DoL, a set of questions on the desired topic is generated from the ontology and sent to TSys. The Test System stores the generated questions and, according to the student's personal characteristics and the available database of already generated test questions, selects a suitable question and sends it to the student's PA. The AO Ambient receives the student's response and verifies its correctness by interacting with specialized ontology, then informs the student and records the information in the eDiary.

Through the CCA notation, the processes of the considered Ambients are formalized as follows:

$$P_{PA} \equiv \big(QO :: < GenerateQuestion, Topic_j > .0|\ TSys :: \big(TestQuestion_k, Topic_j\big).AO :: < Ans\_TestQuestion_k, Topic_j > .0|\ AO :: (Result).0\ \big) \tag{13.1}$$

$$P_{QO} \equiv \big(PA :: \big(GenerateQuestion, Topic_j\big).ONet :: < GiveMeKnowleges, Topic_j, PA_i > .0|\ DoL :: \big(KnowlegesOnTheTopic_j\big).TSys :: < ListGeneratedTestQuestions, Topic_j, PA_i > .0\ \big) \tag{13.2}$$

$$P_{DoL} \equiv \big(QO :: \big(GiveMeKnowleges, Topic_j, PA_i\big).QO :: < KnowlegesOnTheTopic_j > .0|\ AO :: \big(CheckAnsQuest_k, Topic_j, PA_i\big).AO :: < Result, PA_i > .0\ \big) \tag{13.3}$$

$$P_{TSys} \equiv \big(QO :: \big(ListGeneratedTestQuestions, Topic_j, PA_i\ \big).PA :: < TestQuestion_k, Topic_j > .0\ \big) \tag{13.4}$$

$$P_{AO} \equiv \big(PA :: \big(Ans\_TestQuestion_k, Topic_j\big).DoL :: < CheckAnsQuest_k, Topic_j, PA_i > .0|\ DoL :: (Result, PA_i).PA :: < Result > .eDiary < Result, Topic_j, PA_i > .0\ \big) \tag{13.5}$$

$$P_{eDiary} \equiv \big(AO :: \big(Result, Topic_j, PA_i\ \big).0\ \big) \tag{13.6}$$

13.3.4 Future Work Discussions

Integrating ontologies into an interconnected network can greatly enhance educational assessments. By combining ontologies from different disciplines such as history, science, literature, and mathematics, educators can create interdisciplinary test questions that encourage critical thinking and deeper understanding. Including ontologies related to real-world contexts, like cultural or historical events, enables the development of situational questions that challenge learners to apply their knowledge practically. Incorporating linguistic ontologies allows for assessments that test comprehension and translation across multiple languages while integrating emotional intelligence and social behaviour ontologies helps in evaluating learners' understanding of emotional responses and ethical decision-making. Expanding ontologies to cover STEM fields and technical skills can lead to complex problem-solving questions that simulate real-world challenges.

For ontology integration for question generation in Bulgarian, ontologies must have the same structure in terms of annotation properties to each class, identity, data property, or object property. This is necessary for the algorithms created to work on the integrated ontologies. Adding these annotation properties is a rather time-consuming process and requires learners' knowledge of the Bulgarian language grammar. To facilitate the process of creating annotation properties for the grammatical needs of the language, a tool has been created that facilitates and supports ontology designers and developers—OntologyGenerator (Grancharova-Hristova, 2024). The tool allows for quick addition of annotation properties, and can also be used to add identities to any of the created ontology structures.

Another development task is the automatic evaluation of the generated tests. A model based on the mathematical theory of fuzzy sets has been developed, which allows students' knowledge to be evaluated more than once, taking into account data from previous results and for individual learners (Glushkova et al., 2024). In this way, the system can monitor each student's progress and generate a fair grade to store in the eDiary. In turn, eDiary is developed through the application of blockchain technologies. A prototype of this component is presented in (Krasteva, 2024).

The use of the educational platform in hybrid learning as well as for independent lifelong learning is one of the main tasks before the development team. We believe that the use of Blockchain technologies to issue factory-numbered documents such as diplomas and certificates as well as electronic certificates is quite useful in the education sector. Each learner, upon registering in the educational platform, receives his own wallet, in which the received diplomas, certificates, electronic certificates from lifelong learning courses, as well as a resume or portfolio, will be stored. The certificates will be generated by an educational certification system built on Blockchain technologies.

After a training course, the learner solves tests from questions generated by the ontology, receiving points and a corresponding grade based on the knowledge he has demonstrated. The obtained result is recorded in the electronic diary. Upon successful completion of a course, a smart contract is invoked, through which an electronic certificate is issued containing a unique identifier for its storage in the blockchain network. Each certificate is recorded and stored in the electronic wallet of the given learner. All electronic documents that are stored in the e-wallet can be used when applying for a job, as well as in front of all institutions where necessary. Any person or organization can verify the authenticity of the certificate by accessing and verifying the unique identifier from the Blockchain certification system.

Blockchain technologies would contribute to increasing the security, transparency and immutability of certificates. At this stage, we are experimenting to implement the Blockcerts system in the education system built on the Hyperledger Fabric.

13.4 Conclusion

The development and implementation of automatic test generation within the Virtual Education Space (VES) represent a significant advancement in educational technology. This approach harnesses sophisticated semantic technologies and dynamic modelling techniques to greatly enhance the educational process.

Using ontologies to structure knowledge within VES allows for the creation of test questions that are both semantically rich and contextually accurate. The Bulgarian Literature and History Ontology (BuLHO) enables the system to generate questions that are not only relevant but also grammatically and syntactically correct. This alignment with the curriculum ensures that the questions accurately assess students' understanding of the subject matter.

Modelling the test generation process through the Calculus of Context-aware Ambients (CCA) provides the flexibility to simulate various scenarios and dynamically adapt to changing educational needs. This adaptability supports personalized learning experiences by tailoring test questions to individual student profiles and learning progress. Real-time adjustments based on student performance and feedback are facilitated by the dynamic interactions of ambients within VES.

Automating the generation and evaluation of test questions significantly reduces the manual effort involved in question creation, ensuring consistency in assessment. This automation allows educators to concentrate more on interactive teaching rather than administrative tasks, enhancing overall teaching efficiency.

Incorporating annotation properties within the ontology ensures that the generated questions adhere to the grammatical norms of the Bulgarian language. This attention to linguistic detail guarantees that questions are not only accurate but also culturally and contextually relevant, thereby improving the quality of the assessments.

Looking ahead, integrating ontologies from various disciplines, such as history, science, and literature, into a unified network could lead to the creation of interdisciplinary test questions. These questions would encourage students to apply their knowledge across different contexts, fostering critical thinking and a deeper understanding of the material. Additionally, including real-world contexts and situational questions within the ontology framework would further enhance the practical relevance of the assessments. The development of tools like OntologyGenerator also supports the creation and management of ontologies by facilitating the addition of annotation properties for grammatical accuracy. This tool streamlines the ontology development process and ensures consistency across different educational domains.

The approach to automatic test generation within VES provides a robust framework for personalized and context-aware education. By leveraging advanced semantic technologies and dynamic modelling, this method not only improves the quality and relevance of assessments but also supports the creation of adaptive and efficient learning environments.

Acknowledgments This study is financed by the European Union-NextGenerationEU, through the National Recovery and Resilience Plan of the Republic of Bulgaria, project № BG-RRP-2.004-0001-C01 and supported by the project FP23-FMI-002 "Intelligent software tools and applications in research in mathematics, informatics, and teaching pedagogy" at the Plovdiv University "Paisii Hilendarski".

References

Ayeni, O. O., Al Hamad, N. M., Chisom, O. N., Osawaru, B., & Adewusi, O. E. (2024). AI in education: A review of personalized learning and educational technology. *GSC Advanced Research and Reviews, 18*(2), 261–271. https://doi.org/10.30574/gscarr.2024.18.2.0062

Cholakov, G. (2020). Approbation of software agent evaluator in a nonspecific environment for extension of its purpose. In *International conference automatics and informatics (ICAI), Varna, Bulgaria, 2020* (pp. 1–5). https://doi.org/10.1109/ICAI50593.2020.9311346

Cholakov, G. (2021). Secondary effects converted to useful knowledge in e-learning system. In *2021 international conference automatics and informatics (ICAI), Varna, Bulgaria, 2021* (pp. 117–120). https://doi.org/10.1109/ICAI52893.2021.9639599

Dameron, O. (2023). Semantic web methods for data integration in life sciences. In *Biological data integration: Computer and statistical approaches* (pp. 25–51). Wiley. https://doi.org/10.1002/9781394257317.ch2

Ennouamani, S., & Mahani, Z. (2017). An overview of adaptive e-learning systems. In *In 2017 eighth international conference on intelligent computing and information systems (ICICIS)* (pp. 342–347). IEEE. https://doi.org/10.1109/INTELCIS.2017.8260060

Garoufallou, E., & Ovalle-Perandones, M. A. (2019). *Metadata and semantic research*. Springer International Publishing.

Glushkova, T., Ivanova, V., & Zlatanov, B. (2024). Beyond traditional assessment: A fuzzy logic-infused hybrid approach to equitable proficiency evaluation via online practice tests. *Mathematics, 12*(3), 371. https://doi.org/10.3390/math12030371

Gramatova, K., Stoyanov, S., & Popchev, I. (2018). Virtual education space realization as an internet of things ecosystem. *Engineering Sciences, LV, 1*, 5–19. https://doi.org/10.7546/EngSci.LV.18.01.01

Grancharova-Hristova, M. (2024). *Research on the creation of semantic models in the field of Humanities*. PhD Thesis, Plovdiv University, 2024.

Halim, M., Tahiri, A., El Ghzizal, Y., Adadi, N., & Chenouni, D. (2024). Web service-oriented E-learning: Proposition of semantic approach to discover web services related to the E-learning system. *Journal of Internet Services and Information Security (JISIS), 14*(2 (May)), 1–17. https://doi.org/10.58346/JISIS.2024.I2.001

Katsaris, I., & Vidakis, N. (2021). Adaptive e-learning systems through learning styles: A review of the literature. *Advances in Mobile Learning Educational Research, 1*(2), 124–145. https://doi.org/10.25082/AMLER.2021.02.007

Krasteva, I. (2024). *Blockchain-based synchronization of Personal Assistants*. PhD Thesis, Plovdiv University, 2024.

Kusuma, S. F., Siahaan, D. O., & Fatichah, C. (2022). Automatic question generation with various difficulty levels based on knowledge ontology using a query template. *Knowledge-Based Systems, 249*, 108906.

Liu, L., & Özsu, M. T. (Eds.). (2018). *Encyclopedia of database systems* (Vol. 6). Springer, ISBN: 978-1-4614-8266-6.

Oblinger, D. G. (2005). Learners, learning, and technology: The EDUCAUSE learning initiative. *Educause Review, 40*(5), 66–75.

Siewe, F., Zedan, H., & Cau, A. (2010). The calculus of context-aware Ambients. *Journal of Computer and System Sciences, 77*(4), 597–620.

Stancheva, N., Stoyanova-Doycheva, A., Stoyanov, S., Popchev, I., & Ivanova, V. (2017). An environment for automatic test generation. *Cybernetics and information Technologies, 17*(2), 183–196. https://doi.org/10.1515/cait-2017-0025

Stancin, K., Poscic, P., & Jaksic, D. (2020). Ontologies in education–state of the art. *Education and Information Technologies, 25*(6), 5301–5320.

Stoyanov, S., Stoyanova-Doycheva, A., Glushkova, T., Doychev, E. (2018). Virtual physical space – An architecture supporting Internet of Things applications. In *20th International Symposium on Electrical Apparatus and Technologies (SIELA)*. https://doi.org/10.1109/SIELA.2018.8447156

Stoyanov, S., Glushkova, T., Stoyanova-Doycheva, A. G., & Krasteva, I. (2021). The virtual education space: Concept, architecture, application. *Informatics and Education, 36*(9), 47–54.

Stoyanova-Doycheva, A., Doychev, E., Ivanova, V., Valkanova, V., & Stoyanov, S. (2015). DiLibS platform for a virtual education space. In *Proceedings of the 2015 Balkan conference on informatics, Advances in ICT*.

Stoyanova-Doycheva, A., Madanska, S., Grancharova-Hristova, M., Glushkova, T., & Cholakov, G. (2022). Development of ontologies in different domains for a test generation environment. In *8th international conference on higher education advances (HEAd'22)* (pp. 925–933). https://doi.org/10.4995/HEAd22.2022.14205

Todorov, J., Krasteva, I., Ivanova, V., & Doychev, E. (2019). BLISS-A CPSS-like application for lifelong learning. In *IEEE international symposium on innovations in intelligent systems and applications (INISTA), Sofia, Bulgaria, 2019* (pp. 1–5). https://doi.org/10.1109/INISTA.2019.8778363

Yilma, B. A., Panetto, H., & Naudet, Y. (2021). Systemic formalisation of cyber-physical-social system (CPSS): A systematic literature review. *Computers in Industry, 129*, 103458. https://doi.org/10.1016/j.compind.2021.103458

Chapter 14
CAN Bus Workplace with Remote Access Using LabVIEW Web Service

Miroslav Kamenský, Ján Halgoš, and Anatolii Kliuchka

Abstract Progress in the deployment of distance learning has accelerated in recent years. Almost every teacher today needs to solve the issue to some extent. Therefore, it is necessary to simplify the related processes even when a remote access should be achieved for a newly built or installed hardware equipment. Corresponding is the situation at our university institute dealing with the fields of electrical engineering, measurements, technical diagnostics. Work with the CAN bus is required in several subjects taught and remote access to a special CAN system was requested. For this purpose, we used our experience in developing a remote access system based on application pairs. In this article we describe the current structure and functionality of this modular system after latest improvements and inclusion of the didactic CAN workplace. We explain the process of adding remote access to the workplace. Thanks to the modular software system architecture, it is enough to modify one module of paired applications without knowledge of web publishing tools. A user receives access via any standard web browser. Finally, we collected feedback from students on such a teaching aid.

Keywords Remote access · LabVIEW · CAN bus · Distance learning

14.1 Introduction

Many approaches and tools have been developed to support remote web collaboration. In recent years, there has been an expansion in their practical usage for distance and blended learning. Despite cultural differences (Silva et al., 2020), the progress was driven over the whole world especially due to the COVID pandemic and wartimes in some countries (Skvortsova et al., 2022). After the COVID situation stabilized, the commercial company remains to use those means, and the academic field is in a similar situation. We observe traditional academic tools and new

M. Kamenský (✉) · J. Halgoš · A. Kliuchka
Institute of Electrical Engineering, Bratislava, Faculty of Electrical Engineering and Information Technology, Slovak University of Technology in Bratislava, Bratislava, Slovakia
e-mail: Miroslav.Kamensky@stuba.sk; Jan.Halgos@stuba.sk; Anatolii.Kliuchka@stuba.sk

E. Smyrnova-Trybulska et al. (eds.), *E-Learning and Enhancing Soft Skills*,
https://doi.org/10.1007/978-3-031-82243-8_14

applications penetrating from the commercial sphere. The former is represented e.g. by Moodle as an enhanced Learning Management System (Flores-Piñas et al., 2022). The example of the latter is Zoom, especially successful as a teaching aid during school lockdowns, or MSTeams, etc. Also, Massive Open Online Courses—MOOCs are applied in the university environment (Smyrnova-Trybulska et al., 2021). Despite advanced technologies and user-friendly environments, a teacher must still have knowledge and strategies about how to create content to keep the effectiveness of a distance learning tool for remote laboratories (Mokhtar et al., 2014). In the case of specialized laboratories, one still needs to find the way of establishing remote access and its publishing via existing LMS (Learning Measurement System) or institute web. Therefore, the question of simplification of such designs remains.

There are a lot of popular programming and scripting environments suitable for web publishing applications. Graphical programming in LabVIEW became a standard at academic institutes due to easy-to-learn features for a beginner. Today, LabVIEW supports data acquisition applications, automation and control and many other areas. It offers several strategies and tools for network data sharing and remote access distribution. However, finally a big project might arise when trying to create a stable application of desired properties. This paper presents a modular system we gradually built which simplifies the task. The process of creating a new module is explained on a special CAN bus workplace suitable for subjects oriented on industrial automation or automotive diagnostics.

14.2 Choosing LabVIEW for the Application

Graphical programming is a way to simplify software design for non-IT technicians. LabVIEW (from Laboratory Virtual Instrument Engineering Workbench) represents a development environment for graphical programing—in this case for language G from NI (National Instruments). Originally, it was oriented for measurement and data acquisition. One can assume that measurement related subjects are a suitable field of application for this programming environment taking into account also the idea that the studied topic influences software preferences of students (Barot et al., 2020). Electromagnetic compatibility (EMC) belongs also to measurement areas where NI software and hardware tools are popular (Otomański et al., 2021) and where our department is engaged. Anyway, LabVIEW is common for sensing and control applications too (Singh et al., 2015) and gradually it became rich with advanced libraries in areas of industrial automation, visualization etc. In this article we will refer mainly to the version LabVIEW 2014 installed in our laboratories.

Design of a new application in LabVIEW is done via two windows: Front Panel and Block Diagram. In Front Panel we arrange elements of graphical user interface (GUI) while in Block Diagram functionalities are implemented via graphical code. Program code in block diagram form can be illustrative and suitable also for calculations and simulation (León-Martínez et al., 2023).

A traditional tool available within the environment for establishing remote access to a designed workplace is known as Remote Panel. It can be activated simply by choosing the Web Publishing Tool item from the Tools submenu in the Block Diagram main menu. So, it is easy to use technology and by a few clicks the front panel can be embedded into a web page. Anyway, a correct concurrent configuration of a LabVIEW Web Server is assumed. Then, the Internet browser, the LabVIEW run-time engine and a browser plug-in are required on the client machine for executing the front panel in a web page. The Remote Panel tool also handles the connection of several users to the application by enabling one remote control at the time keeping other clients in the position of observers waiting for a full access.

Disadvantage of Remote Panel resides in the requirement of the LabVIEW run-time engine and plug-in, which prevents securing access from mobile devices. Fortunately, other technologies (Tripathi et al., 2012) are also commonly available today allowing establishing remote access even from smart phone Internet browsers. A more daring designer can reach for Web Services supported from LabVIEW version 8.6 which also uses built-in Web Server. It is a set of protocols and standards that allow data to be exchanged between applications with great space for customization by designer. It is available within LabVIEW project and typically consists of G code *.vi files implementing back-end programming tasks, and HTML plus the JavaScript part forming the project front-end with the GUI layout. In HTML code visual elements are defined while the JavaScript part uses DOM (Document Object Model) for accessing HTML elements and adding interactivity.

Web publishing is not the only task when creating access over the network. A set of distributed applications might require data sharing. This is a space for DataSocket usage. It is a traditional technology coming with the LabVIEW package which enables live data transfer via DataSocket Server. It is suitable for simple applications as it does not require a project to be established. A newer technology for data exchange is Shared Variable, however, for this case a LabVIEW project must be established.

14.3 System Design

Based on previous experience at our institute with technologies described above we were improving our strategy for the remote access system design. During past years a modular topology settled for our system where the control of a workplace is realized via a pair of application modules (Kamenský et al., 2023). A local or target module controls the hardware and enables or limits all actions that a user can physically "do", therefore we call it *Do_Module*. The code for remote access is not directly implemented here, *Do_Module* only follows given principles for communication with another module. Such a publishing module is called *Go_Project* as this is the part that provides data to "go" to the remote user. *Go_Project* is a LabVIEW project utilizing a Web Service. As DataSocket variables are used for communication between module pairs, *Do_Module* can be a simple LabVIEW file (*.vi) and stay outside any project. Advantages of such a concept are:

- **Access from any browser**—front-end utilizes web services, therefore access is achievable from any standard and actual Internet browser without special requests for plug-ins or runtime engines.
- **Software security**—only *Go_Project* requires access from outside of the local network. *Do_Module* controlling probably unique and expensive hardware can reside on a local PC with restricted Internet access, if access to *Go_Project* over the local network is provided.
- **Simplified workplace design**—if a teacher wants to create a new workplace accessed remotely, no knowledge about web publishing technologies is required. *Go_Project* is an immutable part of the system. Furthermore, substantial parts of *Do_Module* code can be reused so an inexperienced designer can proceed by merely modifying existing *Do_Module* or a template. A universal template was created. And also to be mentioned, derived *Do_Module* should be able to work as a standalone application too—meaning with local control only without any handshake with *Go_Project*.

To further describe our approach, we will explain parts of GUI which determine other functionalities including claims for data sharing. GUI is composed of a few components: five Button Labels and a Parameter field for user inputs and Graph, Listing, Image and Message for data representation. The *Do_Module* template prepared in LabVIEW has the form shown in Fig. 14.1a. Graph/Listing/Image are aligned as tabs with message text below, buttons are lined up on the left side and parameter input field above. On top of the window, we can see the application

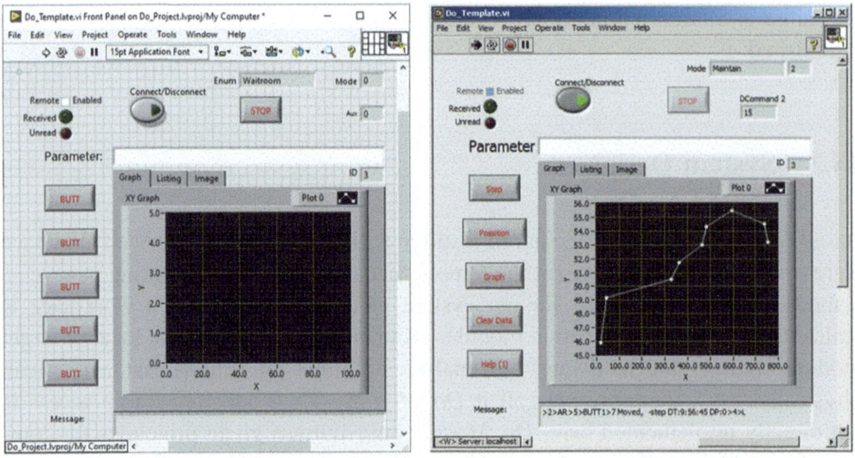

a) b)

Fig. 14.1 Front panel of *Do_Module*: (**a**) before the application starts with unassigned buttons; (**b**) after the application starts with connection established and assigned buttons. (Source: Own work)

control/indication part. It includes parts for enabling remote control and establishing/cancelling connections (to workplace and to remote application), a button for stopping the application and indicators showing the current state. The template has general form, the button labels are not assigned at startup and will be specified after choosing Connect. In Fig. 14.1b we can see the template already adapted for CAN workplace in the connected state and with remote control enabled. Cooperating *Go_Project* should contain corresponding components of workplace control and data representation; however, alignment might be different—see Fig. 14.6a.

Even though the application can run in purely local mode, it was developed with the purpose of building a remote access system. Implementation of remote control is part of the source code and needs to be explained. For communication with *Go_Project* five DataSocket variables are used. Two of them, *Gcom*, *Gvar* are intended for receiving commands and data at the beginning of interaction, which is always initiated by *Go_Project*. Then a reply is formed into the *Dcom* DataSocket variable containing a response command accompanied by data passed via *Dvar*. A special *Dpicture* variable is used if a picture should be transferred within the reply.

Command sent via 16-bit *Gcom* and *Dcom* DataSocket variables specify request of *Go_Project* on the one side and describe content of *Do_Module* reply on the other side. Lower bits refer to parts of GUI while higher bits are used for special requests or alternatives. For example, the lowest bit number 0 in *Gcom* corresponds to button 1 and if it is high, it means that this button was pressed by a remote user and appropriate response of *Do_Module* is expected. *Gvar* has the meaning of an additional string pasted by the user into the Parameter field. The highest *Gcom* bit number 15 means a request for disconnection. Within a response the target module informs *Go_Project* via lower *Dcom* bits about data to be presented in GUI—whether new button labels, Graph or Listing content or Picture are being sent with the command. And again, the highest bit number 15 is informing about disconnection if such a case is happening. The meaning of the mentioned bits is shown in Fig. 14.2. Apparently, there was an intention to form the command syntax as a flag register where the receiver can easily check the appropriate bit to recognize which GUI item is referred. Subsequently, the data of *Dvar* or *Dpicture* variables are read and interpreted in a specified manner on the *Go_Project* side.

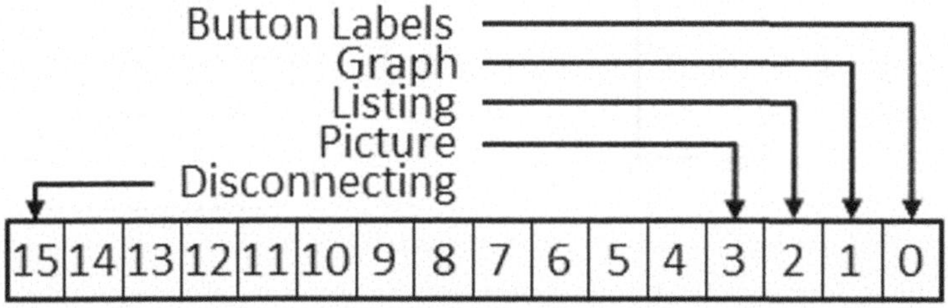

Fig. 14.2 Meaning of the main dedicated bits in the *Dcom* 16-bit command variable used to transfer reply from *Do_Module* to *Go_Project*. (Source: Own work)

14.4 Target Module Structure

The structure of a *Do_Module* can be represented by a diagram depicted in Fig. 14.3. Software is working as a finite state machine. After starting the application, it is waiting for local user interaction in state 0: **Wait-room**. Only local users are allowed to proceed with the Connect/Disconnect button with a pre-selected option for enabling remote control. After establishing the connection an **Initialization** in state 1 is performed where DataSocket variables are opened. Next, button labels are assigned in state 6. *Do_Module* then stays in state 2: **Maintain** waiting for further user action. This can be performed locally by pressing a button or remotely—if previously allowed—via a command received from *Go_Project*.

If a command was received in state 2, it is decoded in state 5: **Decide**, where also data variable *Gvar* is read, and a decision is made about further procedure. In case of a simple action not related to the workplace control—like a request for disconnection or reading of existing data—all steps might be performed and reply command and data prepared in state 5. The reply is then sent out in state 4: **Reply**. If HW or advanced action is requested, then the right state from 6 and above (**Actions**) is chosen within decision in state 5. Finally, the reply is always sent in state 4. After data transfer the application returns to state 2 and continues to maintain the connection. In case of local interaction, state 4 is omitted as no reply is needed. However, the application will remember that some data left unread (see Unread Boolean Indicator in Fig. 14.1) by *Go_Project* and unread data can be remotely requested later using appropriate bits of *Gcom*. This means that *Go_Project* managing remote

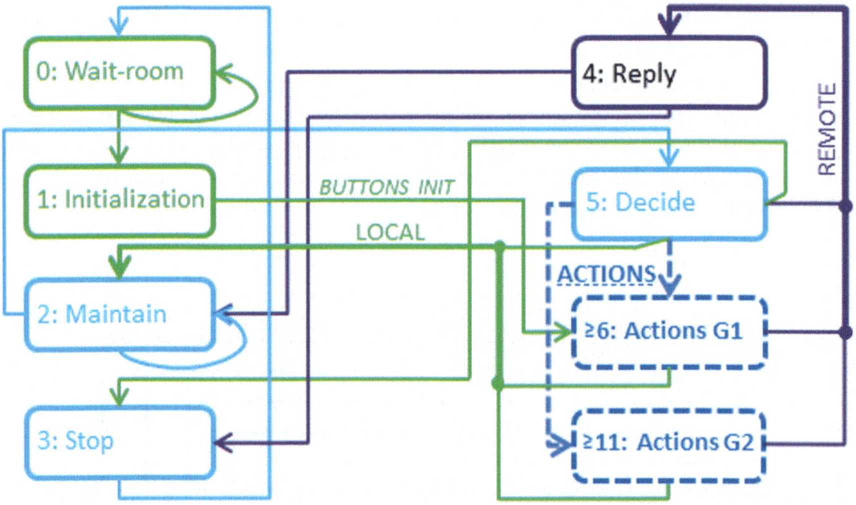

Fig. 14.3 Block scheme of *Do_Module*. **Action** states number 6 and above are divided into two parts **G1** and **G2**. Hence, the diagram can be understood as a general case of a *Do_Module* with one submenu or as a diagram corresponding to *Do_Module* controlling the CAN workplace. (Source: Own work)

connection can refresh even data changes resulting from local activities. For this purpose, *Do_Module* stores and updates a special internal state register (*RemStat*) containing flag bits of unread data—Button Labels, Graph, Listing or Picture.

In Fig. 14.3 the local behavior is drawn in green, remote access related progress is indicated by purple while blue color covers both alternatively. Regardless of this, states can be then divided into two parts. Lower numbers 0 till 5 are the core of the *Do_Module* structure and cannot be omitted. Here, state 0 can be understood as purely local while later also remote action is accepted. Only state 4 is purely related to remote action. If a designer comes with a new workplace, he must understand and most likely also modify mode 5 which decides how the software responds to received command and which corresponding state should follow. The choice is done between states 6 and above, not considering now cases of simple reaction performed within state 5 itself. So, the second group of states are those user states 6, 7 etc. Exact numbering of action states is arbitrary; however, assignment of button labels would be recommended in state 6. Anyway, action states might be further divided into several groups (**Actions G1**, **G2**..) meaning that submenus may exist within the application when new button labels need to be assigned. Content of **Action** states depends on the application and is fully on the workplace designer.

14.5 CAN Bus Workplace

Within subjects related with communication buses, diagnostics especially in automotive area or even with industrial automation a CAN bus didactic workplace might be necessary. One such workplace used at our institute is principally depicted in Fig. 14.4. The pedagogical purpose here is to show live data from a known hardware with ability to influence measured input. More specifically, it is based on the rotary position sensor type RT8CN with CAN output whose measured position is adjustable by means of a stepper motor. CAN data are collected on PC using CAN to USB adapter. Stepper motor having a common shaft with the sensor is controlled from the same PC via digital outputs of the DAQ card and the motor driver circuit.

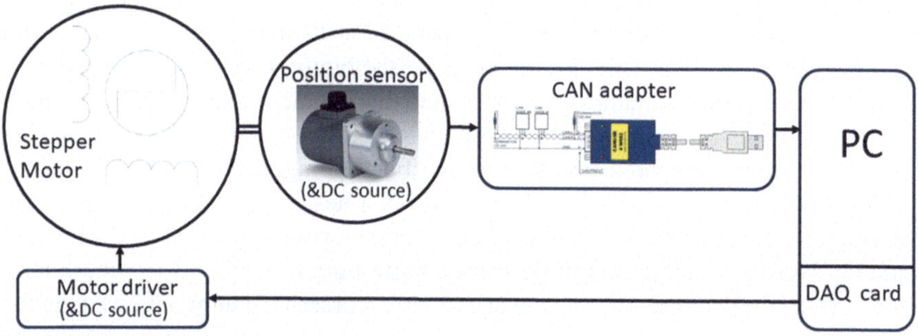

Fig. 14.4 Block scheme of hardware part of designed CAN bus workplace. (Source: Own work)

Originally, students were using the system locally in our laboratory. However, as we built only one unique prototype, it seemed useful to add remote access to it; thus extending working hours of access for students. In advance, we need to think about the desired way of functioning of the workplace. Here we planned five buttons:

- Step—rotates the stepper motor in one sequence of positions (which should be considered as one step in this system); response to GUI is a confirmation message,
- Position—collects 10 packets of measured position from the sensor CAN output and stores the mean value of positions into hidden internal array alongside with timestamp; response are data of 10 undecoded packets shown in Listing indicator,
- Graph—shows the history of positions vs time; response are data for Graph indicator,
- Clear Data—clears internal arrays; response is message and new empty data for Graph indicator,
- Help (1)—jumps into submenu of 3 help pages where user can select whether to observe diagram of system hardware, instructions for software usage or CAN data syntax explanation; response is a new set of button labels, in submenu then help picture for Image indicator or original set of labels when getting back from submenu.

To achieve desired behavior, we had to insert parts of code for data acquisition and digital sequence generation into the *Do_Module* template with some other relevant modifications.

14.6 Design of Target Module

Modification of *Do_Module* template consists in adding appropriate number of states above 6 and correction of decisions in state 5 in reaction to received button press commands. Structure of this new *Do_Module* for the CAN bus workplace is just what we have seen in the block scheme in Fig. 14.3. The basic set of labels is set in state 6 and then states 7, 8, 9, 10 correspond with actions related to expected responses to clicks on buttons no. 1 to 4. Button no. 5 Help(1) then leads to state 11 where a new set of button labels is distributed with storing of the actual menu level into an internal variable. From that point the button press action is served by states 12, 13, 14. Button number 4 is omitted in this submenu and response to button number 5 press is returning to main menu via state 6.

State 5 must be prepared for the intended functioning, see Fig. 14.5. The main part of its code (block diagram in LabVIEW) is a nested case structure where the case number is actual *Gcom* command code temporarily stored in the GCommand variable. Normally, request coming from remote button number 3 has the binary code 1100100b (decimal 100), where active bit 2 represents button 3 pressed, active bits 5 and 6 mean a concurrent request for the content of status register and of other unread items (possible changes resulting from previous local actions). In case of

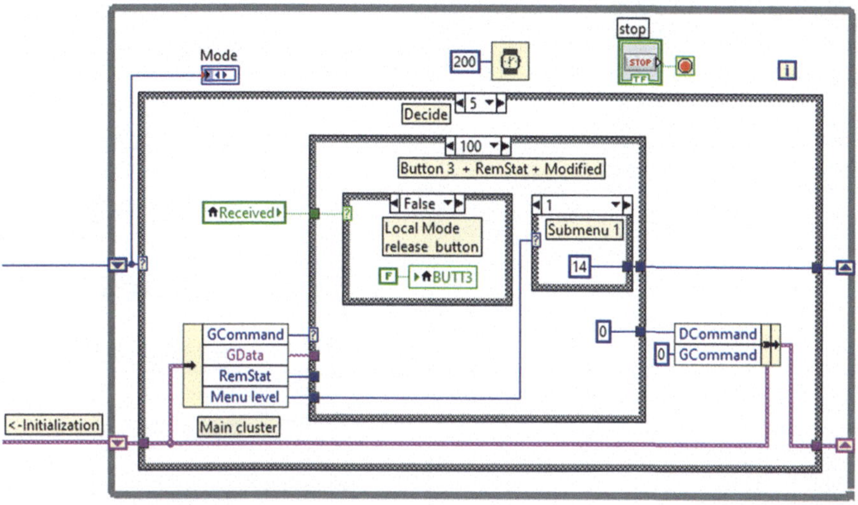

Fig. 14.5 Part of the block diagram of *Do_Module* controlling the CAN bus workplace showing the logic inside state (Mode) 5: **Decide**. (Source: Own work)

simple action, the response is prepared in the DCommand variable in state 5 already. However, if state 6 or above follows—like 9 or 14 here depending on menu level—DCommand will be specified in a subsequent state. Besides DCommand, connection to hardware or internal variables is implemented subsequently in states >6. Finally, the DCommand content will be sent via the *Dcom* DataSocket variable in state 4 accompanied by appropriate data in *Dvar* and *Dpicture*.

The code related to hardware resides mainly in state 7—sending digital sequence to the DAQ output; and state 8—reading data from CAN to the USB adapter. However, this was not a complicated task. Besides this, some modifications were needed also in initialization parts of the *Do_Module* template. It means that initialization of additional variables should be performed, like arrays of stored positions and timestamps, to make them accessible for manipulation via the main cluster brought to the border of the main case structure of the state machine.

14.7 Discussion

The designed workplace was planned for subjects taught in the engineering degree of study which means 4th and 5th grade. However, during the testing phase we offered access to the workplace for students of the subject Electrical engineering 1 taught in 1st grade hence in bachelor studies. Front-end or user interface shown in the browser on the client side is shown in Fig. 14.6a. Alignment of controls is a little different from local Front Panel (Fig. 14.1). Below the title lines of the actual time and progress state message the UI starts with control buttons then followed by parts

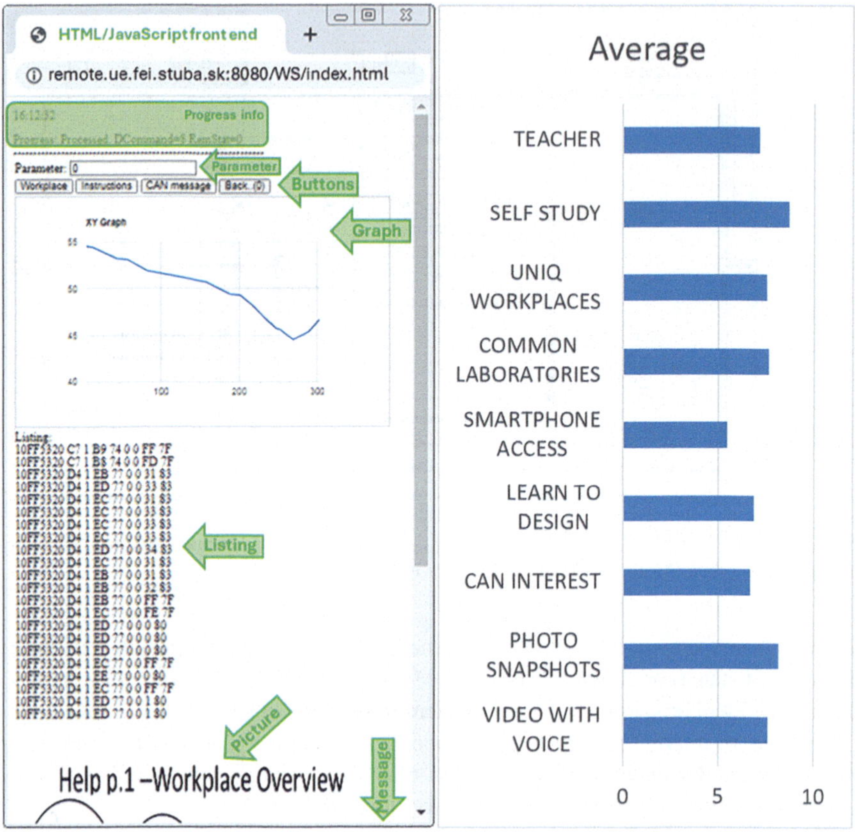

a) b)

Fig. 14.6 (**a**) The front-end of *Go_Module* as shown in an Internet browser (explaining comments are in green); (**b**) results of survey. (Source: Own work)

for presented data: Graph, text lines corresponding to Listing, Picture, and Message line (not shown, a message would appear below the help picture). It does not have form of tabs, but of cells located below each other as separate lines implemented within <table> tags of html code in *Go_Project* (Kamenský et al., 2023). The trend line in XY Graph (X axis is time in seconds and Y position in %) shows that several steps and measurements were performed until the rotary position reached lower limit 45%. Therefore, direction was automatically changed after 250 s—counted from the last Clear Data—to keep the sensor inside middle of range ~45–55%.

A student can decode parts of the CAN message which settles on the position specified by the first two bytes of the CAN frame 0x01D4~46.84% (from full range of 30 turns) expressed also as a ratio 0x77ED~30,701 to 65,535 in next two bytes. The calculated value can be then confronted with Graph. If a help is needed the picture below can describe the workplace parts or the CAN message syntax. In

Fig. 14.6a the application resides in the Help submenu, therefore the buttons let the user select the help window shown in the picture below Listing data.

In Fig. 14.6b results of the survey are presented. We were working with a group of 20 volunteers and asked them for feedback to understand what perspective students of first cycle studies see in the utilization of such a remote tool in their upcoming studies. Our quiz questions were:

1. Do you like if the TEACHER uses such a remote tool?
2. How do you like remote access to a workplace for SELF STUDY?
3. How do you like to have a complete topic taught only remotely in case of UNIQUE WORKPLACES?
4. How do you like to complete COMMON LABORATORIES taught only remotely?
5. How important is SMARTPHONE ACCESS i.e. without the need of a PC?
6. Would you like to LEARN TO DESIGN a remote access to a workplace?
7. How INTERESTing is this workplace of CAN (bus) for you?
8. Do you consider it necessary to allow the application to show PHOTO SNAPSHOTS?
9. Do you consider it necessary to observe the remote workplace as VIDEO from camera WITH VOICE?

Every question was scored from 1 to 10 where 10 points means the positive reply to the question. Collected results are graphically summarized in Fig. 14.6b as averaged scores. Young technicians seem to be in favor of remote access tools. Most positive was the option for self-study, however, the majority of them would accept also common laboratories lead in this way. We can assume that it saves them time and costs for travelling but offers also extra time or access to a hardware at the expense of being remote. Students would be glad to see the workplace at least on camera screenshots, which might be simply implemented via Picture indicator. On the other hand, the access from a smartphone is not such an essential feature like we expected.

14.8 Conclusion

The didactic CAN bus workplace with remote access was described. The software part was designed in LabVIEW based on a prepared universal modular system of applications. The system consists of pairs of modules. This structure simplifies the task of establishing remote access as only the target module needs to be modified when creating a new workplace such as the discussed CAN bus equipment. The designer does not even need to understand web publishing technologies. This is already pre-programmed in the publishing module.

The described CAN bus workplace consists of a rotary sensor whose position is controlled via a stepper motor. The system was tested on a group of 20 students of first cycle studies. The goal of deployment was to collect feedback from students which are in the decision phase and let them influence the teaching methods in their study. Survey results showed that remote access seems preferable not only for self-study but also for common laboratories.

Acknowledgements This work was supported by grant of the Slovak Scientific Grant Agency No. VEGA 1/0045/21 and by the Slovak Research and Development Agency grant APVV-20-0042.

References

Barot, T., Korenova, L., Krpec, R., & Vagova, R. (2020). Utilization of mathematical software in favour of tutoring processes. In E. Smyrnova-Trybulska (Ed.), *Innovative educational technologies, tools and methods for E-learning* (E-learning) (Vol. 12, pp. 279–289). Katowice–Cieszyn. https://doi.org/10.34916/el.2020.12.24

Flores-Piñas, W. V., Flores-Piñas, H., Chiri-Saravia, P. C., & Laura-De La Cruz, K. M. (2022). Moodle in distance education. *PURIQ, 4*, e417. https://doi.org/10.37073/puriq.4.417

Kamenský, M., Hilko, K., Bitter, M., & Králiková, E. (2023). Employing LabVIEW web service in design of application twins with remote access. In *Measurement 2023. 14th International conference on measurement. Smolenice, Slovakia* (pp. 199–202). https://doi.org/10.23919/MEASUREMENT59122.2023.10164593

León-Martínez, V., Peñalvo-López, E., Montañana-Romeu, J., Andrada-Monrós, C., & Molina-Cañamero, L. (2023). Assessment of load losses caused by harmonic currents in distribution transformers using the transformer loss calculator software. *Environments, 2023*(10), 177. https://doi.org/10.3390/environments10100177

Mokhtar, A., Mikhail, G. R., & Seong-Joo, C. (2014). A survey on remote laboratories for E-learning and remote experimentation. *Contemporary Engineering Sciences, 7*(29), 1617–1624. https://doi.org/10.12988/ces.2014.411218

Otomański, P., Pawłowski, E., & Szlachta, A. (2021). The evaluation of expanded uncertainty of DC voltages in the presence of electromagnetic interferences using the LabVIEW environment. *Measurement Science Review, 21*(5), 136–141. https://doi.org/10.2478/msr-2021-0019

Silva, N., Alvarez, I., & Pinto, P. (2020). E-learning implementation – From action to equity in an international higher education. In E. Smyrnova-Trybulska (Ed.), *Innovative educational technologies, tools and methods for E-learning* (E-learning) (Vol. 12, pp. 290–304). Katowice–Cieszyn. https://doi.org/10.34916/el.2020.12.25

Singh, A. K., Chatterji, S., Shimi, S. L., & Gaur, A. (2015). Remote lab in instrumentation and control engineering using LabVIEW. *International Journal of Electronics and Electrical Engineering, 3*(4), 297–304. https://doi.org/10.12720/ijeee.3.4.297-304

Skvortsova, S., Ishchenko, A., Halitsan, O., & Haievets, Y. (2022). Digital education content in the learning environment of educational institutions in the context of distance and blended learning in mathematics. In E. Smyrnova-Trybulska (Ed.), *E-learning in the transformation of education in digital society* (E-learning) (Vol. 14, pp. 89–104). Katowice–Cieszyn. https://doi.org/10.34916/el.2022.14.07

Smyrnova-Trybulska, E., Sekret, I., & Morze, N. (2021). Preliminary analysis of the development and implementation of the MOOC project: A case study. In E. Smyrnova-Trybulska (Ed.), *E-learning in the time of COVID-19* (E-learning) (Vol. 13, pp. 137–150). Katowice–Cieszyn. https://doi.org/10.34916/el.2021.13.12

Tripathi, P. K., Mohan, M. J., & Gangadharan, K. V. (2012). Design and implementation of web based remote laboratory for engineering education. *International Journal of Engineering and Technology, 2*(2), 270–278. ISSN: 2049-3444.

Chapter 15
Virtual Laboratory as a Tool to Empower IT Students

Kateryna Yalova, Mykhailo Babenko, Yurii Babenko, Nataliia Polyakova, and Zubrytska Anna

Abstract The paper presents the results of design research, the final product of which is a virtual laboratory for students in the field of information technology for a course on algorithms and data structures. The work justifies the necessity and relevance of developing and integrating into the educational process tools that enhance IT students' acquisition of professional competencies and skills. Emphasis is placed on the importance of IT students acquiring practical skills as a crucial factor in their value in the IT job market. It is noted that e-learning tools and virtual laboratories, in particular, are effective means of supporting self-paced, instructor-led, and offline studying. The paper describes the main results of each stage of the conducted design research, namely: problem analysis, creation of domain models, building the solution, and preparing the developed virtual laboratory for testing and integration into the educational process. The models that formalize domain knowledge, including its functional and object models, are presented in this work. The proposed models are universal and can be applied during the development of virtual laboratories for academic IT study. The proposed design solutions were validated by introducing the developed virtual laboratory to the educational process of training second-year students of bachelor studies in the specialty of software engineering at the Dniprovsky State Technical University (Ukraine).

K. Yalova (✉)
University of Information Technology and Management, Rzeszow, Poland

Dniprovsky State Technical University, Faculty of Computing Technologies and Energy, Kamianske, Ukraine

M. Babenko · Z. Anna
Dniprovsky State Technical University, Faculty of Computing Technologies and Energy, Kamianske, Ukraine

Y. Babenko
Taras Shevchenko National University of Kyiv, Kyiv, Ukraine

N. Polyakova
University of Information Technology and Management, Rzeszow, Poland

Zaporizhzhia National University, Zaporizhzhia, Ukraine

E. Smyrnova-Trybulska et al. (eds.), *E-Learning and Enhancing Soft Skills*,
https://doi.org/10.1007/978-3-031-82243-8_15

Keywords Virtual laboratory · Distance learning · e-learning · Higher education institution · IT studying

15.1 Introduction

With the advancement of information technologies, access to knowledge has become more open and accessible than ever before. The use of audio-video equipment, tools for recording and broadcasting educational materials, computers and mobile devices, the implementation of virtual and augmented reality technologies, and the widespread application of the Internet of Things allow modernization of the knowledge acquisition process, making it more interactive, interesting, and motivating (Dancsa et al., 2023). Information technologies are effectively used both in traditional classes and for organizing distance learning processes. Distance learning (DL) is a form of education that involves a distance between the participants in the educational process. The history of DL dates back to 1840 when Sir Isaac Pitman organized the process of interaction and teaching his students through the English postal system. In 1999, with the advent of the Learning Management System (LMS) concept, the term «e-learning» emerged, defining a specific type of DL process organized through information and communication technologies and using the Internet for information transmission. The definitions «online» and «virtual learning» became widely used as synonyms for e-learning (Lydia et al., 2023).

As Internet and computer accessibility increased for a wide range of people, private companies and universities began creating and distributing online educational content. Notable projects in this field include OpenCourseWare from the Massachusetts Institute of Technology, Coursera from Stanford University, edX from Harvard University, Udacity and Udemy from their respective private educational organizations, the Netacad and Skillsforall platforms from Cisco, and the Ukrainian educational platform Prometheus (Chrominski et al., 2023), which provide commercial and/or free access to Massive Open Online Courses (MOOCs). MOOCs can be used in self-paced and instructor-led modes and can partially be integrated into academic university courses, fulfilling an informal component of the educational process.

Researchers from various countries have dedicated numerous studies to analysing the effectiveness of e-learning applications. A significant scientific and practical task highlighted in contemporary scientific works is the development of e-learning tools (Ahmed & Hasegawa, 2021; Asiksoy, 2023; Hao et al., 2021; Okoyeigbo, 2020) and their implementation to support various forms of the learning process (Brillantes et al., 2022; Usman et al., 2021; Dancsa et al., 2023; Reeves et al., 2021). The primary scientific method in the field of e-learning involves conducting surveys, processing, analysing, and comparing statistical data from student responses (Chrominski et al., 2023; Glazunova et al., 2023; López-Sánchez et al., 2024; Yalova et al., 2016) and teachers (Botero-Gómez et al., 2024; Hasan, 2024). Experts in pedagogy, psychology, and medicine (Ismail et al., 2016) are also involved in analysing the data regarding the effectiveness of e-learning tools.

15.2 Research Background

The need to support formal academic education—the learning process that takes place within educational institutions—has laid the foundation for the creation and implementation of specialized Learning Management Systems such as Blackboard, Moodle, and Google Classroom (Mucundanyi & Woodley, 2021). These systems enable e-learning within the scope and scale set by educational programs, curriculums and syllabuses of the academic courses. In this context, the source of educational content is academic instructors rather than MOOCs vendors.

During the COVID-19 pandemic, the implementation of e-learning ensured the continuity of educational services for universities worldwide (Usman et al., 2021). For countries like Ukraine, where military actions, emergency and planned power outages, and internet access disruptions are prevalent, the adoption and use of various e-learning tools is an effective means of maintaining high-quality professional technical higher education (Glazunova et al., 2023; Tiahunova & Lavryk, 2023). By providing 24/7 online access to educational content, students can adapt their learning process based on current conditions and safety situations.

An integral component of the educational process in higher education institutions is conducting practical and laboratory classes. These classes aim not only to develop general and professional competencies, but also to acquire practical skills, which are especially crucial for STEM (science, technology, engineering, and mathematics) education (Sellberg et al., 2024).

This is equally true for the professional academic training of students in the field of information technology, where acquiring practical skills is paramount. The value of IT graduates in the job market is largely determined by the quality of their professional skills and their ability to demonstrate these skills in specific practical tasks (Yalova & Yashyna, 2020). The complexity of developing practical skills for IT students lies in the vast number of professional competencies required, fully encompassing the software development process. This includes skills and abilities such as:

- requirements and data domain (DD) analysis;
- designing application architectures, computer networks, user interfaces (UI), and databases;
- writing program code;
- testing and code debugging;
- deploying and maintaining the developed software;
- analysing, organizing, and managing information processes for data collection, storage, processing, and protection.

The issue of developing practical skills for students is one of the main challenges of e-learning, especially in disciplines such as physics, chemistry, and medicine, which heavily rely on data collection from experiments, where the lack of access to laboratory equipment can become a critical problem (Usman et al., 2021). A solution to this situation could be the use of virtual laboratories (VL) of either simulator or training type (Elmoazen et al., 2023). Simulator VLs often use Virtual and/or Augmented Reality and are widely implemented into the educational process in medicine and aviation. The primary goal of training VLs is to explain complex phenomena clearly and simply through graphical visualization of real practical examples. The purpose of using VLs in educational institutions is to provide an interactive virtual learning environment for solving practical problems and developing practical skills, where experiments can be conducted both independently and under the guidance of an instructor.

The aim of this work is to present the results of design research, a type of scientific study that aims to create theoretical models and their practical implementation as the final product of the research (Easterday et al., 2018). The product of this design research is a VL for the course «Algorithms and Data Structures» (ADS), which is a basic course in training students in the field of information technology.

15.3 Virtual Laboratories as Part of Academic e-Learning

There are many definitions for the term «virtual laboratory» (Elmoazen et al., 2023), but the most general and widely used one is: a VL is a specialized computer environment where a user can conduct experiments or learn without real equipment, where practical actions, observations of results, and interactions with virtual equipment are carried out through a computer screen (Sellberg et al., 2024). VLs address the problem of lack of access to real laboratory equipment or materials that may be expensive, dangerous, or difficult to access. They are also useful for education because they allow students to practice and develop skills without risk to health or property. This is an excellent way to provide access to scientific research and experiments to a large number of people, regardless of their location or financial capabilities (Usman et al., 2021).

VLs and practical labs are powerful learning tools that can be applied in various fields of knowledge, ranging from natural sciences such as chemistry and biology to exact and information sciences, including mathematics and programming (Asiksoy, 2023; Garcia-Zubia et al., 2016; Ismail et al., 2016; Reeves et al., 2021; Munawar et al., 2018). These tools offer students a unique opportunity to master material through practical experience without leaving home or the classroom, which is especially important in modern e-learning (Reginald, 2023). VLs, which recreate various phenomena and processes in a virtual environment, have many advantages as an effective means of developing practical skills in e-learning, namely (Herraez, 2022):

1. Safety: In these laboratories, users do not interact directly with laboratory equipment, significantly reducing the risk of injury or damaging scientific equipment.
2. Accessibility and convenience: Users can access the laboratory anytime and from anywhere with an Internet connection.

3. Variety and experimentation: Users can choose the sequence of their actions, try new combinations, or set new values and make corresponding conclusions.
4. Visualization: Most VLs allow the visualization of the studied process through animations and images, demonstrating the steps of the work or explaining a phenomenon.
5. Cost-effectiveness: Using an online service is much cheaper than using laboratory equipment, which carries the risk of damage during research.

Many researchers focus on justifying the use of VLs and evaluating their effectiveness in STEM education (Sellberg et al., 2024). In fields like chemistry or physics, VLs enable users to take measurements, mix different reagents in virtual flasks, observe chemical and physical processes, and study research results without health risks. Studies by authors such as (Reeves et al., 2021; Asiksoy, 2023; Usman et al., 2021) provide examples of creating VLs and evaluate the results of their implementation in specific universities. In mathematics, VLs can include interactive learning models and graphs that help students visualize abstract concepts and solve practical problems (Okoyeigbo, 2020).

Furthermore, VLs can serve as platforms for interaction between students and instructors. They can facilitate assignment distribution and assessment, message exchanges, feedback, online testing, and overall evaluation of student performance.

Numerous studies analyse the relationship between cognitive, motivational factors and the learning process, using frameworks such as Moreno's and Mayer's Cognitive-Affective Theory of Learning with Media (CATLM) to justify the use of multimedia support, VLs, virtual and augmented reality in the educational process. These technologies serve as a means of motivation and as a positive accelerator for the acquisition of knowledge and skills (Makransky & Peterson, 2019).

The studies by Asiksoy (2023), Ismail et al. (2016), Garcia-Zubia et al. (2016), and Lydia et al. (2023) present the results of experiments designed to analyze the impact of VL on the learning process. These experiments involved control and experimental groups, assessing both the subjective attitudes of students toward VL and the objective outcomes in terms of knowledge acquisition. The findings from these experiments arguably confirm the hypothesis that VLs positively influence student learning outcomes. This served as the foundation for developing the presented VL for the Algorithms and Data Structures course.

15.3.1 Virtual Laboratories for IT Learning

There are numerous online educational resources that can be utilized to support IT students, particularly in courses like ADS. Despite using generalized principles of e-learning tools, they often have a narrow focus, for example:

- organizing remote online interactions (Google Meet, Microsoft Teams, Cisco Webex, Zoom, etc.);

- tools for distributing and collaboratively using educational content (Blackboard, Moodle, Google Classroom, etc.);
- tools for creating, distributing, and using test assignments (LearningApps, Online Test Pad, ClassMarker, Google Forms, etc.);
- MOOCs;
- video courses on YouTube channels and social networks.

The main problems with their use for academic purposes are:

- fragmentation of information sources, which requires learning process participants to spend time searching for the necessary topic, code snippet, or educational video;
- mismatch of educational content with the established regulatory documents of the academic course;
- imperfections in the legal regulation regarding the place, role, and methods of recognizing learning outcomes from external knowledge sources.

These factors lay the foundation for the creation and implementation of software tools to support the university educational process. This allows, on the one hand, formatting of a virtual learning environment in accordance with academic regulatory documents, and on the other hand, combining active knowledge acquisition mechanisms into one online resource.

VL for IT students can serve as environments for software development (Okoyeigbo, 2020; Munawar et al., 2018), database and computer networks construction (Yalova & Yashyna, 2020) algorithm analysis and creation (Perhac & Simonak, 2022; Striuk et al., 2020), where students can step-by-step study the informational, hardware, and technical processes of IT technologies, assemble, debug, and test programs, architectures, and algorithms, and track their execution and results. It should also be noted that one of the biggest advantages of using VL is the engagement of students in the actual process of forming practical skills, rather than just mastering theoretical knowledge. It is known that knowledge is much better absorbed through practice than through theory alone (Glazunova et al., 2023).

For disciplines related to programming, particularly ADS, there can be challenges in visualization and practical work with code. Students often struggle with understanding abstract concepts and the principles of algorithm operation without proper visualization and practical experience. This complicates the learning process and reduces its effectiveness because, without visual representation, it is difficult to grasp the functionality and application of ADS in practice. This is why the algorithms and data structures course was chosen as the informational content. Examples of VL implementations for the course for university-level student preparation are provided in the papers of (Perhac & Simonak, 2022; Striuk et al., 2020). The VL presented in (Perhac & Simonak, 2022) includes lab sessions on data structures and search, binary tree traversals, and graph algorithms. The VL described in (Striuk et al., 2020) contains lab assignments solely on graph structures and their processing algorithms. It is worth noting that these developments are the proprietary work of instructors teaching ADS and are not commercialized or open for use by external users.

15.4 Virtual Laboratory for Algorithms and Data Structures Course: Aspects of Implementation

The objective of the ADS course is to develop knowledge related to the design of algorithms using modern data structures, gain skills in software development, and master the methodological foundations for creating algorithms to solve problems in the field of IT. The informational content for the VL was developed based on the curriculum and course cards for bachelor's degree programs in information technology specialties from Dniprovsky State Technical University (DSTU) in Ukraine and the University of Information Technology and Management (UITM) in Poland.

The main stages of the conducted design research, according to the principles of this methodology (Easterday et al., 2018), are:

- problem analysis and definition of research objectives;
- creation of domain models and students learning activity models;
- building the solution;
- preparing the solution for testing and implementation in the educational process.

The stage «building the solution» was implemented following the stages of the software development lifecycle, including requirements analysis, design, software implementation, testing, and deployment.

15.4.1 Requirements Analysis

Requirements analysis is the first stage of building the solution. The specification of requirements for the VL includes descriptions of both functional and non-functional requirements, enabling the determination of what the VL should perform and how it should function. The proposed design solutions were tested by integrating the VL into the existing information portal of DSTU (URL: https://dstu.dp.ua) and providing access to its content through the ADS course page. Below is a fragment of the requirements specification highlighting the main requirements without detailing.

Functional requirements included, for example, the implementation of software mechanisms for:

- online access to educational content;
- non-linear navigation through the educational content;
- organizing dialogue and applying user support tools;
- distributed access to laboratory practice data.

Non-functional requirements deal with more abstract concepts, for example: non-functional requirements included:

- usability: ensuring the simplicity and ease of use of the VL;
- maintainability: ensuring the operational reliability of the VL;
- security: ensuring information security and distributed access to data according to established rights;
- scalability: ensuring the software's ability to expand according to changes in the domain and functional requirements of the VL.

Additionally, architecturally significant requirements were defined, which implementation impacts the architecture of the VL, for example:

- access to the VL should be provided through the DSTU information portal, starting with user authorization;
- educational content and student work results should be stored in a relational database;
- use of free software tools for implementation and data storage mechanisms;
- client-server multi-layer architecture for the software implementation of the VL.

15.4.2 Pedagogical Background of the Virtual Laboratory Implementation

To build a model of VL interaction with students, Edgard Dale's Pyramid of Experience was adapted, which reflects the relationship between the effectiveness of retention and the type of experience. People remember best what they read and hear, even better what they see, and the best what they actively do by interacting with the material. This pyramid can be adapted to the knowledge acquisition process in the field of IT, highlighting specific types of learning activities for developing practical skills (Fig. 15.1).

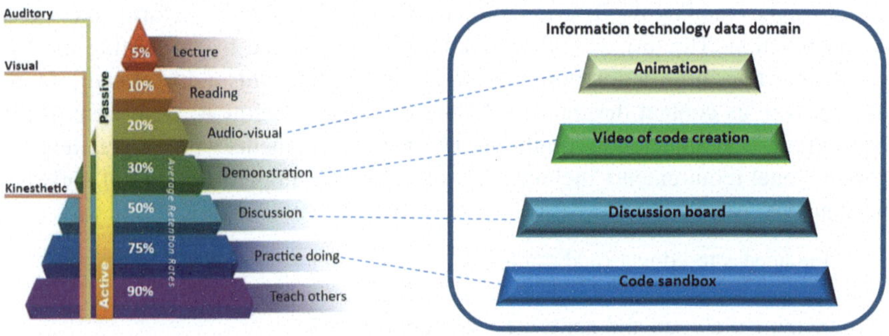

Fig. 15.1 Pyramid of knowledge. (Source: Adopted from the NTL institute of applied behavioural science)

When determining the types of active learning activities for implementation in the VL, CATLM (Makransky & Peterson, 2019) was used, considering that the ways of presenting input learning information should have a positive motivating effect, arouse interest, implement cognitive load with the possibility of self-regulation, and assess self-efficacy. The specification of active learning activities types in the VL was formed as follows:

1. Embedding animations. Visualization of data structures and algorithms can be implemented using short multimedia video clips. Table 15.1 lists interactive online tools that allow creating animations of data structure processing or algorithm steps.
2. Code implementation demonstration. In addition to visualizing algorithm steps and data structure operations, demonstrating the writing of program code within a specific integrated development environment is advisable.
3. Practice doing in the code sandbox. To allow students to create their own code or modify a generalized implementation of a specific algorithm, it is advisable to use «code sandboxes»—tools embedded in the learning course for writing and

Table 15.1 On-line tools for data structures and algorithms visualisation

Name	URL	Description	Ability to save created animation
VisuAlgo	https://visualgo.net/en	Creating animations for data structure visualization, e.g.: Arrays, linked lists, binary heap, hash table, binary search tree, graph structures and sorting, recursive algorithms, algorithms to find the shortest path.	Conditional (only for authorized users)
Algorithm visualizer	https://algorithm-visualizer.net/	Allows visualizing steps for sorting and graph-searching algorithms.	–
Interactive online platform	https://algorithm-visualizer.vercel.app/	Allows visualization for sorting, Pathfind and ConvexHull algorithms.	–
Online visualizer by David Galles from University of san Francisco	https://www.cs.usfca.edu/~galles/visualization/Algorithms.html	It is possible to visualize the process of creating and processing data structures such as stacks, queues, lists, heap-like, and graph data structures. Recursive, indexing, sorting, and geometric algorithms, as well as dynamic programming algorithms, can also be visualized with user-inputted or randomly generated values.	–

Source: Own work

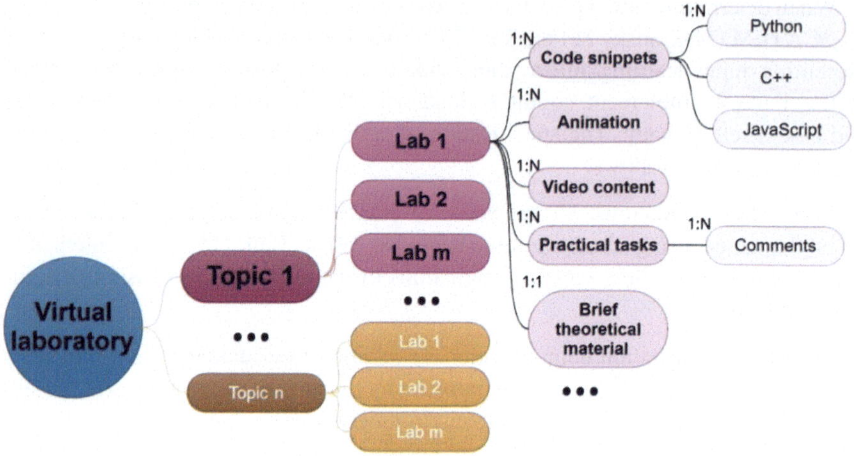

Fig. 15.2 Learning content architecture. (Source: Own work)

interpreting/compiling program code. These code sandboxes enable running the written code and providing error descriptions if any are present.

The architecture of the VL learning content is shown in Fig. 15.2 and presented as a highly branched tree, allowing not only the display of connections between structural units but also the visualization of the multiplicity of these connections.

As seen in Fig. 15.2, it is clear that the VL contains n topics of the ADS course. Each topic, in turn, includes m laboratory sessions, and each laboratory session consists of k code snippets, t animations, j videos, where n and m are the quantities established by the course regulatory documents, and the values of k, t and j are determined by the instructor-creator of the educational content.

The next step in designing the VL is to create a DD's functional model, defining the active participants and their available actions.

15.4.3 Data Domain's Actors and Use Cases

A DD's functional model is created to formalize the data processing processes, describing the business logic of the interaction between actors, answering the questions: what, how, and by whom actions are allowed within the VL. The domain actors are the academic teachers and students, who can interact asynchronously through the laboratory's Discussion board and by reviewing the results of practical tasks of the lab. The overall management and maintenance of the VL's functionality are performed by the administrator. Since access to VL data occurs through the DSTU information portal, the user authentication process occurs outside the VL and is part of the operation and access rights management of the information portal.

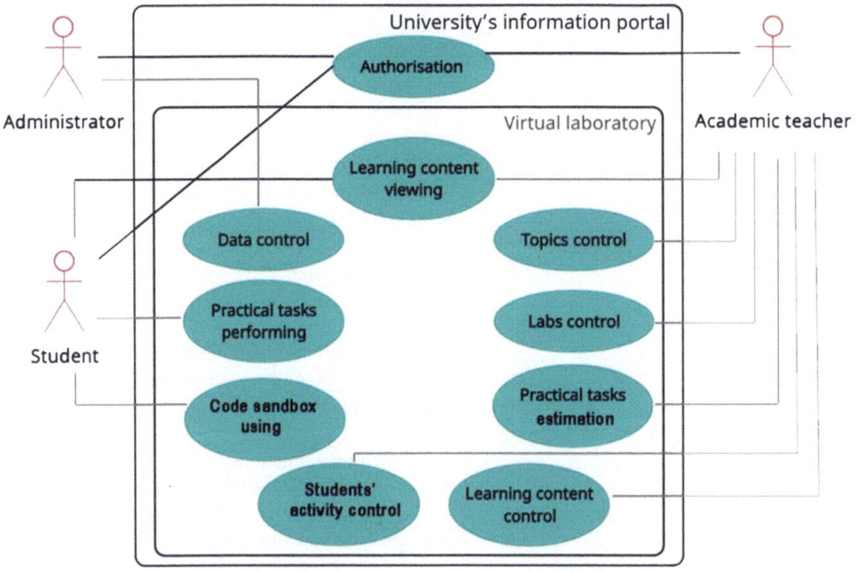

Fig. 15.3 Data domain's use case diagram. (Source: Own work)

A Use Case Diagram is employed to display the possible actions of an actor, aiming to visually represent the connections between the domain's participants and use cases, thus describing the DD at a conceptual level (Fig. 15.3).

Table 15.2 provides a description of the main use case diagram precedents' extends and includes. Inclusions illustrate which other precedents are always used to perform the operations of the basic precedent. Extensions are actions that are performed depending on the results of checking a specific condition.

The data presented in Table 15.2 reflect the main components of the DD functional model, which are used to develop the DD object model, define the data storage architecture, and build a prototype of the UI.

15.4.4 Data Storage

The process of creating data storage consists of stages involving the creation of models at different levels of abstraction: DD object model → fully-attributed data model → physical data model.

The domain object model represents the highest level of abstraction, focusing on its structural characteristics. It captures the composition of entities, their attributes, and relationships significant for solving a specific task. This model is independent of the software implementation tools. The components of the object model include specifications of entities, attributes, and relationships between entities. The specification of the entities includes descriptions of:

Table 15.2 Precedents inclusions and extensions

Precedent	Actor	Description
Authorization	Student, teacher, administrator	Extend: The action is performed on the DSTU information portal page. If authorization is successful, access to VL data is granted; otherwise, a message about the lack of access rights is provided, and the user is prompted to try again
Viewing educational content	Student	Include: navigating the VL; viewing the course structure and the content of each lab; reviewing brief theoretical information; viewing examples of completed program implementations of the selected data structure or algorithm; watching animations demonstrating the operation of specific data structures or algorithms; viewing video recordings showing the program code coding and testing
Completing practical tasks of the lab session	Student	Include: viewing the task; practical implementation of the task; submitting results for evaluation.
Working with code sandbox	Student	Include: writing, testing, and debugging the developed program code; conducting personal experiments; saving the developed code in a source code file.
Managing topics	Teacher	Include: Adding, deleting, editing, and saving topics according to the structural-logical scheme of the course
Managing lab sessions	Teacher	Include: Adding, deleting, editing, and saving the description of labs, and the text of practical tasks for individual self-paced work
Managing educational content	Teacher	Adding, deleting, editing, and saving the text of theoretical educational material, animations, videos, and code snippets for each lab.
Assessing results	Teacher	Include: viewing submitted practical tasks; saving the grade for the task;
Managing the activity journal	Teacher	Include: Viewing multidimensional data presented in the structure: Student—Practical task—Grade—Date of completion.
Managing data	Administrator	Adding, deleting, editing, and saving data entities: Users, topics, labs, practical tasks, student activity logs

Source: Own work

- user roles;
- teachers;
- student groups;
- students;
- course topics;
- labs;
- code snippets;
- video demonstrations;
- animations;
- practical tasks of laboratory sessions;
- comments on the execution of practical tasks;
- student activity journal;

According to the architecture of the educational content shown in Fig. 15.2, a domain object model was created. A fragment of this model is presented in Fig. 15.4 as an Entity Relationship Diagram.

The fully-attributed data model was constructed using data normalization rules. The data from the object model were brought to the third normal form for storage in a relational database created using the MongoDB database management system. At this stage of the VL life cycle, all data and educational content are stored on the server and provided to the user through the processing and execution of queries sent via a browser. The possibility and feasibility of implementing cloud storage for VL data are also being considered.

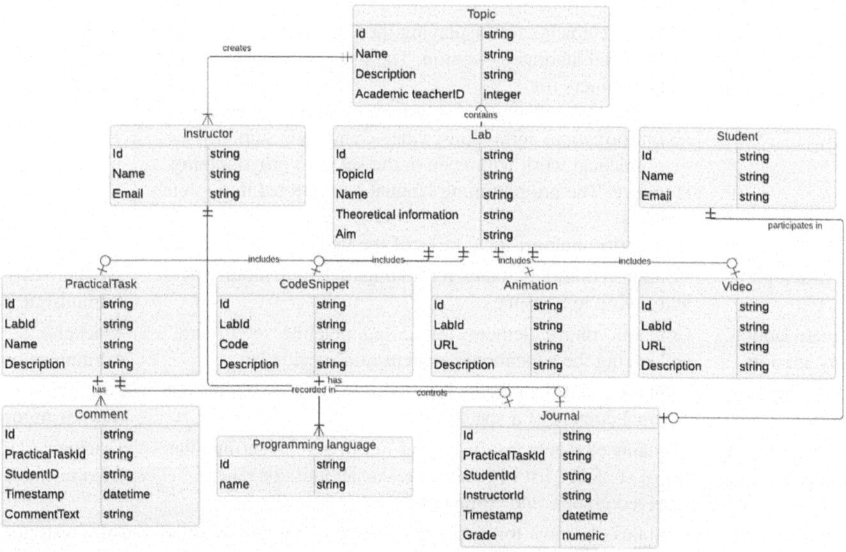

Fig. 15.4 Data domain object model. (Source: Own work)

15.4.5 User Interface and Navigation

The UI consists of elements such as menus, lists, buttons, input or selection fields, messages, and dialogs. These elements allow the user to conveniently interact with the VL and receive responses to their actions. The control element used for confirming transactions on the web form is buttons. The UI was designed with usability principles in mind and possesses the following properties:

- unification and standardization;
- predictability for the user, forming a consistent response from the VL to identical actions performed on different web forms;
- provides user support with reference materials and system messages;
- intuitive for the user as it implements common practices in web interface design;
- ergonomic.

Table 15.3 lists the web pages created to implement the functionality of the VL, and Fig. 15.5 shows an example of a VL web page.

VL is a multi-window web-based application, with navigation facilitated through the use of main, submenu, and context menus. The main menu is visible and accessible from any page within VL, providing users with the ability to quickly

Table 15.3 Web pages of the VL

Web page	Functionality description	User access
Main page	Contains a sidebar menu in the form of a tree with the names of available topics for selection and a field for forming search queries.	Student Teacher Administrator
Labs	Contains elements for displaying the educational content of the selected laboratory session. The page is divided into logical containers for: displaying theoretical information, providing access to text information, animations, videos, and code snippets; practical work with code in the selected programming language. The programming language is selected through tab controls; performing practical tasks of the labs;	Student Administrator
Create topic	Contains control elements for adding, deleting, modifying, and saving topic names.	Teacher Administrator
Create laboratory session	Contains control elements for adding, deleting, modifying, and saving the educational content of a specific lab.	Teacher Administrator
Edit laboratory session	Contains control elements for modifying and saving the educational content of a specific lab.	Teacher Administrator
Student activity log	Contains elements for displaying information regarding the list of students, list of practical tasks, submission dates, received grades, and comments.	Teacher Administrator
Administrator panel	Contains elements for displaying data stored in the database for performing CRUD operations.	Administrator

Source: Own work

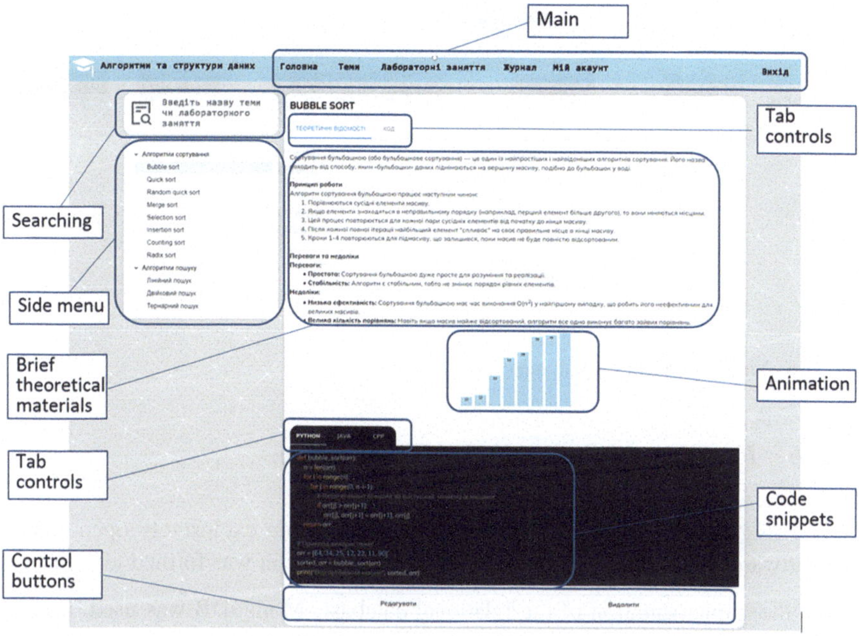

Fig. 15.5 Virtual laboratory's web page example. (Source: Own work)

return to a chosen section. The main menu is arranged as a horizontal menu at the top of the page and contains hyperlinks to the main sections of VL. The composition of the main menu depends on the current user and, based on access rights to the data, displays the names of available web pages.

As an auxiliary navigation tool, a side menu is implemented, presented in the form of a tree that shows the composition of labs and their components. The side menu can be hidden from the page upon user request. To display the relationship between parent and child web pages, a dropdown menu, which opens when hovering over one of the main menu items, is used. These menus allow for the distribution of information across multiple levels and help avoid overloading the main menu with a large number of options. Figure 15.6 shows an example of VL's path map, a logical hierarchical structure of nodes, each representing a page of VL, from a teacher's point of view.

Context menus appear when performing certain actions, such as right-clicking. They provide additional options that depend on the specific context in which the user is operating.

Due to the non-linear navigation, all educational content is simultaneously available to students, enabling them to create their own learning paths, skip certain topics, or return to previously covered materials.

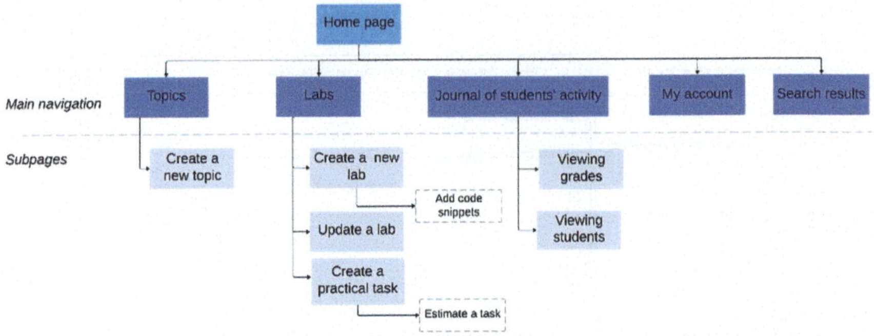

Fig. 15.6 Virtual laboratory's path map. (Source: Own work)

15.4.6 Features of Software Implementation

Considering the requirement to implement the VL using exclusively open source free software, the technology stack for the implementation was formed as follows:

1. For the implementation of the relational database, MongoDB was used. It is an open-source, document-oriented database management system that does not require table schemas and supports storing documents in a JSON-like format. It has a flexible query language, can create indexes for various stored attributes, efficiently stores large binary objects, and can be configured for fault tolerance.
2. For the implementation of the business logic of the VL, Spring Boot was applied. This framework simplifies and accelerates the development of web applications and micro services based on the Spring Framework. It provides software developers with flexible configuration, reliable batch processing, efficient workflows, and numerous tools for developing robust and scalable applications.
3. The UI was developed using React. To ensure a more attractive and easy-to-use design, built-in libraries were used, for example:

   - Material UI: Includes a collection of ready-to-use components like buttons, lists, and input fields that can be used in development once imported into the project;
   - React Quill: A powerful, feature-rich text editor that can be easily embedded into a project and offers a wide range of customization options and features. It supports various formats such as bold, italic, underline, lists, etc., making it essential for creating textual information for displaying theoretical knowledge in labs.
   - React Syntax Highlighter: A useful library for developers creating documentation, educational content, or any application that requires clear code display, one of the goals of the labs. This component allows syntax highlighting in a React project, used for displaying code examples as snippets.

4. For the implementation of the code sandbox:

   - Monaco Editor: A browser-based code editor developed by Microsoft, used in several Microsoft products like VS Code, Azure DevOps, and Visual Studio Online. It is designed for fast and efficient coding, featuring syntax highlighting and autocompletion. It also supports various programming languages and technologies, including JavaScript, TypeScript, HTML and CSS;
   - Piston API: An API offering a convenient way to remotely execute code, allowing developers to test, debug, and run their programs without setting up their own runtime environment. It supports numerous programming languages, such as Python, JavaScript, Java, C++, among others, and works in conjunction with the Monaco Editor.

Testing of the developed software was performed manually for the local version of the VL. The developed VL will be trialed by integrating it into the educational process for second-year bachelor's students in the Software Engineering program at DSTU. To evaluate the effectiveness of the proposed design solutions, an entry survey and a post-course survey are planned.

15.5 Conclusion

The academic preparation of IT students aims to develop general and professional competencies, acquire knowledge, and build practical skills. The rapid development of information technologies complicates this process, as educational content needs to be continuously updated to reflect current trends in the IT job market.

Integrating supplementary e-learning tools, particularly VL, into the learning process engages students in more active learning, supports the acquisition of theoretical knowledge through practice, and provides opportunities for experimentation. The foundation for developing non-commercial academic VLs lies in the need for effective formation of hard skills in basic courses, such as the ADS course. The practical classroom training for this course amounts to 32 h at DSTU and 28 h at UITM, while the volume of knowledge and skills required by regulatory documents is substantial. Therefore, implementing and using supplementary educational tools to support an active learning process is a highly relevant scientific and practical task.

The final product of the conducted design research is a VL for the ADS academic course. During the research, the following practical tasks were successfully implemented:

- analyse DD and create VL requirement specification;
- construction of domain models to formalize knowledge about the structure, characteristics, and relationships between domain entities;
- development of VL multilayer client-server architecture;
- development of a data storage presented as a relational database;

- creating an interface and user dialogues using the principles of web interface usability.
- software implementation of VL with open source programming tools.

The practical significance of the obtained results lies in providing students and teachers with a convenient virtual environment for acquiring practical skills in algorithm design and software development using various data structures. The expected outcome of using the developed VL is to enhance the effectiveness of the process of forming practical skills in students. Approbation and testing of the developed VL effectiveness will be carried out during the offline educational process of second-year students majoring in software engineering at DSTU. The assessment of the proposed laboratory's effectiveness will be conducted during the 2024–2025 academic year, beginning with a student survey administered both before and after the use of the virtual laboratory. A comparison of the survey results will reveal not only the initial students' attitudes and expectations but also any changes in opinions that may occur.

References

Ahmed, M. E., & Hasegawa, S. (2021). Development of online virtual laboratory platform for supporting real laboratory experiments in multi domains. *Education Sciences., 11*, 548. https://doi.org/10.3390/educsci11090548

Asiksoy, G. (2023). Effects of virtual lab experiences on students' achievement and perceptions of learning physics. *International Journal of Online and Biomedical Engineering, 19*(11), 31–41.

Botero-Gómez, V., Ruiz-Herrera, L. G., León, J. A., & Flores, L. V. (2024). Perspectives on the use of virtual tools among university teachers in the context of an emerging economy. *Cogent Education, 11*(1), 1–14.

Brillantes, C. J., Sarona, J. M., Buslon, J. B., & Alieto, E. O. (2022). Prepared for the traditional, trained in the virtual: Impact of online teaching practicum on student teachers' readiness. *Asia CALL Online Journal, 2022*, 155–132.

Chrominski, K., Przybyla-Kasperek, M., Smyrnova-Trybulska, E., Bazarbayeva, A., & Morze, N. (2023). Comperative international research in the area of educational platforms and MOOCs: An opinion of IT students using data mining analysis. In E. Smyrnova-Trybulska (Ed.), *E-learning and artificial intelligence* (pp. 15–30). Katowice–Cieszyn. https://doi.org/10.34916/el.2023.15.02

Dancsa, D., Štempeova, I., Takac, O., & Annuš, N. (2023). Digital tools in education. *International Journal of Advances Natural Sciences and Engineering Researches, 7*, 298–294.

Easterday, M. W., Lewis, D. G., & Gerber, E. M. (2018). The logic of design research. *Learning: Research and Practice, 4*(2), 131–160. https://doi.org/10.1080/23735082.2017.1286367

Elmoazen, R., Saqr, M., Khalil, M., & Wasson, B. (2023). Learning analytics in virtual laboratories: A systematic literature review of empirical research. *Smart Learning Environments, 10*, 1–13.

Garcia-Zubia, J., Cuadros, J., Romero, S., Hernandez-Jayo, U., & Orduna, P. (2016). Empirical analysis of the use of the VISIR remote lab in teaching analog electronics. *IEEE Transactions on Education, 60*(2), 149–156. https://doi.org/10.1109/TE.2016.2608790

Glazunova, O., Hurzhii, A., Korolchuk, V., & Voloshyna, T. (2023). Selection of the digital tools for organizing students' group work in distance education. *Information Technologies and Learning Tools, 94*(2), 87–101.

Hao, C., Zheng, A., Wang, Y., & Jiang, B. (2021). Experiment information system based on an online virtual laboratory. *Future Internet, 13*, 1–19. https://doi.org/10.3390/fi13020027

Hasan, M. (2024). Innovation and sustainability in engineering and technology: revolutionizing the future of education through virtual labs. *SSRN Electronic Journal, 4*, 1–13.

Herraez, A. (2022). Virtual laboratories as a tool to support learning. *Turkish Journal of Biochemistry, 47*(2), 195–200. https://doi.org/10.1515/tjb-2019-0146

Ismail, I., Permanasari, A., & Setiawan, W. (2016). STEM virtual lab: An alternative practical media to enhance student's scientific literacy. *Journal Pendidikan IPA Indonesia, 5*(2), 239–246.

López-Sánchez, J. A., Patiño-Vanegas, J. C., Valencia-Arias, A., & Coronel, A. M. (2024). Model of adoption of virtual tools by university students in the context of an emerging economy. *Frontiers in Education, 9*, 1–13.

Lydia, M. Z., Naidu, V. R., Bhat, A. Z., & Frrag, S. (2023). Impact of online tools on the learning experience of students in higher education. *SHS Web of Conferences, 156*, 1–4.

Makransky, G., & Peterson, G. B. (2019). Investigating the process of learning with desktop virtual reality: A structural equation modelling approach. *Computers and Education, 134*, 15–30. https://doi.org/10.1016/j.compedu.2019.02.002

Mucundanyi, G., & Woodley, X. (2021). Exploring free digital tools in education. *International Journal of Education and Development using Information and Communication Technology, 17*(2), 96–103.

Munawar, S., Toor, S. K., Aslam, M., & Hamid, M. (2018). Move to smart learning environment: Exploratory research of challenges in computer laboratory and design intelligent virtual laboratory for e-learning technology. *Journal of Mathematics, Science and Technology Education, 14*(5), 1645–1662.

Okoyeigbo, O. (2020). Design and implementation of a java based virtual laboratory for data communication simulation. *International Journal of Electrical and Computer Engineering, 10*(6), 5883. https://doi.org/10.11591/ijece.v10i6.pp5883-5890

Perhac, P., & Simonak, S. (2022). Interactive system for algorithm and data structure visualization. *Computer Science Journal of Moldova, 30*, 28–48.

Reeves, S. M., Crippen, K. J., & McCray, E. D. (2021). The varied experience of undergraduate students learning chemistry in virtual reality laboratories. *Computers and Education, 175*(2020), 104320. https://doi.org/10.1016/j.compedu.2021.104320

Reginald, G. (2023). Teaching and learning using virtual labs: Investigating the effects on students' self-regulation. *Cogent Education, 10*(1), 1–15. https://doi.org/10.1080/2331186X.2023.2172308

Sellberg, C., Nazari, Z., & Solberg, M. (2024). Virtual laboratories in STEM higher education: A scoping review. *Nordic Journal of Systematic Reviews in Education, 2*, 58–75. https://doi.org/10.23865/njsre.v2.5766

Striuk, A., Rybalchenko, O., & Bilashenko, S. (2020). Development and using of a virtual laboratory to study the graph algorithms for bachelors of software engineering. *CEUR, 2731*, 1–10. https://doi.org/10.31812/123456789/4462

Tiahunova, M., & Lavryk, V. (2023). Virtual laboratory as an effective solution during distance learning. *Systems and Technologies, 66*(2), 125–131. https://doi.org/10.32782/2521-6643-2023.2-66.14

Usman, M., Suyanta, & Huda, K. (2021). Virtual lab as distance learning media to enhance student's science process skill during the COVID-19 pandemic. *Journal of Physical Conference Series, 1882*, 1–9. https://doi.org/10.1088/1742-6596/1882/1/012126

Yalova, K., & Yashyna, K. (2020). Virtual educational laboratory for databases discipline. *Advances in Intelligent Systems and Computing, 1080*, 623–636.

Yalova, K., Zavgorodnii, V., Romaniukha, M., & Sorokina, L. (2016). Challenges and prospects in development of e-learning system for IT students. *Engineering Education and Life-Long Learning, 1*(26), 25–43.

Chapter 16
Examination of Teachers' Competencies for Online Teaching

Bülent Döş

Abstract In recent years, the landscape of education has undergone significant transformation, particularly with the rapid expansion of online learning. This shift has underscored the critical importance of teachers' competencies in effectively delivering online education. Teachers' proficiency in online teaching not only impacts their ability to deliver content effectively but also influences their attitudes and behaviors towards the learning process. Thus, understanding and developing these competencies is essential for enhancing the overall quality of education in the digital age. In this research, teachers were asked questions about their qualifications. For example, they were asked about their competence in teaching online or their competence in creating online content. 56 teachers attending a certificate program participated in the research. The average age of the teachers is 28, with 82% of the participants being women and the rest men. According to the research results, the majority of teachers do not consider themselves competent in creating content digitally. They also stated that they did not want to provide online education due to this inadequacy. It has become clear that while online education is valuable, it must be delivered in a highly qualified and efficient way to maximize its impact.

Keywords Online education · Teacher digital competencies · Digital content · Quality of online learning

16.1 Introduction

The rapid expansion of technology in recent years has resulted in significant changes across various industries, with education being one of the most profoundly impacted. The introduction and spread of online learning platforms transformed the traditional classroom setting, providing unparalleled flexibility and access to educational resources. However, this transition has brought unique problems, particularly

B. Döş (✉)
Gaziantep University, Nizip Faculty of Education, Educational Sciences, Gaziantep, Türkiye
e-mail: bdos@gantep.edu.tr

E. Smyrnova-Trybulska et al. (eds.), *E-Learning and Enhancing Soft Skills*,
https://doi.org/10.1007/978-3-031-82243-8_16

regarding the skills teachers need to effectively conduct online education. Teachers' digital competence is a multifaceted and essential aspect of modern education (Kiryakova & Kozhuharova, 2024). In the digital age, a teacher's role extends beyond merely conveying knowledge; it also involves the development, curation, and facilitation of engaging online content that may satisfy the unique requirements of learners in a variety of circumstances. Modern society requires teachers with digital skills to effectively use digital technology in the teaching and learning process (Reisoglu & Çebi, 2020). In order to help students improve their digital competencies, teachers must also build their own digital competencies (Redecker, 2017).

Teachers' competencies in online education have grown increasingly crucial as educational institutions worldwide adapt to mixed and entirely online learning settings (Archambault et al., 2022). These competencies encompass a wide range of skills, including technological literacy, instructional design, digital content creation, and the ability to foster a sense of community and engagement in virtual classrooms. Lázaro Cantabrana et al. (2019) considers teachers' digital competence as the ability to apply and transfer their knowledge, strategies, skills, and attitudes about educational technologies in practical, real-world situations in practice. It is essential for instructors to adapt their abilities and methods to the constantly evolving digital terrain. The capacity of the instructor to use digital tools and platforms, create engaging and interactive lessons, and assist students in a virtual setting are all critical to the success of online learning. Without these skills, the quality of online learning could decline, which would cause students to become disengaged and achieve suboptimal learning results.

In this regard, researching teachers' online teaching competencies is not only appropriate but also necessary to comprehend the opportunities and difficulties associated with digital education. There's growing worry that many teachers do not have the abilities needed to produce and deliver high-quality online education, as evidenced by recent studies. (Bragg et al., 2021) Despite the growth in online learning, research on student, faculty, and institutional readiness for online learning highlights the need for educators and institutions to better understand how to effectively design, develop, and deliver high-quality instruction online (Keefe et al., 2020). This competence gap is particularly concerning given the increasing reliance on online education to address global challenges, such as the COVID-19 pandemic, which has forced many educational institutions to adopt remote learning strategies at an unprecedented scale (Ali, 2020). The effectiveness of these strategies largely depends on the preparedness and proficiency of teachers in utilizing digital tools and methodologies.

This study investigates the competencies of teachers participating in a certificate program, specifically focusing on their proficiency in online teaching and digital content creation. Many of the publications examined emphasize the need for teachers to improve their digital competence through specialized training, notably in the pedagogical use of technology, particularly in practical and experience-based training (Dias-Trindade & Ferreira, 2020). Thus, although teacher training programs are an effective means of encouraging teachers to develop digital competence for pedagogical purposes, this component appears to be inadequately integrated into the actual programs (Tomte et al., 2015). The findings of this study are expected to help instructors improve their digital competencies through training. This includes not only technical competencies but also a deep understanding of online pedagogy and the ability to adapt teaching strategies to the digital environment (Archambault et al., 2022). The development of these competencies should be a priority for educational institutions, as it directly impacts the quality and success of online learning initiatives. For pupils to acquire the skills needed to fully utilize new technologies, teachers must possess a strong level of digital competency (OECD, 2019). According to Ilomaki et al. (2014), digital competence is a reflection of the digital abilities and knowledge that are politically recognized as essential for full participation in our increasingly technological society.

In conclusion, the examination of teachers' competencies for online teaching is crucial for understanding the current state of digital education and identifying areas for improvement. As the demand for online education continues to grow, so too must the efforts to equip teachers with the skills and knowledge they need to succeed in this new environment. The findings of this study provide valuable insights into the challenges faced by teachers in the digital age and offer a starting point for discussions on how to enhance the quality of online education through targeted professional development and support.

The purpose of this study—The purpose of this research is to reveal the perceptions of teachers enrolled in a teaching certification program towards online education.

The questions of this research are given below.:

- Is online education valuable according to teachers' perceptions?
- Do teachers want to teach online? Do they feel competent in this regard?
- Do teachers feel competent in preparing e-content?
- What are the advantages and disadvantages of online education according to teachers?

Based on the research questions above, the following hypotheses could be formulated, focusing on teachers' perceptions, competencies, and the advantages and disadvantages of online education:

Hypothesis 1: According to teachers' perceptions, online education is considered less valuable compared to face-to-face education.

Hypothesis 2: A significant portion of teachers do not wish to teach online and do not feel competent in this regard.

Hypothesis 3: Most teachers do not feel competent in preparing e-content.

Hypothesis 4: When evaluating the advantages and disadvantages of online education, teachers are more likely to emphasize disadvantages such as limited technological access and reduced interaction over advantages like flexibility and time-saving.

16.2 Research Design

In this study, phenomenological design was preferred among qualitative research designs because we examined teachers' experiences, perspectives and perceptions. Phenomenology is a qualitative research design that aims to understand how individuals experience and make sense of a particular phenomenon. This design is suitable for examining in depth teachers' perceptions of online education and how these perceptions are formed.

A data collection tool was prepared by the researcher to answer the research questions. Data collection was applied to students who attended a certificate program completely online for 1 month at the end of the online training. 4 open-ended questions were asked to the participants. (Sample question: What are your thoughts on creating materials to be used in online training? Do you consider yourself competent in creating e-content?). Questions were collected via Google Forms. An attempt was made to reach all students who attended the course via the Whatsapp group. The reason for constructing this research qualitatively is to deeply understand the general perspective on online education, identify the problems and form the basis for future studies planned accordingly.

The study included 56 teachers as participants, of whom 46 were female (82%), and 10 were male (18%) with an average age of 34. Almost half of the participants were preschool teachers, while others taught subjects such as accounting, sociology and law. The participants had an average of 5 years of teaching experience.

16.3 The Results of the Study

16.3.1 Findings Related to Value of Online Education

Analysis of the first research question results shows that the perceptions of the participants (Table 16.1).

This analysis indicates that participants generally have positive views about online education. Most participants believe that online education is economical and has a future. However, some participants highlighted that online education is less efficient compared to face-to-face education and is limited due to technological infrastructure deficiencies. A few participants expressed skepticism about the future of online education. Participants believe that online education will be far more effective if it is made interactive and content-qualified in terms of content. They believe that today's online learning activities are inefficient. Respondents report that it lowers the price of transportation, lodging, and other standard schooling expenses. The economic advantages and cost-effectiveness of online learning are highlighted by this theme. Many respondents think that the future of online education is bright and that it is worthwhile. They list benefits like time and resource savings, implying that online learning is a useful strategy with a lot of promise. This theme affirms the

Table 16.1 Value of education according to the participants

Theme	Description	Representative quotes	Frequency
Online education is economical	A significant number of respondents highlighted the economic benefits of online education, noting it reduces costs related to transportation, accommodation, and other expenses associated with traditional learning	"Online education is economical."	22
Online education is valuable and has a future	Many respondents believe that online education is valuable, has a future, and is an effective method to save time and resources	"Precious. It has a future."	18
Concerns about effectiveness	A number of participants expressed concerns regarding the effectiveness of online education, particularly in terms of student engagement, the quality of learning, and the lack of direct interaction with teachers	"I am against online education. Education ignores the interaction between students and teachers."	15
Conditional value	Some respondents see online education as valuable under certain conditions, such as when paired with adequate infrastructure or when it is used for specific situations (e.g., emergencies)	"It's valuable if it's done interactively."	12
Preference for face-to-face education	Several respondents indicated a preference for face-to-face education over online education, citing reasons such as better interaction, effectiveness, and overall learning experience	"I think face-to-face education is more efficient."	10
Online education as a complementary tool	Some teachers believe online education is a useful complementary tool that can be effective when combined with traditional methods, especially in hybrid models or for specific types of content	"I think it is useful training, it also saves time."	7

usefulness and significance of online learning in the long run. Concerns have been raised about the effectiveness of online education, notably in terms of student engagement, learning quality, and the lack of direct teacher connection. This theme highlights the drawbacks and difficulties associated with distance learning.

16.3.2 Findings Related to the Perception Towards Providing Online Education

Analysis of the second research question results shows that the perceptions of the participants (Table 16.2).

The analysis reveals that the majority of the respondents are not willing to provide online education primarily due to feelings of incompetence and inadequacy in this area. Most respondents stated that they prefer to provide face-to-face education instead of online. If sufficient and qualified online education is provided to the teacher candidates and the tools to be used are easy to understand, use and learn, the opinions of the respondents about providing online education may change. It appears that many educators favor the more direct, personal connections that may be made in

Table 16.2 Perception towards providing online education

Theme	Description	Representative quotes	Frequency
Preference for face-to-face education	Many respondents prefer face-to-face education over online education, citing reasons like better interaction, effectiveness, and overall learning experience	"I think face-to-face education is more efficient."	15
Willingness to teach online	Some respondents are open to teaching online but with certain reservations, such as needing more technological skills or only under specific conditions like emergencies or blended learning	"I would like to, but even if I am not good enough right now, I would improve myself in this regard."	14
Technical challenges and limitations	Several respondents highlighted technical challenges or limitations, such as inadequate resources, insufficient experience, or concerns about student engagement in online settings	"I would not teach online unless I was in a very difficult situation."	12
Confidence in online teaching ability	A few respondents feel confident in their ability to teach online, citing their comfort with technology or previous experience	"Of course, I would like to give online education. Since I can use technology comfortably, I could spend the lessons much more efficiently."	11
Conditional support	Respondents who have a mixed or conditional stance on online education, seeing it as useful only in specific situations or with certain age groups	"Yes, I would like to. Those who explain it well face to face will also understand it online."	10
Potential benefits of online education	Some respondents recognized the potential benefits of online education, particularly in terms of flexibility, time efficiency, and utilizing technology for effective learning	"I think I may have some shortcomings, but I can fix this with a short training course."	9

a physical classroom because of the emphasis on improved engagement and learning outcomes. This choice might also be a result of worries about how difficult it would be to recreate these relationships in an online setting where there is less opportunity for in-person engagement and physical presence. Willingness to teach online indicates a cautious openness to online education among educators who understand its promise but lack confidence in their ability to provide it successfully. The statement that greater technology abilities are required points to a digital literacy deficit that might be filled with focused professional development. Furthermore, the fact that their readiness is conditional implies that, although they recognize the benefits of online learning, they are still unsure about its comparability to in-person instruction, especially in non-emergency scenarios. Their willingness to accept online learning can be strongly impacted by technological obstacles, particularly if they feel unsupported or underprepared. Concerns regarding student participation also highlight the challenges associated with developing a dynamic and engaged virtual classroom, which can reduce the efficacy of distance learning. In order to overcome these obstacles, educators must have better access to technology as well as resources and training that will enable them to successfully manage the intricacies of online learning. The positive attitude towards online teaching suggests that with the right skills and experience, educators can effectively manage and excel in a digital learning environment. These confident educators could serve as mentors or leaders in professional development initiatives, helping to build a broader base of online teaching expertise within educational institutions. The recognition of online education's benefits indicates that, despite reservations, some educators see the advantages of digital learning environments. These benefits, such as flexibility and time efficiency, are particularly valuable in today's fast-paced and diverse educational landscape.

16.3.3 Findings Related to Being Competent on Online Education

Analysis of the third research question results shows that the perceptions of the participants (Table 16.3).

The majority of respondents express feelings of inadequacy and lack of knowledge about e-content development. They also do not feel confident in this area. Only a small portion of the respondents stated that they prepare online content. The respondents feel inadequate and insecure about online content development, especially due to reasons such as technical knowledge, technical difficulties, and lack of education. This research demonstrates a significant deficiency in educators' digital literacy and emphasizes the necessity of focused professional development. Their willingness to participate in online education may be significantly impacted by the widespread feeling of inadequacy, which could result in lower-quality digital content and fewer successful teaching outcomes. Improving the general efficacy of online

Table 16.3 Being competent on online education

Theme	Description	Representative quotes	Frequency
Lack of competence and confidence	Many respondents expressed a lack of confidence in their ability to create e-content for online education, often citing insufficient skills or experience with technology	"I don't see myself as competent in e-content creation."	17
Desire for improvement and development	Some teachers acknowledged their current limitations but expressed a strong desire to develop their skills and improve their competencies in online content creation	"I want to develop myself in this area if I become a teacher in the future."	10
Partial competence	A number of respondents indicated that they have some level of competence in using specific tools or adapting existing materials but feel they need further development in other areas of digital content creation	"I am good at using existing materials but not at creating new ones."	9
Technical challenges	Some educators highlighted technical challenges, including lack of resources, skills, or opportunities to create digital content, which hinder their ability to produce effective online materials	"I think creating materials for online education would be inadequate because I believe we need to touch and feel the material."	8
Practical strategies and creativity	A subset of teachers shared practical strategies and creative approaches they use to engage students in online settings, such as using interactive tools or creating multimedia content	"I would send timed polls during lessons to keep students engaged and track attendance."	6

education requires addressing this lack of confidence through assistance and instruction. The desire for growth indicates that, while teachers may currently feel unprepared, they are eager to learn and adapt. Their drive is a sign that they have what it takes to develop and flourish in the online learning environment. Educational institutions should take advantage of this openness by offering instructors easily accessible and useful training opportunities that enable them to acquire the skills required to provide high-quality online content. This partial competence highlights the varied levels of digital skills among educators. It suggests that while some teachers are comfortable with certain technologies, there is a need for more comprehensive training that covers a broader range of digital content creation tools and techniques. The technological difficulties mentioned by the participants highlight how crucial it is to guarantee that educators have access to the resources and tools they need to produce digital content. Institutional support, such as improved technology infrastructure, expert help, and opportunities for practical experience with digital tools, may be necessary to address these issues. In order for teachers to participate in online education in an effective manner, these obstacles must be removed. The presence of practical strategies and creative thinking among some teachers indicates that a foundation for effective online teaching practices already exists.

16.3.4 Findings Related to the Advantages and Disadvantages of Online Education

Analysis of the third research question results shows that the perceptions of the participants (Table 16.4).

The table presents a comprehensive analysis of the numerous topics that surfaced from the feedback provided by respondents regarding online education, providing an understanding of the perceived benefits and difficulties associated with this method of instruction. The number of responses indicates the significance or prominence of each subject, each capturing a unique aspect of online education as experienced and expressed by the participants. Respondents highlighted how the flexibility of online learning allows them to manage their time more effectively, balancing educational commitments with personal and professional responsibilities. This flexibility is particularly beneficial for those with busy or unpredictable schedules, enabling them to engage with learning materials at their own pace and in their own time. A significant portion of the participants highlighted the financial advantages of distance learning. Online education is a more cost-effective choice for many people due to the decrease in expenditures associated with lodging, transportation, and other fees involved in attending in-person classes. Because of this financial benefit, online education is a more accessible and inclusive choice for people who reside distant from educational institutions or have low financial resources. Some respondents had concerns regarding the efficacy of online education in comparison to conventional face-to-face learning, despite the convenience and financial advantages. This theme highlights a widespread belief that online learning might not offer the same degree of contact, engagement, or educational impact as traditional classroom settings. One major negative is that there isn't any direct, face-to-face connection between students and teachers, which could result in a less engaging and productive learning environment. The availability of the technology infrastructure directly affects how successful online learning is. The concerns raised by respondents include outdated hardware, limited digital materials, and bad internet access, all of which can seriously impede the learning process. These technical difficulties can worsen student inequality, especially for those who do not have access to high-quality technology, further hindering the overall educational experience. The reduced opportunity for social interaction is one of the main disadvantages of online learning, according to respondents. In-person instruction is highly valued because it allows more personal

Table 16.4 Advantages and disadvantages of online education

Theme	Description	Representative quotes	Frequency
Convenience and flexibility	Many respondents appreciate online education for its convenience and flexibility, emphasizing how it allows them to manage their time more effectively and comfortably	"It is more comfortable, more flexible.", "online education is easier, more flexible, and time-saving."	25
Economic benefits	Respondents highlighted the economic advantages of online education, noting that it significantly reduces costs associated with transportation, accommodation, and other expenses typically incurred with traditional in-person learning	"It is economical, especially for those who have to travel.", "no need for transportation costs."	15
Lack of effectiveness	Respondents expressed concerns about the effectiveness of online education compared to face-to-face learning. They noted that online education often lacks the personal interaction and engagement that are crucial for effective teaching and learning	"Online education is not as effective as face-to-face.", "it is less efficient and lacks personal interaction."	12
Technological challenges	Respondents pointed out that the effectiveness of online education heavily depends on the quality and reliability of technological infrastructure	"It depends on the technological infrastructure.", "there are internet issues that make it less efficient."	8
Social interaction	Respondents emphasized that face-to-face education offers superior opportunities for social interaction and communication, which are often lacking in online settings	"Face-to-face education allows for better social interaction.", "online lacks the interaction and communication."	10
Reduced discipline	Respondents noted that while online education provides a more relaxed and comfortable learning environment, this ease can negatively impact discipline and the seriousness of the educational experience	"Online education is more relaxed, but this affects discipline.", "the comfort of online education can hinder seriousness."	6

relationships between professors and students than online learning can. The development of interpersonal and collaborative skills—two essential elements of the educational experience—may be hampered by the lack of in-person connection and communication that can make online learning feel solitary. Another benefit that was often emphasized was the time-saving nature of online education. Online education is considered more convenient and accessible for individuals with busy schedules due to the elimination of commute time and the opportunity to engage in classes from home. This effectiveness raises the appeal of online learning by enabling students to devote more time to other obligations or interests. Although some people view the laid-back atmosphere of online learning as a plus, it can be difficult to maintain seriousness and discipline. A lack of focus and distractions brought on by the informal nature of online learning environments have the potential to undermine the structure and rigor that are frequently required for learning to be effective. As a result, respondents expressed concerns that reliability and student engagement may suffer due to the ease of access to online learning.

16.4 Conclusions

Many people have questioned online education, particularly during the Covid-19 pandemic, when it became an unavoidable necessity for practically everyone. With the sudden emergence of the pandemic, problems such as inadequate infrastructure, lack of knowledge and skills, and concerns about efficiency have also emerged. Many institutions and organizations providing online education have had to provide superficial and unqualified training due to these issues. Such problems have led to the emergence of negative perceptions that online education is of poor quality and inefficient. It is especially important to reveal the perceptions and thoughts of teachers who were asked about the value of online education, its advantages and their own competencies. According to the research results, most teachersbelieve that online education is valuable and will have a significant place in our lives in the future. In the report published by Allen & Seaman, 2011, participants also made positive statements about the value of online education (Allen & Seaman, 2011). However, they stated that these trainings are inefficient and unqualified due to the inefficient processing of online training. For this reason, teachers indicated that the value of online training will increase especially if prepared in a more qualified, efficient and scientific way. Sun and Chen (2016) stated that professional development for teachers should emphasize promoting effective online collaboration for students, setting high expectations, adjusting teaching to conform to the online environment, and creating appropriate online teaching strategies, etc. In addition, online faculty also need adequate training in the technologies applicable to online teaching: how to use the online system and course software, and what to do and whom to contact when certain technology problems occur.

Although the majority of participants found online education valuable, they stated a preference for face-to-face education while teaching. They explained that this preference stemmed from a lack of confidence in their competency to provide online education and a deficiency in technical knowledge and skills. Kiryakova and Kozhuharova (2024) found that very few teachers use technology regularly in their teaching practice. In contrats Saha et al. (2022) observed that while more than three-fourth of participants preferred online teaching during the pandemic, only about 10.05% of selected teachers preferred online teaching after the pandemic. Analyzing the data obtained from the research question regarding whether the participants perceive online education as valuable led to the following conclusion: although the participants find online education valuable, they are reluctant to use it due to their lack of technical knowledge and skills.

In this sense, it can be concluded that teachers will be more likely to provide training if they are offered an easy-to-use, effective and efficient online training platform. Mayantao and Tantiado (2024) state that the more teachers use digital tools such as Powerpoint/Canva, video editing tools like Capcut and Excel, spreadsheets, the higher their confidence level in technology. Antón-Sancho et al. (2021) report that teachers at the Latin America University in Spain predominantly use PowerPoint, Google Forms, and Google Classroom in their lessons. Dos (2014) used some digital learning materials and tools in his thesis, including Adobe Captivate, Camtasia for presentation, Adobe Presenter and Moodle Content Creator for content creation, Moodle Quiz Creator, Hot Potatoes and Adobe Captivate for quizzes. Therefore,it can be concluded that teachers need to know and use tools for creating presentations, developing content, designing quizez, and facilitating communication in a digital environment. They should be expected to use Canva, Adobe Presenter for creating presentations; Camtasia, Google Docs for content creation; Quizizz, Kahoot for quizzes, and social media platforms such as Zoom, Google Meet, Twitter, Facebook, and Whatsapp for communication.

Providing technical information is considered especially important for delivering effective online training. A small number of participants indicated that they felt competent in online education and could provide it effectively.

One of the most important issues in online education is the development and use of online learning materials, which is vital for enhancing teacher competencies. Most teachers feel quite inadequate when it comes to creating online learning materials. One of the main reasons for this may be that many free and widely used material creation platforms are only available in certain languages. For example Powtoon, StoryboardThat, Prezi, Bandicam, Quizizz do not support the Turkish language which poses a significant challenge for those who do not know any language other than their mother language.

Garzón et al. (2020) found that the digital content creation dimension showed alarming results, as it was close to zero. This indicates that teachers continue to use ICTs as a minimum support tool but lack the self-efficacy to create and share their own digital content with other users. The requirement for software knowledge, the presence of paid resources, and insufficient training in developing online education may contribute to teachers' feelings of inadequacy in this area.

Regarding the advantages of online education, teachers noted that it especially benefits them in in terms of time management. The fact that online courses are shorter than face-to-face courses, allowing teachers to instruct whenever and wherever they choose. Additionally, teachers who have to travel long distances to school or face heavy traffic can provide instruction from home, eliminating transportation costs, which has emerged as an advantage of online education. However most participants expressed concerns that the lack of interaction-both teacher-student and student-student reduces the effectiveness of online learning compared to traditional environments. Other challenges includetechnological problems and difficulties in classroom discipline. While time savings and cost reduction are highlighted as advantages of online education, the lack of interaction and concerns about content quality come to the fore as disadvantages.

Based on the research data, the following general conclusions can be drawn. Teachers acknowledge that online education is valuable but they express concerns regarding quality and knowledge gaps. Most teachers are reluctant to teach online lessons due to a lack of confidence and perceived competence in this area. They feel completely inadequate in preparing e-content. Online education presents both advantages and disadvantages; flexibility and cost savings are viewed as benefits, while lack of interaction and lack of technical knowledge are seen as disadvantages.

References

Ali, W. (2020). Online and remote learning in higher education institutes: A necessity in light of COVID-19 pandemic. *Higher Education Studies, 10*(3), 16–25.

Allen, I. E., & Seaman, J. (2011). *Going the distance: Online education in the United States, 2011*. Sloan Consortium.

Antón-Sancho, Á., Vergara, D., Lamas-Álvarez, V. E., & Fernández-Arias, P. (2021). Digital content creation tools: American university teachers' perception. *Applied Sciences, 11*(24), 11649.

Archambault, L., Leary, H., & Rice, K. (2022). Pillars of online pedagogy: A framework for teaching in online learning environments. *Educational Psychologist, 57*(3), 178–191.

Bragg, L. A., Walsh, C., & Heyeres, M. (2021). Successful design and delivery of online professional development for teachers: A systematic review of the literature. *Computers & Education, 166*, 104158.

Dias-Trindade, S., & Ferreira, A. (2020). Digital teaching skills: DigCompEdu CheckIn as an evolution process from literacy to digital fluency. Icono 14. *Revista De Comunicación y Tecnologías Emergentes, 18*(2), 162–187. https://doi.org/10.7195/ri14.v18i1.1519

Dos, B. (2014). Developing and evaluating a blended learning course. *The Anthropologist, 17*(1), 121–128.

Garzón, A. E., Martínez, T. S., Ortega, M. J. L., Marín, J. A., & Gómez, G. G. (2020). Teacher training in lifelong learning—The importance of digital competence in the encouragement of teaching innovation. *Sustainability, 12*(7), 2852. https://doi.org/10.3390/su12072852

Ilomaki, L., Paavola, S., Lakkala, M., & Kantosalo, A. (2014). Digital competence – An emergent boundary concept for policy and educational research. *Education and Information Technologies, 21*(3), 655–679. https://doi.org/10.1007/s10639-014-9346-4

Keefe, L., Rafferty, J., Gunder, A., & Vignare, K. (2020). Delivering high-quality instruction online in response to COVID-19: Faculty playbook. *Online Learning Consortium.*. https://eric.ed.gov/?id=ED605351

Kiryakova, G., & Kozhuharova, D. (2024). The digital competences necessary for the successful pedagogical practice of teachers in the digital age. *Education Sciences, 14*(5), 507. https://doi.org/10.3390/educsci14050507

Lázaro Cantabrana, J. L., Usart Rodríguez, M., & Gisbert Cervera, M. (2019). Assessing teacher digital competence: The construction of an instrument for measuring the knowledge of pre-service teachers. *Journal of New Approaches in Educational Research, 8*, 73–78. https://doi.org/10.7821/naer.2019.1.370

Mayantao, R., & Tantiado, R. C. (2024). Teachers' utilization of digital tools and confidence in technology. *International Journal of Multidisiplinary Research and Analysis., 7*(5). https://doi.org/10.47191/ijmra/v7-i05-16

OECD. (2019). *Education at a glance 2019: OECD indicators*. OECD Publishing. https://doi.org/10.1787/f8d7880d-en

Redecker, C. (2017). *European framework for the digital competence of educators: DigCompEdu (No. JRC107466)*. Joint Research Centre (Seville site); Publications Office of the European Union.

Reisoglu, I., & Çebi, A. (2020). How can the digital competences of pre-service teachers be developed? Examining a case study through the lens of DigComp and DigCompEdu. *Computers and Education, 2020*(156), 103940. https://doi.org/10.1016/j.compedu.2020.103940

Saha, S. M., Pranty, S. A., Rana, M. J., Islam, M. J., & Hossain, M. E. (2022). Teaching during a pandemic: do university teachers prefer online teaching? *Heliyon, 8*(*1*), e08663.

Sun, A., & Chen, X. (2016). Online education and its effective practice: A research review. *Journal of Information Technology Education: Research, 15*, 157–190. Retrieved from http://www.informingscience.org/Publications/3502

Tomte, C., Enochsson, A. B., Buskqvist, U., & Karstein, A. (2015). Educating online student teachers to master professional digital competence: The TPACK-framework goes online. *Computers & Education, 84*, 26–35. https://doi.org/10.1016/j.compedu.2015.01.005

Chapter 17
Digitalization of Social Services for Vulnerable Groups in Territorial Communities of Ukraine

Serhii Gunko, Maryna Lekholetova, and Viktor Liakh

Abstract The processes of regional digitalization are one of the important tasks of state policy in Ukraine. This primarily concerns the digital transformation of territorial communities in communication and information infrastructure; digitalization of public services; e-democracy tools and their implementation in social spheres; organizational and resource support for digitalization; and information security.

The authors of the article conducted an analysis of social service digitalization programs in which Ukrainian communities participate (community informatization program, smart community call center, mobile suitcase for Administrative Service Centers (ASC), community chatbot, and participatory budget).

The geographical scope of the study covered four regions of Ukraine (Vinnytsia, Volyn, Dnipropetrovsk, Odesa). The communities involved in implementing digitalization programs for social services were selected through open competitions, during which factors such as the state of digital infrastructure, the infrastructure for providing public services (including social services), the motivation of local government bodies to adopt new technologies, and the level of development of civil society institutions were considered.

To achieve the goal, the authors used methods of analysis, synthesis, and systematization of scientific sources to define basic concepts; the comparison method to identify the features of digitalizing social services for vulnerable groups; and the analysis of the regulatory and legal framework to determine the legal foundations for digitalizing social services for vulnerable groups in Ukrainian communities.

It has been established that the implementation of social service digitalization programs used in Ukrainian communities, including work with vulnerable groups opens opportunities for open and transparent public administration focused on the interests of people and aimed at developing an information society. This includes the

S. Gunko · V. Liakh
East Europe Foundation, Kyiv, Ukraine
e-mail: SHunko@eef.org.ua; vliakh@eef.org.ua

M. Lekholetova (✉)
Borys Grinchenko Kyiv Metropolitan University, Kyiv, Ukraine
e-mail: m.lekholetova@kubg.edu.ua

E. Smyrnova-Trybulska et al. (eds.), *E-Learning and Enhancing Soft Skills*,
https://doi.org/10.1007/978-3-031-82243-8_17

use of information technologies and digital solutions; conducting economic activities based on effectiveness and efficiency principles; creating conditions for the development of informatization and digitalization infrastructure; and ensuring sustainable innovative development of the community as a whole.

The study highlights the advantages of implementing social service digitalization programs for both local self-government bodies and community residents, including vulnerable groups.

Keywords Digitalization of social services · Information and communication technologies · Community informatization program · Smart call center (IVR) · Mobile suitcase for administrative service centers (ASC) · Community chatbot · Participatory budget · Vulnerable groups

17.1 Introduction

Currently, one of the key tasks of Ukraine's state policy is the digitalization of regions and the establishment of social interaction between cities, towns, villages, government agencies, and non-governmental organizations through the implementation of information technologies and digital solutions. Consequently, there is a need for the digitalization of social services that can be provided in each territorial community. The demand for social services is formed according to the needs of the community residents, and the digitalization of these services can significantly impact the efficiency of service delivery and the quality of social services, which is crucial for vulnerable population groups. This has become particularly relevant during the full-scale invasion by the Russian Federation, when the protection of rights and interests, as well as access to quality social services, became critical for the population in Ukrainian communities. This is especially true for those population groups who are in difficult life circumstances or have specific needs and requests.

These conditions have led to new challenges for communities, local governments, and social service professionals, particularly regarding the need to develop the digitalization of executive and local government bodies, expand digital infrastructure in communities affected by Russia's full-scale invasion of Ukraine, and implement digital tools and solutions for accessing public services.

The EGAP Program, implemented by the Eastern Europe Foundation and Innovabridge Foundation with funding from Switzerland, in partnership with the Ministry of Digital Transformation of Ukraine, aims to use advanced information and communication technologies to improve governance quality, enhance government-citizen interaction, and promote social innovations in Ukraine (EGAP, 2022).

Thus, there is a need to conduct research to analyze the digitalization programs of social services in which Ukrainian communities participate. In particular, community informatization programs, smart community call centers, mobile service centers, community chatbots, and participatory budget. The implementation of these programs in Ukrainian communities expands opportunities for residents, including vulnerable population groups, to receive quality social services according to their needs.

17.2 Literature Review

The digitalization of social services is currently highly relevant for the development of communities. An increasing number of scholars are focusing their research on the study of digitalization and management, the development and understanding of digital competence, educational technologies (Punie et al., 2013; Daniel & Uvali-ć-Trumbić, 2012), digital access and engagement in social media tools (He et al., 2020; Anderson et al., 2020), the use of social media as a means of digital transformation of the communication system between Ukrainian government bodies and citizens (Vovk & Markovets, 2022), the digital competence of clients of territorial social service centers (Karkach & Semigina, 2023), and modern approaches to teaching digital skills to the elderly (Blažič & Blažič, 2020).

Providing effective social services to vulnerable population groups is now at the forefront of social work practitioners' attention (Denysiuk et al., 2023). There is an increasing need to direct state policy in the field of social work to address the issues faced by individuals with disabilities (Dulia et al., 2024); to develop and implement programs aimed at enhancing digital competencies, including ICT, for social workers employed in government services and non-governmental organizations, particularly those working with women combatants (Faidiuk et al., 2022); and to provide psychosocial support to victims of war-related sexual violence (Shved et al., 2024), among other areas.

Researchers note that the digitalization of social services will contribute to the establishment of interaction processes between the government and the community, with new governance processes fostering greater citizen participation in decision-making (Gunko et al., 2023; Bingham et al., 2005). The assessment of local budget transparency is also important, where the key task is to assist local governments in implementing solutions and practices of budget transparency and engaging citizens in the budget processes at the local level (Liakh, 2018).

Examining the state of legal regulation and issues arising from the provision of social services in electronic form, particularly pension services and unemployment benefits, Yehorova-Lutsenko (2020) notes that a key medium-term task for advancing Ukraine toward a unified digital space with the European Union could be the merger of types of social administrative services and the mechanisms of their provision with those offered in European Union countries. The principles underlying the provision of social services include: targeting and individualized approach; accessibility and openness; humanity; maximum efficiency in the use of budgetary and extrabudgetary funds by service providers; legality; social justice; ensuring confidentiality; compliance with quality standards, and responsibility for adherence to ethical and legal norms (Gunko et al., 2024).

The advantages of digitizing social services include: improving the efficiency and transparency of aid distribution to those in need; digital tools can help track and manage the work of service providers, thereby enhancing coordination between organizations and the government (Ihnatenko & Sadzaglishvili, 2023). At the same time, it is important to ensure that the digitalization of social services does not exclude vulnerable population groups who may not have access to technology (Preiß, 2019). For example, in researching the issues faced by refugees, scholars emphasize that digitalization should be inclusive, not only contributing to community input during the development of new technology but also ensuring that displaced persons have access to the same services as the host country (Martin-Shields, 2021).

The analysis of the above-mentioned studies, the new challenges faced by social service providers and vulnerable population groups due to the martial law in Ukraine, highlights the need for analyzing the digitalization programs of social services used in Ukrainian communities, including in working with vulnerable population groups. It also emphasizes the advantages of implementing social service digitalization programs for both local governments and community residents, including vulnerable groups. These tasks formed the basis of our research.

17.3 Analysis of the Selection Process for Participation in the Competition for Implementing Digitalization Projects in Ukrainian Communities Based on Applications from Local Governments

In the first stage of the study, in order to identify the main problems and needs related to the digitalization of social services in Ukrainian communities and to determine which tools are lacking for effective work with the population, including vulnerable groups, we have clarified the essence of key concepts, considered the experience of digital transformation in the public services sector in European countries, and analyzed the competitive selection process and its results based on requests from local government bodies.

The organizational and legal foundations for providing social services aimed at preventing difficult life circumstances, overcoming or minimizing their negative consequences for individuals/families in difficult life situations are regulated by the Law of Ukraine "On Social Services" (2019) (Verkhovna Rada of Ukraine, 2019). In the law, the concept of "vulnerable groups of the population" is defined as individuals/families who are at the highest risk of falling into difficult life circumstances due to the impact of adverse external and/or internal factors. Digitalization of social services is viewed in scientific literature as government management through the interaction of government bodies and society using advanced information and communication technologies under conditions of accessibility, openness, and transparency (Chuba, 2022); organizing simplified access for citizens to administrative services (Yehorova-Lutsenko, 2020); and creating a unified information space for the social protection sector (Kulyna & Nalukova, 2020).

Estonia remains a recognized leader in the digitalization of public services in the EU, with 100% of state public services already converted to digital (online) format. At the local level, digital tools have been created for transitioning local services to an online format. Specifically, to simplify the submission of applications, streamline the work of municipal officials, and ensure transparency, the SPOKU system has been introduced for receiving and processing requests for local government bodies and for providing a self-service environment for residents (SPOKU, n.d.). Digitalization is actively supported by the Estonian government and local government associations. The exceptional importance of digitalization is recognized at the European Union level, as evidenced by the adoption of the EU Digital Decade 2030 Strategy in 2021 (European Commission, n.d.).

Comparing foreign experience in e-government with the domestic system of providing administrative services through e-government, researchers A. Kolosok, T. Martyniuk, and O. Melnyk emphasize that in Ukraine, this process is extremely difficult to establish due to: the large number of regulatory documents governing this process; the need for a significant number of individuals to service it; and the existence of a vast number of people with specific needs who receive social services (housing subsidies, privileged categories of the population, internally displaced persons, combatants, and other individuals in difficult life circumstances) (Kolosok et al., 2023). In today's reality, it is necessary to develop and implement a digitalization strategy for social services in communities, taking into account the problems and needs of social service recipients, including vulnerable population categories, as well as the experience of foreign countries.

The development and implementation of digitalization programs for social services in Ukrainian communities involves determining the needs of the community's population for social services. According to the Law of Ukraine "On Social Services" (2019), this includes collecting, summarizing, and analyzing information about the presence of vulnerable groups of the population in the community, individuals/families in difficult life situations who cannot overcome them independently, and their needs for social services, based on which management decisions regarding the organization of such services are made (Verkhovna Rada of Ukraine, 2019).

The geographical scope of the study was limited to the Volyn, Vinnytsia, and Dnipropetrovsk regions of Ukraine, where the following communities were selected as experimental sites: Viitivetska, Stryzhavska, Zymnivska, Smidynska, Solonyanska, and Mezhivska. These communities were selected through open competitions in each region.

The conditions of the competition were communicated on all information resources of the Eastern Europe Foundation, through regional state administrations, and directly to each community in the region.

The selection committee considered the following criteria:

- The community's experience in implementing innovative technologies and digital tools or other projects aimed at the systematic development of the community;
- The presence of or readiness to introduce positions for a digital leader (CDTO), IT specialist(s), and a community engagement specialist;
- Readiness for co-financing some joint projects;
- The presence of civil society organizations striving for change and community development, as well as real examples of cooperation (joint projects) between local governments and the community;
- Willingness to use digital tools for organizing local government activities and to promote democracy by increasing community influence in decision-making.

For each competition, an online information session is held on a ZOOM platform, where community representatives can ask questions and receive answers regarding the competition conditions, future joint project plans, etc.

After receiving applications for participation, the selection committee evaluates them to create a shortlist of communities for interviews. At this stage, the information provided by the community is assessed according to the proposed form, the content of examples of successfully implemented projects, interaction with civil society institutions, the community's digital readiness (already implemented tools and experience in using them), etc. The shortlist usually contains three times the number of "vacant" places. In the second phase of the EGAP program, 8 pilot communities were selected in each target region based on the results of two selection processes (Table 17.1).

In addition to the ZOOM web platform for conducting the selection process, other tools were also used. Specifically, Google Forms were utilized to collect basic information about the current capabilities and previous projects implemented in the candidate communities (communities filled these out independently), and the "Decentralization" portal, which was created in 2014 for communicating the reform of local self-government and territorial organization of power in Ukraine (Decentralization, n.d.); for analyzing the structure of the local budget, the State Budget

Table 17.1 Information about respondents

Programs used for digitalizing social services	Participating communities	Implementation period
Community Informatization Program	Viitivetska, Stryzhavska, Smidynska	2024
Smart Community Call Center	Viitivetska, Stryzhavska, Solonyanska, Mezhivska	2023-ongoing
Mobile Service Center Kit (digital suitcase)	Viitivetska, Stryzhavska, Zymnivska, Smidynska, Solonyanska, Mezhivska	2022-ongoing
Community Chatbot	Viitivetska, Stryzhavska, Zymnivska, Smidynska, Solonyanska, Mezhivska	2021-ongoing
Participatory Budget	Viitivetska, Stryzhavska, Zymnivska, Solonyanska, Mezhivska	Annual activity

Portal for Citizens was used (Open Budget, n.d.); and for tracking local budget expenditures, the Single Web Portal for Public Spending was used (Spending, n.d.).

Specified digitalization programs for social services were developed and implemented in collaboration with EGAP pilot communities.

In the second stage of the competition, interviews are conducted with community teams, including the head of the community, their deputy responsible for Center for Administrative Services (ASC) operations and/or digital development of the community, the head of ASC, the head of the IT department, and other community representatives as designated by the head. During the interviews, the level of motivation to implement innovative technologies, the presence of "political will" for systemic changes, the desire for community development, and the ability to work as a team in implementing digital tools and managing change are assessed.

During this stage of the study, in-depth research was conducted by experts on the existing digital infrastructure, the operation of ASC, and the automation of local government operations in the selected communities. Following expert recommendations, strategic sessions are held together with each community for short- and medium-term planning of joint activities.

To assess the effectiveness of implementing digital tools in communities, a list of indicators was developed, taking into account the quantitative data on the use of each tool. For example, for the digital mobile office of the Center for Administrative Services (CNAP), information is collected and accumulated on the number of public services provided using the mobile office and the number of trips made (considering that multiple services may be provided to several applicants during a single trip). For the Smart Call Center, data is collected on the number of incoming calls as well as the number of connections made with a specialist at the institution. It is assumed that if the applicant receives complete information through the automated system (auto-reply), they do not select the option to "connect with a specialist" in the voice menu. The effectiveness indicators for the Participatory Budget include the number of submitted projects and the number of community residents who voted for at least one project, compared to the total population of the community. To evaluate the effectiveness of the chatbot implementation, the relative number of community residents who connected to the chatbot and the number of interactions are assessed.

17.4 Digitalization Programs for Social Services Implemented in Ukrainian Communities

Based on the previous research phase (analyzing the selection process for participating in the competition to implement digitalization projects in Ukrainian communities following the requests from local governments), we analyzed the digitalization programs for social services in which Ukrainian communities are participating (Fig. 17.1). We also characterized the challenges identified by expert groups during the practical implementation of digital services.

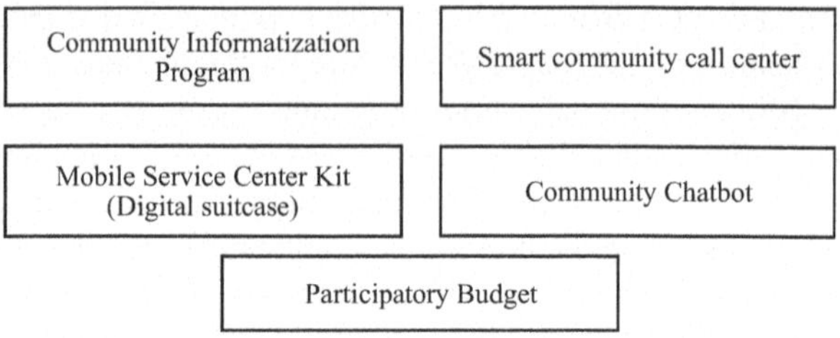

Fig. 17.1 Digitalization programs for social services implemented in Ukrainian Communities. (Source: Own work)

We will further examine the essence of the digitalization programs mentioned above in more detail.

Community Informatization Program This is a document approved by the local council, outlining plans for informatization and digital transformations in various areas of community life in the short and medium term. It includes an action plan with clear objectives and specific implementation deadlines and serves as a tool for choosing the most optimal solution from possible alternatives, ensuring the planned result with minimal resource use.

Within the framework of the program, the community addresses the transition to open and transparent public governance focused on people's interests and the development of an information society, including the use of information technologies and digital solutions. It also promotes economic activities based on efficiency and effectiveness, creates conditions for the development of informatization and digitalization infrastructure, and ensures sustainable innovative development of the community as a whole. The program's formation and implementation involve a wide range of community representatives in addressing life-sustaining and sustainable development issues through informatization and digital solutions.

The process of forming and implementing the community's informatization program consists of several key stages, namely:

- Initiating the development of the informatization program
- Creating a working group and discussing urgent issues
- Developing and publicly discussing the program's concept
- Approving the program's concept
- Preparing the program draft
- Conducting an expert review and coordinating the program draft with the general state customer of the National Informatization Program and approving the program
- Including the program in annual social and economic development programs
- Organizing the implementation of the program's measures and tasks and monitoring their execution
- Making changes to the program
- Monitoring the program's implementation and preparing interim, annual, and final reports

Smart Call Center for Communities A software complex that automatically processes calls from users and provides them with the necessary information without the need for a staff member's participation. Implementing such a call center in the operations of the Center for Administrative Services (ASC) can significantly reduce the load on operators and shorten the time users wait for answers to their queries. This, in turn, will ensure faster and better service delivery, reduce queues, and improve customer satisfaction.

The most effective solution for the ASC could be the introduction of a call center with an answering machine based on a cloud (online) service. This implementation option allows for reduced costs on equipment and infrastructure since there is no need to purchase and maintain an on-site server. Thus, the implementation of a cloud-based answering machine call center can simplify implementation and reduce the financial costs of the ASC.

Algorithm for the operation of a cloud-based answering machine call center:

1. The client calls the call center
2. The answering machine responds
3. The administrator provides additional consultation
4. A specialized expert processes the client's request

The process of implementing a call center into the ASC operations can be divided into several key stages, namely: needs and requirements analysis; solution selection and preparation; designation of responsible persons and operators (workstations); preparation of answering machine responses; system setup and preparation of workstations; information campaign and launch of the new service; updating and actualizing information; adding new functions and services.

Mobile Service Center Kit (Digital Suitcase) A mobile remote workstation for an ASC administrator. The mobile remote workstation allows the administrator to conduct off-site receptions at the place of residence or stay (location) of the service applicant or another address specified by them, within the relevant administrative-territorial unit.

The list of categories of applicants who can receive services through a mobile remote workstation and the procedure for the administrator's work at the mobile remote workstation is determined by the body that decided to establish the ASC.

The mandatory category of applicants who are provided services through a mobile remote workstation includes persons with disabilities in group I and others who, according to the conclusion of a medical consultative commission, are unable to care for themselves and require constant external assistance.

Currently, there are two options for servicing client categories using a Mobile Office:

- Servicing at the residence or stay of citizens. This model may involve servicing either a specific group of citizens (e.g., people with disabilities, hospitalized citizens, elderly people, etc.) or all residents of the area without restrictions.

– Servicing citizens at a permanent address of a designated service point. That is, in the case of servicing citizens in a remote settlement or city district by the Mobile Office, it is advisable to designate a permanent location for receiving citizens according to a set schedule (for example, in a cultural center, the elder's office, etc.).

Community Chatbot A program that uses a simple mechanism to provide users with the necessary information automatically and in a familiar interface, such as Viber, Telegram, or another messenger or website-embedded application. Chatbots are usually used as an additional communication channel and source of information for users or as an independent unique service.

A classic chatbot works based on pre-defined (programmed) request/response templates or step-by-step algorithms for providing results according to the request selected by the user. This operating mechanism can be considered the simplest form of artificial intelligence.

Chatbot operation algorithm according to the request selected by the user (steps 2 and 3 can be repeated several times, depending on the service's complexity):

1. The chatbot provides the user with a list of possible requests
2. The user selects the desired request from the options provided and gives additional data
3. The chatbot processes the request or information received from the user
4. The chatbot provides the user with the result or response to the request

Participatory Budget A tool for direct community influence on the distribution of local budget funds, which creates opportunities for residents to submit their initiatives and projects for implementation, as well as select projects that they consider the most important and relevant.

To make the Participatory Budget operational, the following steps are necessary: the local government decides to allocate local budget funds for implementing public initiatives and announces the start of the public budget cycle; community representatives submit projects that undergo moderation by the relevant local government units (primarily examining compliance with legislation and feasibility); projects that pass moderation are published for public discussion and voting; based on the voting results, a specially created commission determines the winners to be implemented in the next budget period; contractors are selected on a competitive basis, and the winning projects are brought to life; based on the analysis of the completed Participatory budget cycle, the amount and conditions for the next cycle are determined.

Participatory Budget implementation algorithm:

1. Allocation of local budget funds (choosing a financing model).
2. Designation of responsible persons.
3. Preparation and approval of the Public Budget Regulation.
4. Connection to the platform.

5. Approval of the order on conducting the Public Budget in the current (next) year.
6. Decision-making on the formation of a Coordinating Body (Coordination Council).
7. Selection of criteria for project division and evaluation.
8. Organizing training for responsible persons.
9. Conducting an information-promotion campaign for residents (if necessary, conducting a Public Budget school).
10. Announcement of the start of the Public Budget cycle.
11. Moderation of submitted projects.
12. Voting.
13. Determination and announcement of the winners.
14. Project implementation and reporting.
15. Conducting an information campaign.

Based on the analysis of scientific sources and digitalization programs for social services, we can conclude that the implementation of digitalization in Ukrainian communities provides broad opportunities for the development and quality of social services delivery to the population, improves the lives of vulnerable groups, enhances the digital competence of specialists and service recipients.

At the same time, it is important to characterize the challenges identified by expert groups during the practical implementation of digital services, namely:

- **Low level of digital readiness among residents of small communities.** The communities where the research was conducted are located in rural areas, where a stable culture of using digital tools has not yet been established. As a result, community residents use the relevant devices (smartphones, computers, etc.) significantly less. This negatively affects the widespread adoption of digital tools, as well as the motivation of local authorities to support and update them. To overcome this challenge, opportunities must be created for mass digital literacy education. A good example would be the reformatting of existing libraries into digital multifunctional spaces.
- **Low level of digital readiness among local government employees.** Digital tools require not only implementation but also constant administration and support. Therefore, for their successful implementation, specialized training for responsible staff must be planned. During the implementation and piloting of such tools, the EGAP team developed so-called "Template Projects" for each tool. These projects include a set of necessary methodologies, algorithms, descriptions, and templates for local policies—everything required for the successful implementation and further use of the tool.
- **Resistance to change.** In cases where the implemented digital solutions replace the non-digital routines familiar to most officials, time and additional effort are required to convince the responsible executors of the greater efficiency of the digital tool, after which they can become ambassadors of digitalization.

17.5 Conclusion

Based on the conducted research, several advantages of implementing digitalization programs for social services have been identified, benefiting both local government authorities and community residents, including vulnerable populations.

The implementation of a community informatization program helps local government authorities create conditions for the innovative development of the economy and the social sector, ensures socio-economic and cultural growth, and promotes the development of an open information society. Additionally, it facilitates effective governance and the development of e-democracy tools. For community residents, the implementation of the informatization program contributes to creating and maintaining a full-fledged, comfortable living environment, enhances the availability and quality of public and social services, and fosters the establishment of direct democracy institutions and the satisfaction of informational needs.

The advantages of implementing a smart call center in the Center for Administrative Services (ASC) for local government authorities include:

- Significant reduction in the workload on ASC employees: the auto-attendant can handle and respond to a large number of inquiries simultaneously, reducing the need for multiple operators.
- Improved efficiency: the auto-attendant promptly provides information and performs standard tasks, speeding up inquiry processing and reducing client wait times.
- Continuous availability: the auto-attendant can operate 24/7 without interruptions, allowing clients to receive answers to their inquiries even outside ASC working hours.
- Increased accuracy and consistency: the auto-attendant always provides standardized responses, eliminating the possibility of errors and inconsistencies in the information provided.
- Enhanced client satisfaction: the quick and accurate responses from the auto-attendant contribute to improved client satisfaction, as their inquiries are processed immediately.

A positive example of the implementation of Mobile Service Center can be seen in the work of ASCs in four territorial communities at the end of 2020, supported by the EGAP Program. For instance, the Mobile Service Center introduced in the ASC of the Illinetska city territorial community allows for the provision of administrative services at home for vulnerable citizens. Now, residents of the Illinetska community who are over 80 years old, people with disabilities in Group 1, and those who cannot move independently due to health conditions, including those receiving inpatient treatment in municipal medical institutions, can receive the necessary administrative services without spending time, effort, and health on visiting the ASC (EGAP, 2020).

Key advantages of a chatbot over a website or social media page include its accessibility on a smartphone, a familiar and user-friendly messenger interface, the ability to create interactive dialogues and integrate services for personalized information delivery, the capability to send personalized messages to each user (depending on their needs), and the possibility of user authentication and the establishment of two-way communication.

The advantages of implementing a participatory budget for local government authorities include:

- The ability to better understand the needs of the community
- Increased community trust through budget transparency
- Uniting the community around the idea of collective development
- Public support for maintaining the local area
- Engaging citizens in the decision-making process
- Greater motivation to pay taxes that contribute to the local budget

For community residents, implementing a participatory budget provides the opportunity to influence the allocation of budget funds, realize their ideas for community development, enhance the community's tourism and economic attractiveness, establish sustainable dialogue, and foster the development of civil society.

Acknowledgments The research was conducted within the framework of the scientific project at the Department of Social Pedagogy and Social Work of the Faculty of Psychology, Social Work, and Special Education at Borys Grinchenko Kyiv Metropolitan University. The theme of the research is "Socialization of Vulnerable Population Groups in the Context of Territorial Community Development in Ukraine," with registration number 0121 U112043. The project duration is from June 2021 to June 2026.

References

Anderson, V., Gifford, J., & Wildman, J. (2020). An evaluation of social learning and learner outcomes in a massive open online course (MOOC): A healthcare sector case study. *Human Resource Development International, 23*(3), 208–237. https://doi.org/10.1080/13678868.2020.1721982

Bingham, L. B., Nabatchi, T., & O'Leary, R. (2005). The new governance: Practices and processes for stakeholder and citizen participation in the work of government. *Public Administration Review, 65*(5), 528–539.

Blažič, B. J., & Blažič, A. J. (2020). Overcoming the digital divide with a modern approach to learning digital skills for the elderly adults. *Education and Information Technologies, 25*, 259–279.

Chuba, N. V. (2022). E-gov-ernance and adaptation of the civil service of Ukraine to EU standards. *Public management., 2*(30), 101–109. https://doi.org/10.32689/2617-2224-2022-2(30)-13

Daniel, J.S., & Uvalić-Trumbić, S. (2012, September 12). Open Educational Resources (OER): The coming of age of ICT in education? *Speech to eLearning Korea*. Retrieved from http://sirjohn.ca/wordpress/?page_id=29. Accessed 8 Sept 2024.

Decentralization. (n.d.). Portal "Decentralization". Retrieved from https://decentralization.ua

Denysiuk, O., Dulia, A., Vainola, R., Epel, O., Zhukov, V., Karpenko, O., Klishevych, N., Kotelevets, A., Lekholetova, M., Liutyi, V., Liakh, T., Ostrolutska, L., Petrochko, Zh., Sapiha, S., Sylantieva, I., Spirina, T., Sulitskyi, V., Faidiuk, O., & Shved, O. (2023). *Current issues in social work: A study guide.* O.V. Epel, T.L. Liakh, I.V. Sylantieva. RIK-U, 420. ISBN 978-617-8276-43-0.

Dulia, A., Lekholetova, M., & Liakh T. (2024). Peculiarities of social work with persons with disabilities as a result of the war. Naukovyi Chasopys Dragomanov Ukrainian State University. 98. 42–46. ISSN 2311-5491.

EGAP. (2020). *Program – Mobile suitcases.* Retrieved from https://egap.in.ua/project/mobilni_valizi. Accessed 8 Sept 2024.

EGAP. (2022). Retrieved from https://egap.in.ua. Accessed 8 Sept 2024.

European Commission. (n.d.). *Europe's digital decade: Digital targets for 2030.* Retrieved from https://commission.europa.eu/strategy-and-policy/priorities-2019-2024/europe-fit-digital-age/europes-digital-decade-digital-targets-2030_en. Accessed 8 Sept 2024.

Faidiuk, O. V., Lekholetova, M. M., & Liakh, T. L. (2022). Professional self-improvement of social workers in working with female combatants: The potential of the MOOC. In *E-learning in the transformation of education in digital society* (pp. 178–190). https://doi.org/10.34916/el.2022.14.13

Gunko, S., Liakh, T., & Karpenko, O. (2023). Public space as a tool of community cohesion. *Humanitas, 2*, 218–224. https://doi.org/10.32782/humanitas/2023.2.32

Gunko, S., Liakh, T., Vainola, R., & Lekholetova, M. (2024, May 24). Features of providing social services in rural communities. In *Society, integration, education. Proceedings of the international scientific conference, volume I* (pp. 652–664). https://doi.org/10.17770/sie2024vol1.7899

He, T., Huang, C., Li, M., Zhou, Y., & Li, S. (2020). Social participation of the elderly in China: The roles of conventional media, digital access and social media engagement. *Telematics and Informatics, 48*, 101347.

Ihnatenko, K., & Sadzaglishvili, S. (2023). The digitalization of social services in response to the war in Ukraine. *Social Work and Education, 10*(3), 276–289. https://doi.org/10.25128/2520-6230.23.3.2

Karkach, A., & Semigina, T. (2023). Digital competence of older people: Specifics, proficiency criteria and development programme. *Humanitas, 2*, 50–56. https://doi.org/10.32782/humanitas/2023.2.8

Kolosok, A., Martyniuk, T., & Melnyk, O. (2023). Digitalization of social service delivery. *Humanitas, 5*, 134–143. https://doi.org/10.32782/humanitas/2023.5.17

Kulyna, H., & Nalukova, N. (2020). Informatization of social providing as a main element in the formation of digital economy. *World of finance., 1*(62), 95–106. https://doi.org/10.35774/sf2020.01.095

Liakh, V. (2018, June 19). Methodology for assessing the transparency of local budgets. *East Europe Foundation.* Retrieved from http://auc.org.ua/sites/default/files/5_lyah_methodology_presentation_kk.pdf Accessed 8 Sept 2024.

Martin-Shields, C. (2021). Digitalization in displacement contexts: Technology and the implementation of the global compact on refugees. *Reference paper for the 70th anniversary of the 1951 refugee convention.* (pp. 1–19).

Open Budget. (n.d.). *State web portal of the budget for citizens.* Retrieved from https://openbudget.gov.ua

Preiß, C. (2019). A global information society? Why digital services for refugees often fail to achieve their objectives. *German development institute/Deutsches Institut für Entwicklungspolitik (DIE).* (The Current Column of 13 May 2019).

Punie, Y., Brecko, B., & Ferrari, A. (2013). *DIGCOMP: A framework for developing and understanding digital competence in Europe.* Publications Office of the European Union. Retrieved from https://bit.ly/3OHWAyX. Accessed 8 Sept 2024.

Shved, O., Liakh, T., Spirina, T., & Lekholetova, M. (2024). Psychosocial support for victims of sexual violence during the war in Ukraine: Challenges for social work. *Journal of Human Rights and Social Work, 9*, 464–473. https://doi.org/10.1007/s41134-024-00336-w

Spending. (n.d.). Unified web portal for the use of public funds. Retrieved from https://spending.gov.ua/new

SPOKU. (n.d.). SPOKU – A system for receiving and processing requests for local governments and a self-service environment for residents. Retrieved from https://spoku.ee/ulevaade. Accessed 8 Sept 2024.

Verkhovna Rada of Ukraine. (2019). *On social services: Law of Ukraine* No. *2671-VIII.* Retrieved from https://zakon.rada.gov.ua/laws/show/2671-19#Text. Accessed 8 Sept 2024.

Vovk, N., & Markovets, O. (2022). Digitalization of public administration: EU experience and prospects for Ukraine. The formation and peculiarities of the implementation of the European Union's Eastern policy. (pp. 71–86). https://doi.org/10.30525/978-9934-26-224-1-5.

Yehorova-Lutsenko, T. P. (2020). Digitalization of the social services system in Ukraine. *Law and Innovative Society, 2*(15), 80–85. https://doi.org/10.37772/2309-9275-2020-2(15)-13

Chapter 18
Tools for Teaching the R Programming Language to Bachelors of Computer Science in the Period of Distance Learning

Olha V. Korotun, Tetiana A. Vakaliuk, and Artur M. Makhno

Abstract The article is devoted to the issue of teaching the elective discipline 'Programming in R' for bachelor of computer science in the period of distance learning, which has become relevant for Ukrainian students and teachers during the full-scale invasion. The educational process at universities must continue even if many participants are far from each other. Defining tools for statistical analysis and data visualisation arise as the IT industry becomes increasingly popular. The authors emphasise the need to include the R programming language in the educational process and describe its widespread use in large companies to solve various problems. A model of studying the discipline 'Programming in R' by a bachelor of computer science in the period of distance learning has been developed since this form of organisation and construction of the educational process provides other options for communication between participants, studying the educational material and its assimilation than full-time education. Thus, this model includes the following blocks: the purpose, objectives and stages of studying this discipline, types of activities and teaching methods, teaching aids, and the topics of content modules. When studying this discipline, using the described online compilers and the RStudio environment is suggested. In order to deepen their knowledge and skills in the R language, bachelors of computer science are encouraged to take one of the presented courses of their choice.

Keywords R programming language · ICT · Tools · Bachelors of computer science · Distance learning

O. V. Korotun · A. M. Makhno
Zhytomyr Polytechnic State University, Zhytomyr, Ukraine

T. A. Vakaliuk (✉)
Zhytomyr Polytechnic State University, Zhytomyr, Ukraine

Institute for Digitalisation of Education of the National Academy of Educational Sciences of Ukraine, Kyiv, Ukraine

Kryvyi Rih State Pedagogical University, Kryvyi Rih, Ukraine

Academy of Cognitive and Natural Sciences, Kryvyi Rih, Ukraine

E. Smyrnova-Trybulska et al. (eds.), *E-Learning and Enhancing Soft Skills*,
https://doi.org/10.1007/978-3-031-82243-8_18

18.1 Introduction

The R programming language is becoming increasingly popular and widespread in modern technology, as many IT companies use it to create business applications. R is becoming a popular programming language among data analysts and scientists, as it helps to import and clean data and perform efficient analyses. According to the PYPL index, which determines the popularity of a programming language based on the number of queries in Google search, as of November 2023, R was among the ten most promising programming languages (PYPL, 2023). The key features of this language include the following: a subject-oriented language specialising in statistical data analysis; an interpreted language (you can run the code without any compiler); free and open source; efficient data processing; provides extensible graphical capabilities (provides libraries that help create high-quality interactive graphics); integration with other technologies, community support.

The R programming language is gaining relevance and popularity in higher education institutions, as it allows students to acquire the necessary theoretical knowledge and practical skills in statistical data analysis, forecasting, data visualisation, and machine learning.

The study **aims** to identify the means of teaching the R programming language to bachelors of computer science in distance learning form. The **objectives** of the study are to describe the capabilities of R language tools, to present the use of selected R language tools by a bachelor of computer science in data analysis and visualisation, and to experimentally test the constructed model of studying the elective discipline 'Programming in R' by bachelors of computer science.

18.2 Literature Review

Let us highlight the areas of human activity in which the R programming language has been implemented. Let us start with a description of its use in medicine; for example, work (Malyutina et al., 2023) discusses the use of the DRDA (Dose-Response Data Analysis) package for performing dose-response data analysis, which is designed to facilitate the analysis of dose-dependent data by implementing efficient and accurate functions. With the help of DRDA, it is possible to fit the constructed models using the least squares method, perform fit tests and select them; (Kumwichar, 2023) emphasises the need to study data analysis for research purposes and use the R language for statistical analysis, for convenience, the R language is used in the Jupyter Notebook environment, and this study also collected feedback from students and their teachers on the use of R, identified problems that arose during its study, and showed how they could be solved.

For example, airlines are doing everything possible to improve their efficiency and productivity (Lee, 2023). Measuring Airline Productivity and Efficiency is a data analysis performed using R that identifies and explains the sources of airline efficiency and helps achieve these goals through state-of-the-art measurement techniques. The results provide airline managers with critical information needed to create better policies and avoid inefficient practices.

For accounting, (Calderon et al., 2023) describes the possibility of using the R programming language, which includes working with various data sets and analysing economic data.

For the analysis of geographic data, the programming language and environment for working with R are used because it is possible to work with various spatial and non-spatial data on one platform (Wimberly, 2023). The application of fundamental methods of processing and visualising tabular, vector and raster data is presented through a series of practical examples, and there are also case studies that combine different types of data to solve more complex problems. Key features of the book include a focus on developing practical tasks for processing and integrating multiple geospatial data sources in R and an example-based approach that teaches R programming and data science concepts through real-world processes related to climate, land cover and land use, and natural hazards.

The use of R in education to teach data science disciplines is discussed in (Estrellado et al., 2020). This is a kind of reference book for learning data science in education. The book answers the following questions: what does a data scientist do in education, how to start learning R, a popular open-source statistical programming language, what does a data analysis project in education look like, etc. The book uses a 'learning by doing' approach and offers eight step-by-step instructions showing data analysis from start to finish, along with code that the reader can practice with.

Teaching mathematics through programming has cognitive benefits, improves practical skills, and improves thinking and problem-solving (Amimour et al., 2021). For this purpose, we use R, a free software environment and one of the most popular programming tools, to solve various mathematical problems by programming. We also describe the use of R to solve some problems posed in complex numbers and consider some exercises from a mathematics textbook.

The basic concepts and tools necessary for developing statistical thinking, descriptive, inferential, and exploratory statistical methods can be learned in a statistics textbook (Heumann & Schomaker, 2023). The text is primarily intended for undergraduate students in data analysis, business administration, social sciences, medicine, politics, and macroeconomics. It contains many examples, exercises, solutions with computer code in the statistical programming language R, and additional material that will allow the reader to adapt the methods to their applications quickly.

Teaching Statistics and Data Analysis with R (Tucker et al., 2022) describes the attitudes and experiences of 672 undergraduate students using a custom-designed online textbook as part of a 10-week introductory statistics course. Students expressed negative attitudes and concerns about R at the beginning of the course. However, most developed more positive attitudes after reading the course materials, regardless of demographic characteristics or previous programming experience.

Work with the Roxygen package is described in (Vidoni & Codabux, 2023). This package allows you to create documentation in the source code for the functions and objects of the package being developed because, with it, users will know how to use the created package. Roxygen is aimed at improving standards in R documentation.

Thus, from the literature analysis, we can draw the following conclusions: R is one of the programming languages used for statistical data analysis by scientists in various fields (medicine, economics, education, etc.) (Vidoni & Codabux, 2023; Marcinkowska & Roszak, 2024), using different packages of this language, different categories of data in the data set, different language instructions, etc. Many articles and books (textbooks, manuals) have been written on the peculiarities of using the R language, which can be used to develop a data analysis course for students of higher education institutions.

18.3 Research Methodology

The following research methods were used in the study: analysis of scientific papers on the use of the R programming language and its capabilities in various fields; analysis and comparison of openly available R language learning tools that allow students to acquire theoretical knowledge and practical skills in mastering the R language for data analysis and visualisation.

A pedagogical experiment was also conducted to determine the level of knowledge of a bachelor of computer science who studied the elective discipline 'Programming in R' according to the model.

The experiment was implemented at the Zhytomyr Polytechnic State University, the participants of which were applicants for higher education in the speciality 122 'Computer Science' of the educational programme 'Computer Graphics and Game Development' of academic groups of the 3rd year of study, the total number of applicants in these groups was 95 people.

18.4 Results

The elective discipline 'Programming in R' will help bachelors of computer science acquire theoretical knowledge and practical skills in the R programming language on the following topics: statistical analysis, manipulation, visualisation and data forecasting.

The objectives of this course are to master the basics of R programming and learn how to use R to solve various data analysis problems.

To achieve these objectives, students must complete the following learning stages:

1. Familiarity with the R language: At this stage, a bachelor of computer science should become acquainted with the R language's basic concepts and terms, as well as its syntax and semantics.
2. Mastering the basics of R programming: at this stage, a bachelor of computer science should learn how to write simple programs in R using loops, conditional statements, and functions.
3. Using R to solve problems: At this stage, a bachelor of computer science should learn to use R to solve various problems, such as data entry and processing, statistical data analysis, data visualisation, and machine learning.

To achieve these tasks effectively, a bachelors of computer science should perform the following activities:

- read and process the study material and complete all tasks offered within the study material;
- perform practical tasks of laboratory work to consolidate the acquired knowledge and skills;
- solve independent tasks to test the acquired knowledge and skills.

The following teaching methods were used to support and motivate bachelors of computer science in the process of learning R:

- interactive teaching methods, such as exercises and laboratory work, which allow bachelors of computer science to be actively involved in the learning process and better absorb the material;
- providing feedback: the instructor should provide bachelors of computer science with feedback on their work so that they can understand their strengths and weaknesses and improve their performance;
- creating opportunities for collaboration: collaboration with other students can help students understand the material better and get additional support.

Let us present a model of studying the elective discipline 'Programming in R' a bachelor of computer science (Fig. 18.1).

In the course of studying this discipline, bachelor of computer science is introduced to the following topics: the concept of 'data analysis' and the connection between statistics and data analysis in the R programming language (data types, data measurement levels, data matrices, frequency tables, description of the central tendency, quartiles and interquartile range, variance and standard deviation); data manipulation in R; implementation of fundamental issues of probability theory and data analysis and classical distributions; descriptive statistics in R; inferential statistics in R; covariance and correlation, linear regression in R; central limit theorem in R; classification and clustering in R; data visualisation in R.

This course was conducted remotely due to the COVID-19 pandemic and the war in Ukraine, bachelors of computer science were offered to use interactive tools, such as online compilers and RStudio software development environments (IDEs), when learning R (Fig. 18.2), which will help students learn this programming language quickly and easily.

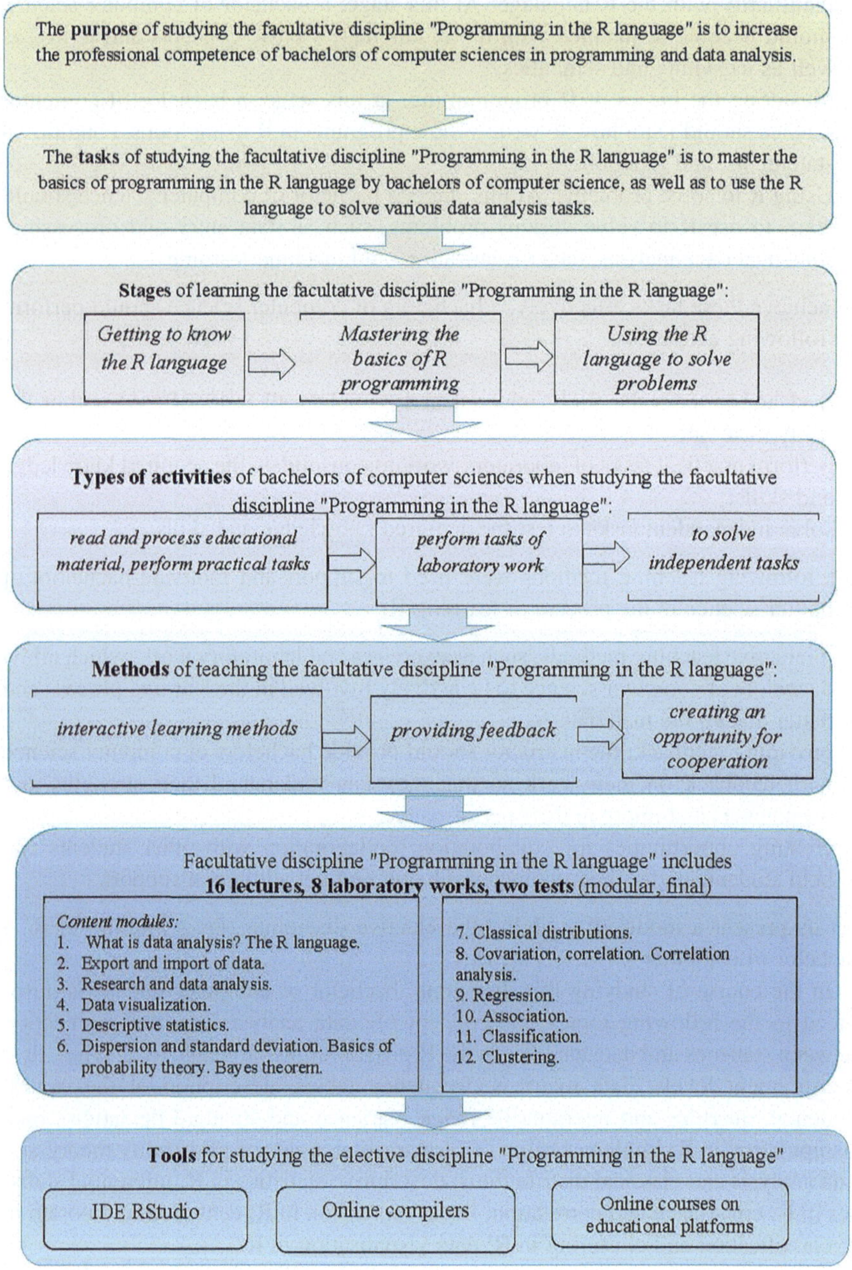

Fig. 18.1 Model of study by bachelors of computer sciences of the elective discipline "Programming in R"

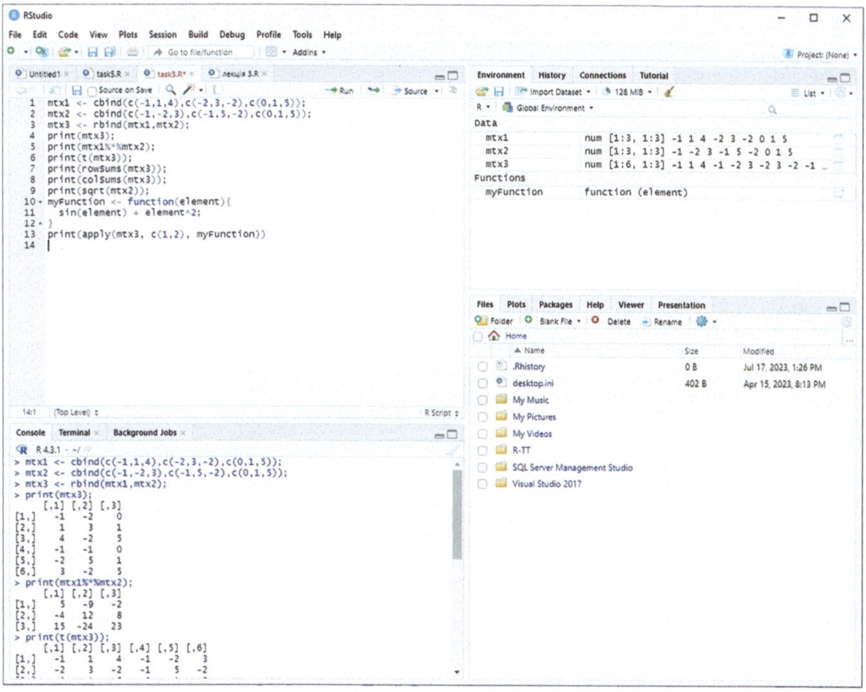

Fig. 18.2 Programming environment RStudio

Let us take a closer look at which online compilers are best for learning this language. They do not require any time to search, download, install, and configure the environment and are very easy to launch and use with a web browser and an active Internet connection.

1. https://rdrr.io/snippets/. Rdrr includes more than twenty-five thousand R packages. When working with this compiler, students can read and study the documentation about R packages and view them using the corresponding menu items. The user can find a description of the required R package, consisting of a specific set of functions.

The compiler page offers a list of recently updated packages. The R documentation for this compiler is divided into the following sections: base (essential functions), datasets (built-in datasets), graphics (functions for drawing graphs), grid (functions for working with tables), stats (statistical functions), etc. All R packages are divided into the following repositories: the central repository (CRAN packages) for packages written in R (Comprehensive R Archive Network, CRAN); the repository of packages for analysing biological data (Bioconductor packages); R-Forge repository (R-Forge packages), which complements CRAN and provides additional opportunities for R package developers, although CRAN is the principal repository, R-Forge offers a more flexible version control and package development system, R-Forge

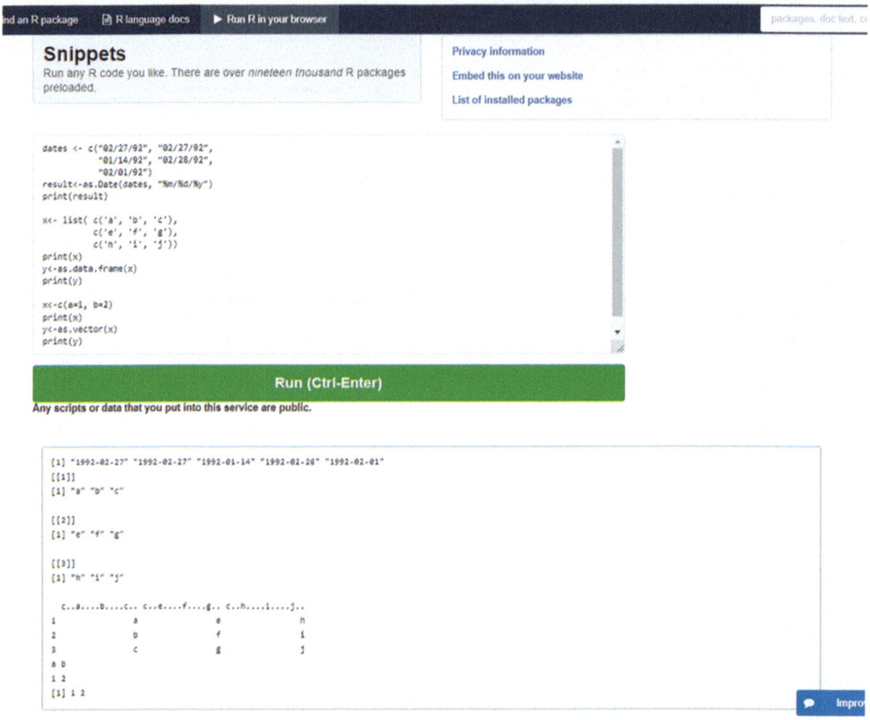

Fig. 18.3 Online compiler Rdrr

often hosts packages that are still under development and not ready for inclusion in CRAN, which allows developers to receive feedback from the community and improve their packages before their official release; GitHub repository (GitHub packages) is designed to store the finished code of a package, its history of changes and improvements, and allows developers to work in a team. Users can test their R programs, and the compiler's workspace is divided into two parts: the upper one has a window for writing code, and the lower one is a window for viewing the results.

To summarise, this compiler has a simple and intuitive interface, does not contain many menu items, but provides extensive information about packages. It allows visualisation of the results obtained during data analysis, does not require registration, and is available in all browsers used by students and teachers in the educational process (Fig. 18.3).

2. https://www.programiz.com/r/online-compiler/. Programiz is an online compiler for many programming languages, including R. It allows you to write code in R and run it online, selecting the programming language you want to learn on the compiler's main page. For each programming language presented, we offer theoretical material with small videos and practical examples, which we recommend that beginners use to help them learn the required language faster.

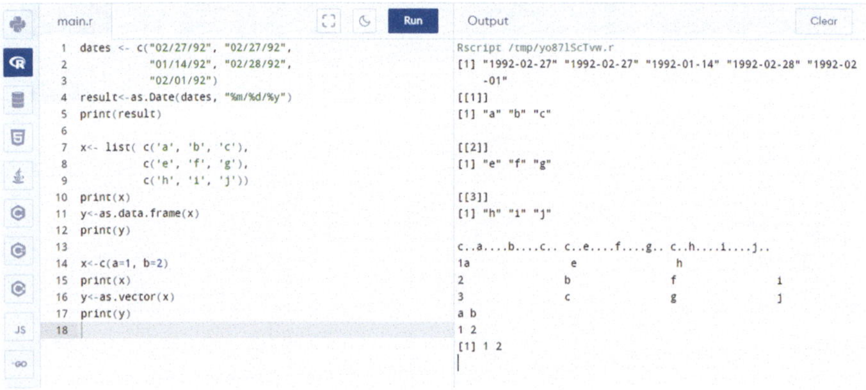

Fig. 18.4 Online compiler Programiz

The structure of the R tutorial is designed to guide the user through the step-by-step process of learning the programming language. It consists of the following sections: introduction (getting started with R, working with comments, variables, constants, data types, etc.), thread management (logical expressions, loops, functions), data structures (strings, vectors, matrices, arrays, lists, data sets, etc.), data visualisation (functions for building different types of graphs and customising them), data manipulation (reading and writing data, finding measures of central tendency, percentiles, etc.). The compiler also provides a link to the official R documentation. Here, students can practice creating R projects in real-time.

Here are some other features of this compiler: syntax highlighting, code auto-completion, a programme window and a result window contained in one workspace, and support for many R language functions is provided. The menu of this compiler is not complicated, easy to understand and use, and can be quickly mastered. You can start working immediately, writing code to solve a specific task (Fig. 18.4).

3. https://www.mycompiler.io/online-r-compiler. MyCompiler is an online compiler that allows you to write, edit, compile, and run code in the programming language of your choice. It currently supports 16 programming languages.

This online compiler is also suitable for students just starting to learn R and trying to write program codes in it. To facilitate this work, the compiler provides auto-completion and syntax highlighting. Figure 18.5 shows the window of the MyCompiler for R compiler; for users' convenience, the workspace is divided into two parts: the left one contains the code, and the right one shows the result obtained after its launch. MyCompiler for R allows you to include various libraries in your code: ggplot2, dplyr, tidyr, and data.table, among others. It also provides integration with the GitHub repository, aimed at importing and exporting code snippets from GitHub. Unfortunately, this online compiler has no sharing or collaborative coding option; it is only used to run code snippets and view the results. The user's workflow consists of the following steps: selecting a programming language,

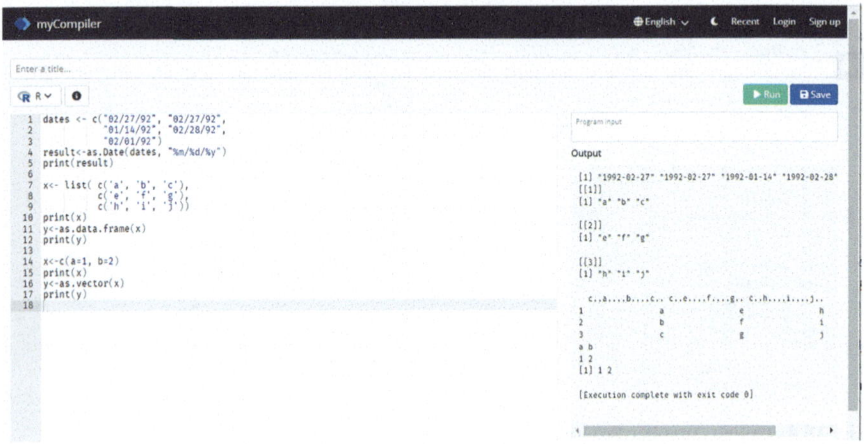

Fig. 18.5 Online R Language IDE MyCompiler

writing the required code snippet, running it, and viewing and interpreting the results.

4. https://www.online-ide.com/online_r_compiler. Online-IDE is an online compiler based on the ACE code editor that both beginners and more experienced developers can use. It is designed for programming in the following languages: C, C++, Python 3, R, Java, PHP, Ruby, Go, Lang, and Bash. Therefore, it can also be used in the educational process to learn the R language, allowing the user to create code, run it, and view the results (Fig. 18.6). This compiler also provides the user with a variety of features for editing code (syntax highlighting, auto-completion, detection of errors in the code), it provides integration with services such as GitHub, Bitbucket, etc., supports work with large projects, and allows users to establish cooperation to work on one project.

Suppose we describe the functionality of the Online-IDE compiler. In that case, we should start with the ability of users to open existing R code files (.r format), create new files, save them in the desired format on a PC, and share them with other users by receiving a link to this code fragment for 24 h, 1 month, or 6 months. Code files are not stored on the server. The code is sent to the server for execution and cleared after completion. The shared code will be stored on the server until the selected validity period expires. The workspace of this compiler is divided into the upper part, where the code is written, and the lower part, where the result is viewed; the code is run by pressing the “Run” button. Such a compiler will be convenient for introducing into the educational process when learning the R language, testing it, demonstrating the code to other students and working together on projects. Online compilers are handy tools for lecturers to quickly present new course material and

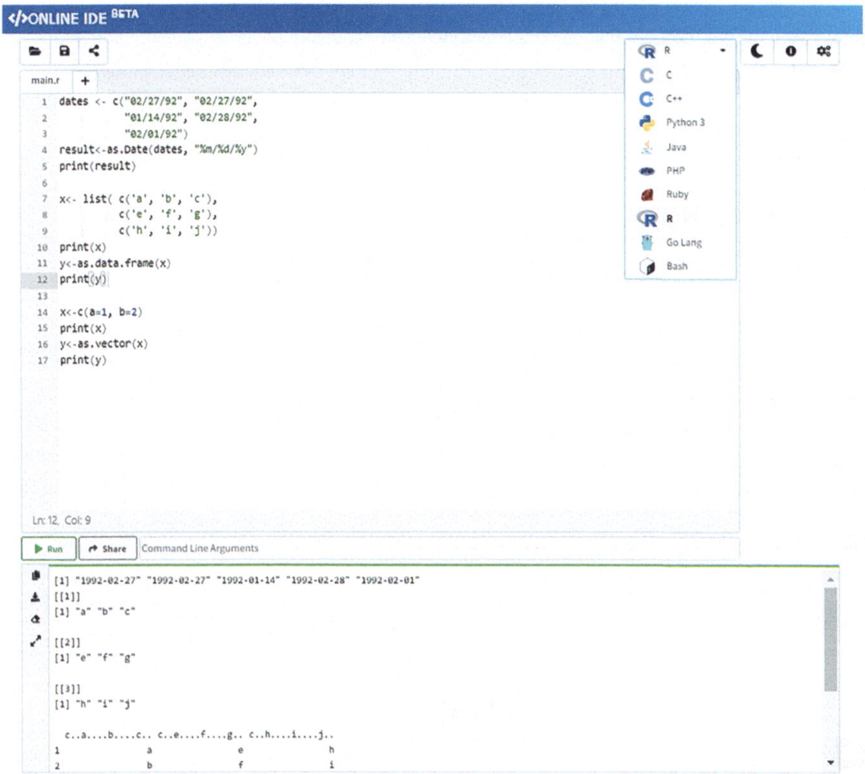

Fig. 18.6 Online compiler Online-IDE

demonstrate results, which aims to improve the efficiency of lecture presentation and student learning during the lecture.

Let us describe how online compilers can be used in R programming language lectures:

- Explain new material and demonstrate functions or libraries. The lecturer can use the online compiler to show students how to use new functions or libraries they are learning.

For example, when studying the topic 'Variance and standard deviation', you can use the online compiler to demonstrate an example of using the functions var() and sd() and explain how they work (Fig. 18.7).

Alternatively, the lecturer can show students how to use the lm() function and interpret the results to build a linear regression model (Fig. 18.8).

- Problem-solving. The lecturer can use the online compiler to show students how to solve a specific problem using R. For example, let us determine the density of

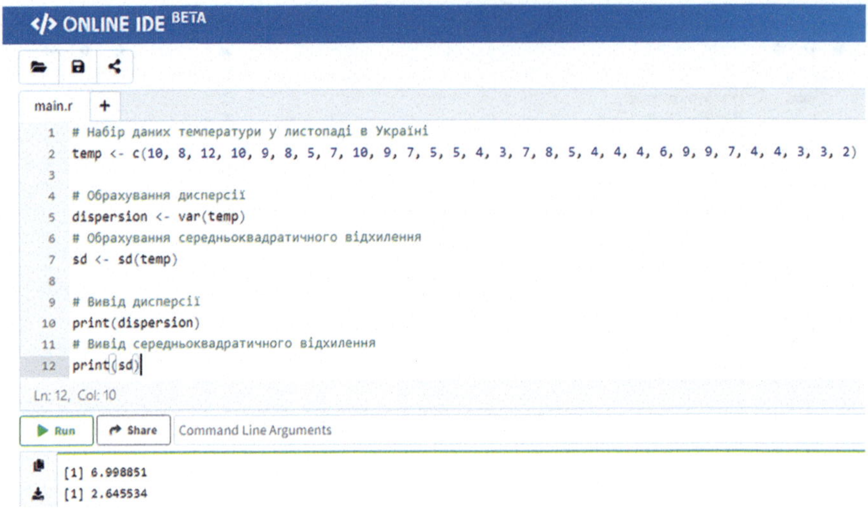

Fig. 18.7 Calculation of variance and standard deviation in the online compiler Online-IDE

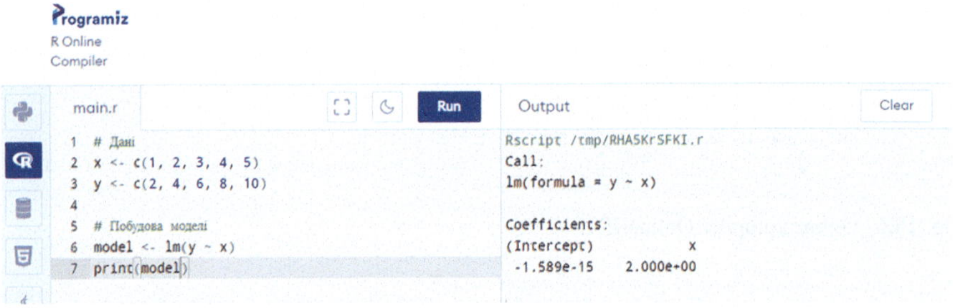

Fig. 18.8 Building a linear regression model in the online compiler Programiz

Table 18.1 Cost of production and market share of the companies

Enterprise	A	B	C	D	E
Cost of production (variable x)	8	3	1	4	2
Market share (variable y)	3	5	6	7	8

the relationship between companies' cost of production and market share, given a dataset of the positions occupied by the top 5 companies in an industry (Table 18.1).

To do this, let us calculate the rank correlation coefficients (Fig. 18.9).

We conclude that all the calculated rank correlation coefficients show an inverse relationship, i.e., as the company's production cost increases, its market share decreases.

Provide students with the opportunity to practice writing code. The lecturer can give students small tasks they must complete using an online compiler and discuss

Fig. 18.9 Calculation of rank correlation coefficients in the online compiler OneCompiler

Tools for collaboratively writing code in R	**RStudio Team Collaboration** is a part of RStudio IDE and allows users to easily share their projects with other users.
	GitHub is a popular website for managing code repositories, allowing users to create shared code repositories.
	GitLab is similar to GitHub.
	Bitbucket is similar to GitHub and GitLab.
	RStudio Cloud allows users to create collaborative RStudio projects in the cloud.
	RPubs allows users to publish their R projects to the Internet.

Fig. 18.10 Tools for collaboratively writing code in R

the results in class. For example, the lecturer may assign students to build a linear regression model for several data sets.

Laboratory work on the R programming language could be performed in RStudio, and a file with completed tasks could be sent to the lecturer, or tools for collaborative writing of R code could be used, and the file with the tasks could be shared with the lecturer. It was suggested that the bachelor of computer science use such tools for collaborative code writing in R (Fig. 18.10).

In order to deepen their knowledge and skills in R, bachelors of computer science were offered to choose one of the MOOCs and take it. This task was set during the students' independent work, and they were asked to choose one of the following courses.

1. Let us start the review of courses with the MOOC Prometheus, which contains both free and paid courses. Moreover, paid courses in various fields, such as IT, business, English, public service, etc. This platform offers a free course called 'Data Analysis and Statistical Inference in R', which requires a registered user status to access its training materials. Completing this course will provide students with the knowledge to understand data. The course is based on mathematical statistics and the modern programming language R, which is the toolkit that data science professionals need to be proficient in. The course covers the basics of obtaining results based on statistical inference using R. If we analyse the topics of this course, it is an introductory course for data analysts.

This course consists of five weeks of training, including short and informative videos and text files. Let us take a look at each of the weeks: In Week 1, the student will learn what data analysis is, get acquainted with the basic concepts of data analysis, what the data analysis process consists of, what categories data is divided into, as well as the levels of their measurement, learn about data matrices and frequency tables, central tendency, and how to visualise data in R; Week 2 is devoted to descriptive statistics and data cleaning with R; Week 3 is a review of the basics of probability theory (trials, events, probability of events), Bayes' theorem, classical distributions (binomial, standard), concepts of covariance and correlation, and linear regression, in Week 4, inferential statistics, the central limit theorem, and the construction of confidence intervals are presented; in the last week of the course, Week 5, statistical hypothesis testing for the mean and proportion is discussed, and a step-by-step plan for hypothesis testing is provided, as well as the errors that can occur. At the end of the course, the student has to pass an exam consisting of 20 questions, and if the student scores more than 60% for the entire course, including the points for laboratory work, tests and the exam, he or she can receive a certificate of completion of this course.

During the course, students will be able to conduct primary statistical analysis of data, formulate hypotheses and evaluate them based on the collected data.

2. The MOOC Alison offers over 4000 free online courses in nine different categories, including R, which are numerous depending on the user's initial skills. The platform offers two types of courses: the first type provides for a certificate (average duration of 2–3 h), and the second type provides for a diploma (average duration of 8–10 h).

Let us look at a few courses on the R programming language on this platform. The "Introduction to R" course will provide students with an introduction to the basics of R, namely installing R and RStudio, workspaces and basic data types, and working with variables, data structures and operators, data frames and tables. This course consists of four modules. Through this course, the student will significantly improve their statistics, data analysis, and programming skills in R.

Upon completion of these certified courses, you can receive an official diploma, for which you need to pass 80% of the course or more. Alison offers three types of diplomas upon completion of the courses: a digital certificate—a PDF certificate that can be downloaded for a fee; a certificate—a physical version of the certificate with an official brand and security mark sent to the user with free shipping; a framed certificate—a physical version of the certificate with an official brand and security mark in a stylish frame sent with free shipping.

Consider the R Programming for Data Science course, which is an introductory course that starts with a demonstration of the importance and benefits of the R programming language and the key topics to learn that are necessary for further understanding and application in data analysis. This course will teach the student how to work with and process data in the R environment. It will use the dplyr package, one of the main packages in R, to get acquainted with the concept of 'big data', the complexity of the R language and the RStudio software that will be used to write the program code. The course includes many practical exercises and example demonstrations designed to help you better understand the language. Completing this course will allow students to use R in real-world analysis, analytical tasks, and research projects to support fact-based decision-making.

The course consists of three modules: the first module is devoted to the basics of the R programming language, an overview of Data Science concepts, installation of R and RStudio, studying data types, working with variables, vectors, matrices, data structures, etc.; the second module is devoted to managing streams and functions in R (logical operators, conditional operators, functions, etc.); the third module is devoted to data manipulation and visualisation in R (data retrieval and cleaning, graphing using various packages, etc.).

3. The MOOC Udacity has a list of R courses, including the course 'Programming for Data Science with R'. This is a free intermediate-level course for learning data analysis with R. This course consists of 10 lessons, including one final project. The course begins with explaining EDA (exploratory data analysis) and its goals. In the following lessons, you will learn about the basics of R programming and working with variables. After that, you will learn about linear regression models and predict diamond prices using other variables in a diamond dataset. Finally, there is one last lesson in which you have to use R and apply exploratory data analysis techniques to investigate the relationships of one variable with multiple variables. Each lesson comes with a set of assignments.

In most cases, free courses need more quality content, but this course is full of knowledge and covers some in-depth concepts with tests, practice sets, and projects. At the end of the course, the user will not receive a certificate but must register.

Another free course on the Udacity educational platform for learning data analysis and visualisation is the Data Analysis and Visualisation course. This course uses the R programming language. It starts with an introduction to R programming and then teaches students some of its complex concepts.

After that, the user will learn about the basics of data visualisation and ggplo. Each lesson includes various quizzes that will help users test their knowledge.

This course covers logistic regression, linear regression, and regularisation. Overaiturse is more practical in nature. The user will learn by working on problem sets. Unfortunately, it also does not require a certificate and registration is required.

4. The MOOC Coursera offers many courses on the R language. Access to the course materials is usually free, but you will have to pay for the certificate.

One of the courses is called 'R Programming'. To check out this course for free, click the 'Register for free' button. A new pop-up window will appear, asking you to select the months of your subscription. However, on the same page, in the lower left corner, there is an option to 'Audit the course'. Select this option, and you will be redirected to the course materials for free.

This course has a 4 week training plan. The first week covers the basics of the R program and practical exercises. In the second week, you will learn about control structures, functions, dates, and time in R programming.

Week 3 covers looping functions and debugging tools in R. The last week is dedicated to modelling and profiling. You will learn the R profiler.

The course tests are locked for the free course.

Data Analysis with R Programming co—This course has a 5-week learning plan. In the first week, you will get acquainted with R and Rstudio. Next week, we will focus on other R programming topics such as vectors, lists, dates, times, general structure, and operating systems. The user will also learn about the tidyverse library.

Week 3 is more practical, describing how to work with data in data frames, clean up data, transform data, and organise data using R programming.

Week 4 covers data visualisation using R programming, and Week 5 will show you how to use R Markdown in RStudio. This course will help me understand how to analyse and visualise data using the R programming language.

Introduction to R Programming for Data Science—this course also has a 5-week learning plan. The first week is dedicated to the basics of R programming. In the second week, the user will learn about vectors, factors, lists, arrays, and matrices in R programming; in the third week—about conditions, loops, functions, regular expressions, and date format in R; and the fourth week of training includes reading text files in R, how to write and save to files, and web scraping in R.

The last week is devoted to the capstone project, but if you register for this course for free, you do not get access to it.

To summarise the above, let us compare the elective discipline 'Programming in R'curriculum modules and the reviewed online courses that will help deepen the knowledge of the Bachelor of Computer Science in data analysis and the R programming language (Fig. 18.11).

The choice of a single course depends on the needs and budget of a Bachelor of Computer Science student. If you are looking for a free course, you can choose one of the first two options. If you want more detailed information and additional materials, you can choose the third option.

Thus, R is one of the essential tools that a bachelor of computer science should have to work in the field of data science. It allows them to competently and correctly analyse data, which in turn allows them to make quality forecasts, make informed decisions, assess risks, etc.

We will describe the steps of a pedagogical experiment to determine the level of knowledge of a bachelor of computer science who studied the elective discipline

Fig. 18.11 Comparison of modules of the work program of the educational discipline "Programming in the R language" and online courses

'Programming in R'according to the model presented above and those students who studied according to the classical programme of studying any programming language (including R).

Table 18.2 The results of the pedagogical experiment

	Empirical frequencies		Empirical proportions		Accumulated empirical frequencies		
Level of knowledge	CG	EG	CG	EG	CG	EG	Difference
High	5	6	0,11	0,13	0,11	0,13	0,02
Sufficient	18	17	0,37	0,36	0,48	0,49	0,01
Average	16	14	0,33	0,29	0,81	0,78	0,02
Low	9	10	0,19	0,22	1	1	0

In the first step, control and experimental groups were formed: the control group (CG) consists of bachelors of computer science who studied the discipline according to the traditional method; the experimental group (EG) consists of bachelors of computer science who studied the R language according to the presented model.

To study the discipline 'Programming in R', students need knowledge of programming and mathematical statistics, so the division of Bachelor of Computer Science into CG and EG was based on the grades of previously studied related disciplines, namely: 'Fundamentals of Programming' and "Probability Theory and Mathematical Statistics". When conducting a pedagogical experiment, it is mandatory to check for equivalence of bachelor's of computer science samples in CG and EG, using the Kolmogorov-Smirnov consistency criterion.

The experiment involved 95 bachelors, and the experimental data fully satisfy the restrictions of the Kolmogorov-Smirnov criterion: the samples are large enough $n_1 = 48$, $n_2 = 47$; the levels of knowledge are ordered in descending order; the number of levels is greater than 3. Let us move on to the formulation of statistical hypotheses: H_0—bachelors of computer science from CG and EG do not differ in the level of knowledge in the disciplines 'Fundamentals of programming' and 'Probability theory and mathematical statistics'; H_1—bachelors of computer science from CG and EG significantly differ in the level of knowledge in the disciplines 'Fundamentals of programming' and 'Probability theory and mathematical statistics'. Let us demonstrate the table for calculating the criterion when comparing the levels of knowledge in CG and EG (Table 18.2).

The maximum difference between the accumulated empirical shares is 0.02. Let us calculate the empirical value of the Kolmogorov-Smirnov criterion using the formula:

$$\lambda_{emp} = 0,02 \cdot \sqrt{\frac{48 \cdot 47}{48 + 47}} = 0,09 \qquad (18.1)$$

The critical value of the Kolmogorov-Smirnov criterion is 0.14 ($\lambda^*_{kr} = 0.14$), then $\lambda_{emp} < \lambda_{kr}$. According to the Kolmogorov-Smirnov statistical criterion, there is reason to assert that there are no differences between the samples, so the null hypothesis (H_0) is accepted.

The next step of the pedagogical experiment is to determine the level of knowledge of a bachelor's degree in computer science in CG and EG. We will test the

Table 18.3 Students' marks for the control test

	Final assessment (number of students)	
	CG	EG
Low	10	6
Average	18	12
Sufficient	13	19
High	7	10
Total	48	47

Table 18.4 Table of empirical frequencies

	'there is an effect'		'there is an effect'	
Groups	number of students	%	number of students	%
CG	21	43%	27	57%
EG	29	62%	18	38%

hypotheses using Fisher's criterion*: H0—the assumption that there is no significant difference between the excellent and good level of students' knowledge of the discipline in the control and experimental groups, that is, the new teaching methodology did not affect the results; H_1—the assumption that there is a significant difference between the excellent and good level of students' knowledge of the discipline in the control and experimental groups, that is, the new teaching methodology affected the results. For this purpose, we will use students' grades in the control test (Table 18.3).

Let us present the actual table of empirical frequencies for two attribute values: if a student received an A or B grade, then 'there is an effect'; otherwise—'no effect' (Table 18.4).

Only the fractions corresponding to the observations for which there is an effect are used in the calculations. Applying Fisher's criterion* is possible because the experimental data fully satisfies the constraints. Using the formula (18.1), we calculate the empirical value of φ*:

$$\varphi^*_{emp} = (\varphi_1 - \varphi_2)\sqrt{\frac{n_1 \cdot n_2}{(n_1 + n_2)}} \approx 0,377 \cdot 4,87 = 1,837 \tag{18.2}$$

The Fisher's criterion* critical value is 1.64 ($\varphi^*_{kr} = 1.64$). The result is $\varphi^*_{emp} > \varphi^*_{kr}$, the empirical value of the Fisher's criterion* is in the zone of significance, which indicates differences in the two samples. According to the terms of the statistical hypotheses, H_0 is rejected, and H_1 is accepted. Therefore, in terms of the degree of homogeneity of the attribute' the level of students' knowledge based on the grades obtained for the final test in the elective discipline 'Programming in R', there is a discrepancy between the two samples. The reliability of the differences in the characteristics of the CG and EG according to Fisher's statistical criterion* at the significance level is less than 0.05.

Also after completing the elective course 'Programming in R', students were asked to answer four questions:

1. Are you interested in this area of study?
2. What advantages of learning a programming language have you identified for yourself?
3. Did you enjoy studying this discipline?
4. Leave your feedback about the discipline.

This discipline was studied by three groups of bachelors of computer science, the following results were obtained:

- first question: 84% of students are interested in this area of study and see further prospects for applying the knowledge and skills they have acquired in their future professional activities, but 17% of students do not see themselves;
- the second question summarises the answers received: free software and its user-friendly interface, easy-to-learn programming language, working on the main topics of analysis (descriptive statistics, basic types of data distributions, correlation, covariance, regression, classification and clustering) and data visualisation;
- the third question received mostly favourable responses (87%), but some students did not like the course, mainly those who would not associate their work with this area in the future;
- The fourth question received mostly positive responses, but some topics students would like to add are data association and data summarisation from the section on unsupervised learning and the use of convolutional neural networks.

After analysing the answers, we concluded that studying such a discipline is mandatory for future bachelor's of computer science and is interesting for students, so it was included in the list of mandatory educational components. We also slightly expanded the topics for study and increased the number of hours for laboratory work.

18.5 Discussion

The results of this study demonstrate the effectiveness of teaching the R programming language to bachelor's students in computer science using the proposed model with special tools.

These findings align with other research on incorporating programming into mathematics and data-oriented science curricula. For example, Amimour et al. described modernizing secondary-level mathematics teaching through R programming and found it enhanced student learning (Amimour et al., 2021). In medical fields, Kumwichar used Jupyter notebooks and R to teach epidemiological data analysis to graduate students and reported positive outcomes (Kumwichar, 2023).

The importance of R programming is also highlighted in other domains like accounting (Calderon et al., 2023) and geographic data science (Wimberly, 2023).

Integrating R into these curricula enables students to develop valuable statistical analysis, data manipulation, and visualization skills that are increasingly in demand.

Proposals for data science courses and integration of R in other fields include: Heumann et al. wrote an introductory statistics and data analysis textbook with examples in R (Heumann & Schomaker, 2023); Marcinkowska & Roszak discuss using interactive R Shiny apps for medical data visualization to promote learner independence (Marcinkowska & Roszak, 2024); Tucker et al. examined student experiences in an introductory statistics course that taught R programming (Tucker et al., 2022).

The positive outcomes of this study and related work suggest that including an R programming course, either as a compulsory or elective subject, would benefit bachelor's programs in computer science and other data-intensive fields. Exposing students to R equips them with programming skills directly applicable to real-world data analysis problems.

However, challenges remain, such as determining the optimal curriculum structure, supporting students with diverse programming backgrounds, and providing adequate computational resources.

18.6 Conclusions

To summarise, let us outline the features of using the R programming language in teaching bachelor of computer science, namely: R is free for use in education; various types of software (online, stationary on a PC) for using R are available; it is easy and understandable to learn; there is much free literature on R; to expand the solution of applied problems, there are many different packages in R; focused on statistical data analysis; contains many different tools for data visualisation and presentation; allows solving problems of artificial intelligence; popular in use in various spheres of human life (economics, business, management, medicine, etc.).

When studying the R programming language, the goal is defined, which includes the acquisition of theoretical knowledge and practical skills by a bachelor of computer science in the programming language for statistical analysis, manipulation, visualisation and forecasting of data, as well as the task, namely, acquaintance with the language, its basics of mastering and using in data analysis. Methods of learning R include interactive teaching methods, providing feedback, and organising cooperation between participants in the process. The tools include the RStudio IDE, online compilers, and online courses.

References

Amimour, A., Benbessai, L., Becharef, K., & Mokhenfer, H. (2021). Modernize the teaching of mathematics at the secondary level through R programming. *Annales Universitatis Paedagogicae Cracoviensis | Studia Ad Didacticam Mathematicae Pertinentia, 13*, 169–183. https://doi.org/10.24917/20809751.13.8

Calderon, T. G., Hesford, J. W., & Turner, M. J. (2023). A framework for integrating "R" programming into the accounting curriculum. In *Advances in accounting education: Teaching and curriculum innovations* (Vol. 26, pp. 209–232). Emerald Publishing Limited. https://doi.org/10.1108/s1085-462220220000026012

Estrellado, R. A., Freer, E. A., Mostipak, J., Rosenberg, J. M., & Velásquez, I. C. (2020). *Data science in education using R*, 292 p.

Heumann, C., & Schomaker, M. (2023). *Shalabh introduction to statistics and data analysis: With exercises, solutions and applications in R* (2nd ed.). Springer, 584 p. https://doi.org/10.1007/978-3-031-11833-3

Kumwichar, P. (2023). Enhancing learning about epidemiological data analysis using R for graduate students in medical fields with Jupyter notebook: Classroom action research. *JMIR Medical Education, 9*, e47394. https://doi.org/10.2196/47394

Lee, B. L. (2023). *Productivity and efficiency measurement of airlines: Data envelopment analysis using R*. Elsevier, 259 p.

Malyutina, A., Tang, J., & Pessia, A. (2023). DRDA: An R package for dose-response data analysis using logistic functions. *Journal of Statistical Software, 106*(4), 1–26. https://doi.org/10.18637/jss.v106.i04

Marcinkowska, J., & Roszak, M. (2024). Interactive RShiny reports – Independence and autonomy in medical data visualization. *International Journal of Research in E-Learning, 10*(1), 1–24. https://doi.org/10.31261/IJREL.2024.10.1.03

PYPL PopularitY of Programming Language, Nov. 2023. https://pypl.github.io/PYPL.html

Tucker, M. C., Shaw, S. T., Son, J. Y., & Stigler, J. W. (2022). Teaching statistics and data analysis with R. *Journal of Statistics and Data Science Education, 31*(1), 18–32. https://doi.org/10.1080/26939169.2022.2089410

Vidoni, M., & Codabux, Z. (2023). Towards a taxonomy of Roxygen documentation in R packages. *Empirical Software Engineering, 28*, 106. https://doi.org/10.1007/s10664-023-10345-4

Wimberly, M. C. (2023). *Geographic data science with R: Visualizing and analyzing environmental change* (1st ed.). Chapman and Hall/CRC. https://doi.org/10.1201/9781003326199

Chapter 19
Developing Digital and Pedagogical Competences in Postgraduate Management Students

Hennadii Karimov, Marina Romaniukha, Ivan Karimov, and Nuzhna Svitlana

Abstract Business management education is a trend in the development of a knowledge-based society within the European Union. In both Ukraine and Europe, business management education is occurring in a highly dynamic context. Developing postgraduate students' teaching and digital skills places significant emphasis on the university's learning environment, the digital competence of scientific advisors and the content of relevant academic disciplines. The article explores methods for improving Management students' teaching and digital skills through mandatory teaching practice. The authors present empirical data from a number of student surveys assessing their learning experience in a typical institution of higher education of Ukraine and summarize students' learning needs. The article identifies a number of growth opportunities, including informal learning, teachers' digital competence and updates to the content of academic disciplines.

Keywords Postgraduate student · Management · Teaching practice · Digital and teaching competences · Informal learning · Teachers' digital competence · Methodology of pedagogical activity

19.1 Introduction

Entrepreneurship is recognized as a key competence in the development of a knowledge-based society in the European Union, with entrepreneurial capacity being one of the primary policy objectives in Europe (Joensuu-Salo et al., 2023). Since the early 2000s, entrepreneurial or business management education has been shaped by major trends such as internationalization, digitalization and interactivity

H. Karimov · M. Romaniukha (✉) · I. Karimov
Dniprovsky State Technical University, Kamianske, Ukraine

N. Svitlana
Dnipro State Agrarian and Economic University, Dnipro, Ukraine
e-mail: nuzhna.s.a@dsau.dp.ua

E. Smyrnova-Trybulska et al. (eds.), *E-Learning and Enhancing Soft Skills*,
https://doi.org/10.1007/978-3-031-82243-8_19

(Ratten, 2023). The COVID-19 pandemic further highlighted the importance of digitalization, encouraging entrepreneurial education management to adapt to rapidly changing contexts and integrate new learning technologies. This shift allowed viewing the pandemic as a catalyst for advancing entrepreneurship education (Ratten & Jones, 2021), reframing the crisis as an opportunity to emphasize the importance of entrepreneurship education for society. In the context of Ukraine, the post-COVID educational environment has experienced considerable growth. However, the entrepreneurial and educational environment has recently become even more unpredictable and changeable due to war started by Russia. This conflict has resulted in internal displacement, the relocation of businesses and educational institutions. For safety reasons, many at-risk educational institutions have had to switch to remote learning, while others have faced interruptions to traditional in-person classes due to air raid alerts, causing inevitable pauses in the learning process.

The system for the attestation of highly qualified scientific personnel in Ukraine requires the modernization in the sphere of higher education and science due to the country's accession to the Bologna Process. Significant positive changes can be achieved by implementing the tasks defined by the participant states in the Bologna process. The introduction of the credit-transfer system, in particular in higher education, was also applied to the training of higher education applicants for the degree of Doctor of Philosophy. Part 6 of Article 5 of the Law of Ukraine "On Higher Education" establishes that the amount of the educational component within the educational-scientific program for training a doctor of philosophy is 30–60 credits according to the European Credit Transfer and Accumulation System (ECTS) (Ministry..., 2021).

This paper examines ways to enhance the digital and teaching competences of postgraduate Management students through teaching practice.

It aims to address the following **research questions**:

- common challenges in training postgraduate students in Ukraine;
- students' assessment of training at Dniprovsky State Technical University;
- students' demand for the improvement of educational services;
- potential growth areas for developing digital and teaching competences in postgraduate students.

The study utilized a combination of theoretical and empirical research methods, including the analysis and synthesis of scientific literature on the research problem to summarize the context of training postgraduate Management students in Ukraine, and a sociological survey involving postgraduate, master's and bachelor's students to identify opinion trends regarding training at Dniprovsky State Technical University. The paper concludes with findings on the problem studied and discusses research limitations.

19.2 Context of Research into the Postgraduate Students' Teaching Practice

The main challenge currently faced by educators worldwide is the need to learn and retrain in order to improve their digital skills for effective online learning implementation (Umryk & Strutynska, 2022). While digital technology offers new opportunities for educators, it also presents challenges such as an extended workload to manage student interactions beyond scheduled teaching hours (Truss & Anderson, 2023). Interestingly, the majority of the students do not prefer fully remote or fully online learning modes, leading to a demand for a range of options, rather than a single choice. This increasingly diverse student environment makes it more challenging to meet the varied demands for different learning scenarios (Ratten, 2023).

Before focusing on the specific challenges facing postgraduate education in Ukraine, it is worthwhile acknowledging the crucial difference in the nature of postgraduate training between European countries and Ukraine. While doctoral education in Europe presents a diverse landscape, its primary focus is on research-specific knowledge and related skills (Hasgall et al., 2019). In contrast, postgraduate studies in Ukraine emphasize preparing a new generation of teachers capable of carrying out professional activities within the context of information and communication technologies (Voinova, 2023). According to Part 1 of Article 5 of the Law (Law…, 2024) the third (educational-scientific) level is recognized as a level of higher education and involves the acquisition of theoretical knowledge, abilities, skills and other competencies sufficient for the production of a person's new ideas, solving complex problems in the field of professional and/or research and innovation activities, mastering the methodology of scientific and pedagogical activities.

Postgraduate education in Ukraine operates within a specific regulatory framework, particularly in terms of teacher training. Unlike pedagogical universities, where the specified competence is developed consistently through all years of study, postgraduate teaching practice in other higher education institutions is typically limited to a course on teaching methodology, followed by obligatory teaching hours (Klimova, 2021; Ignatyuk, 2019). Currently, the development of science in Ukraine is legislatively focused on informatization. The national doctrine for the advancement of the education system emphasizes the capacity of a new generation of teachers to carry out professional activities using information and telecommunication technologies, thus integrating informatization into the educational environment (Voinova, 2023). This significant emphasis on enhancing the digitally-supported teaching skills of young researchers places increased demands on their scientific advisors. Digital competence has become a key instrument for modern scientific research and academic competition. Naturally, there is both positive experience and an urgent need to further improve the digital competence of scientific advisors to ensure effective advisor-student collaboration (Vakaliuk et al., 2024).

Despite extensive research, further solutions are needed to effectively enhance the digital and teaching competences of postgraduate Management students.

19.3 Learners' Assessment of University Training

The following Sects. 19.3.1 and 19.3.2 explore students' perceptions of the skills of scientific and pedagogical personnel, including those with a Doctor of Philosophy degree. These teaching practices are largely reflected by postgraduate students during their obligatory teaching hours.

19.3.1 Postgraduate Students' Assessment of University Training

In accordance with the Resolution of the Cabinet of Ministers of Ukraine No. 261 on the Procedure for training candidates for higher education degrees of Philosophy Doctor and Doctor of Science in higher educational institutions (scientific institutions), a postgraduate student is supposed to acquire the following competencies in accordance with the National Framework of Qualifications:

– in-depth knowledge within the specialty (the estimated amount of such educational component is at least 12 ECTS credits);
  - general scientific (philosophical) competences (the estimated amount of such educational component is four to six ECTS credits);
  - universal researcher's skills (the estimated amount of such educational component is at least six ECTS credits);
  - language competences (the recommended amount of such educational component is six to eight ECTS credits) (Poriadok …, 2016).

In Dniprovsky State Technical University the educational and scientific program "Management" for the third (educational and scientific) level of higher education envisages the final result of training as integral competence confirming "the ability to produce new ideas, solve complex problems in the field of management and administration, which involves deep rethinking of existing and creating new comprehensive knowledge and/or professional practice, applying the latest methodologies of scientific and pedagogical activity, conducting own scientific research of some scientific novelty, theoretical and practical significance" (Koreniuk et al., 2024).

From December 2023 to January 2024 the Laboratory of Sociological Research at Dniprovsky State Technical University (DSTU) conducted a survey on student satisfaction with educational services. A total of 42 students in third-level educational and scientific programs, spanning both humanities and technical fields, participated in the survey. The results show that all postgraduate students (100%) expressed satisfaction with the progress in the chosen educational programs (see Fig. 19.1). However, some students noted certain subjects, such as philosophy, as "unnecessary". DSTU, the primary institution for this study, represents a typical higher educational institution offering both technical and humanities programs.

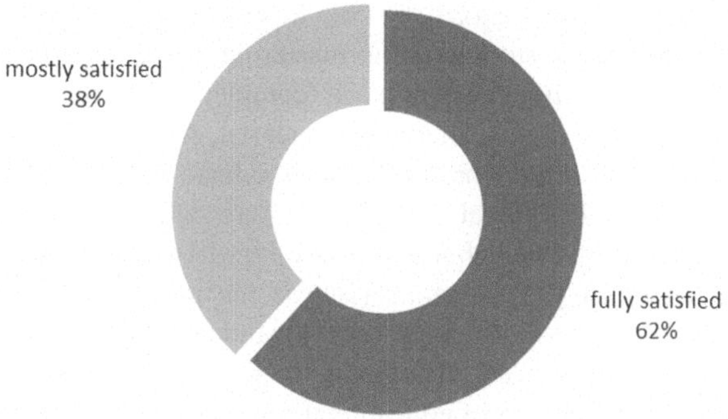

Fig. 19.1 PhD students' satisfaction with the progress in the selected program of study. (Source: Own work based on (Rezultaty …, 2024))

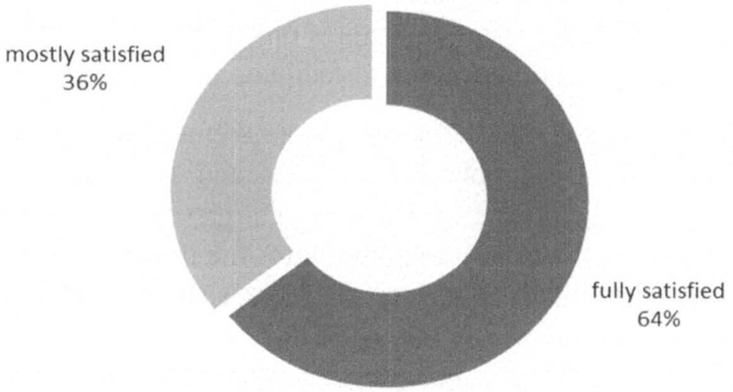

Fig. 19.2 PhD students' satisfaction with the quality of teaching. (Source: Own work based on (Rezultaty …, 2024))

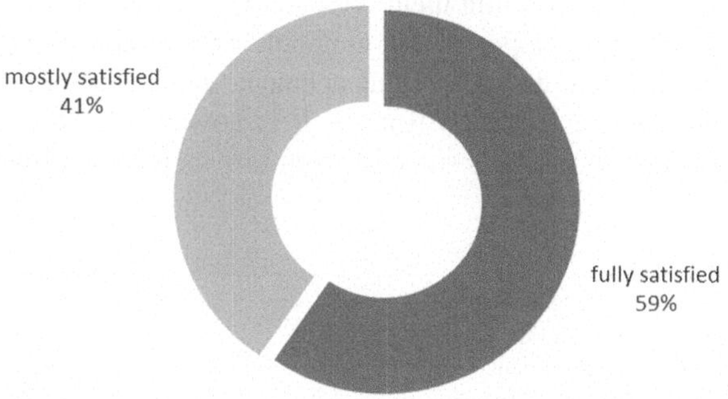

Fig. 19.3 PhD students' satisfaction with the methods of teaching. (Source: Own work based on (Rezultaty …, 2024))

In addition, the lack of technology in the educational process and the small number of selectable subjects influenced the level of satisfaction with the progress in the educational program.

Simultaneously, the majority of PhD students (100%) are satisfied with the quality of teaching (see Fig. 19.2), though some of them note the lack of attempts by teachers to enliven the learning process and the large amount of information in just one lecture, some methodological materials are missing.

At the same time, (100%) PhD students are satisfied with the teaching methods (see Fig. 19.3), although some of them note how timely distance education during the war is.

Besides, the majority of PhD level students (98%) believe that the time spent in classrooms is sufficient for mastering the disciplines, and the amount of time allocated for independent work (completion of scientific works/projects, individual tasks, processing the theoretical material, etc.) is sufficient. Also, 98% of students are satisfied with the work of the postgraduate department.

Among PhD level students, 95% are satisfied with the equipment and personnel of classrooms and laboratories (see Fig. 19.4), although 4.8% note that there is a lack of technology. A separate issue is the process of communication with teachers, which satisfies 100% of PhD level students. War is the main problem complicating the learning process. The main suggestions for improving the level of satisfaction with the provision of educational services can be grouped as follows:

- more communication
- possibility to listen to lectures online
- full digitization in order to facilitate the learning process.

Evidently, the areas demanding improvement are, firstly, revitalization and digitization of teaching tools and methods, and secondly, "redundant" (non-specialized) subjects. The first area is the subject of present research, so let us move to the second one. We support the standpoint that "the underestimation of the role of the cultural component hides a serious threat: a specialist who receives only pragmatically oriented knowledge is not able to go beyond the boundaries of a narrow specialization, and therefore is not able to become a representative of intellectual and spiritual elite, which is supposed to analyze societal challenges, predict risks and develop strategies to confront them" (Pidhotovka …, 2017, p. 44).

One of the important learning outcomes is the ability to develop and teach special disciplines in the area of management, thus the future doctor of philosophy can be considered a potential teacher. It is now worthwhile to consider the generalized opinions of master's students regarding their educational practices.

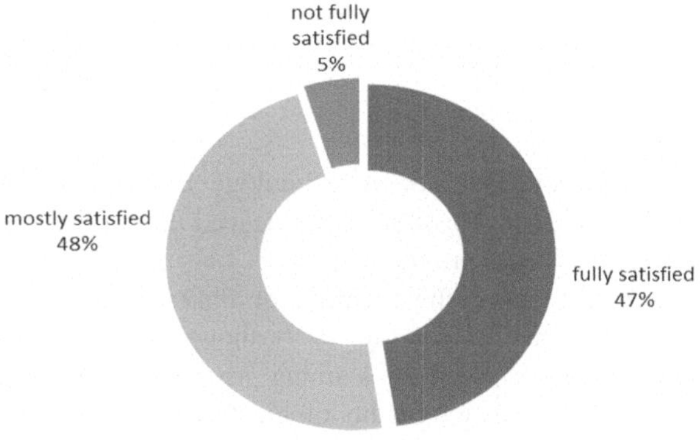

Fig. 19.4 PhD students' satisfaction with the equipment, equipment of the classrooms and laboratories. (Source: Own work based on (Rezultaty …, 2024))

19.3.2 Master Students' Assessment of University Training

"As a university graduate, philosophy doctor is entrusted with the mission of nurturing the next generation of students and scientists" (Pidhotovka …, 2017, p. 44), thus today's graduates of the master's level of higher education shape further development of the educational environment. That is why it is master students' expectations and vision of future teachers that help develop the competencies by graduate students in accordance with the National Framework of Qualifications—the acquisition of universal skills in the use of modern information technologies for the organization and conduct of educational classes (CMU, 2012).

In order to study the opinion of higher education applicants, the sociological laboratory of DSTU conducted an anonymous electronic survey using a specially developed toolkit. It was offered to 146 applicants receiving a second level education (Master's course).

The results of the survey of higher education applicants regarding their satisfaction with the provision of educational services indicate that although the majority of master's degree applicants (98%) are satisfied with their studies in the chosen educational program (see Fig. 19.5), there are directions for improvement (answers of applicants):

- understanding that they chose the wrong specialty;
- incomplete understanding of some subjects;
- lack of technology in the educational process;
- small number of selective subjects;
- many unnecessary subjects;
- lack of practical classes in the crucial disciplines for the future;
- obsolescence of educational material.

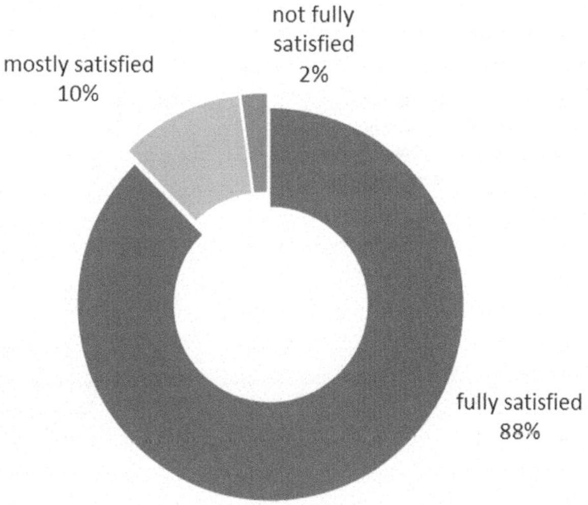

Fig. 19.5 Master students' satisfaction with studies according to the selected educational program. (Source: Own work based on (DSTU, 2024))

At the same time, the majority of master students (99%) are satisfied with the quality of teaching (see Fig. 19.6).

Students' reasons to be dissatisfied:

- teachers do not want to enliven the teaching process and monotonously retell the learned material;
- there is a lack of practical collaborative tasks related to immersion in a real working environment;
- small number of classroom classes;
- not all teachers communicate in distance format (DSTU, 2024).

At the same time, the majority (99%) of master students are satisfied with the teaching methods (see Fig. 19.7).

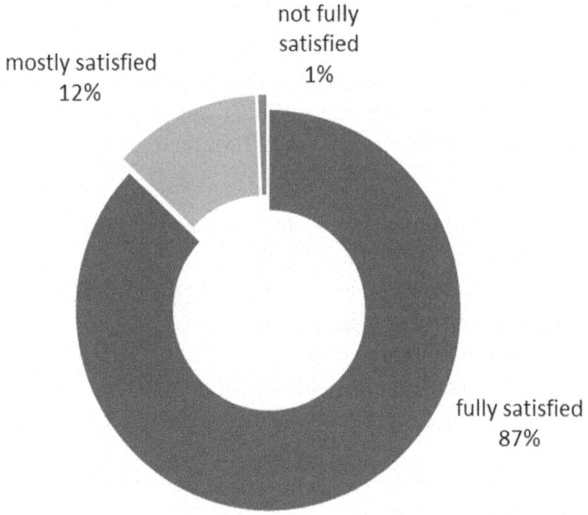

Fig. 19.6 Master students' satisfaction with the quality of teaching. (Source: Own work based on (DSTU, 2024))

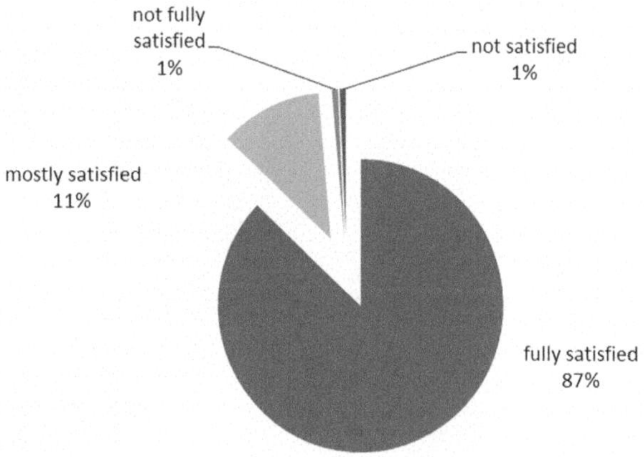

Fig. 19.7 Master students' satisfaction with teaching methods. (Source: Own work based on (Rezultaty …, 2024))

Among the reasons for dissatisfaction, the following were noted:

- few technologies and modern teaching methods;
- I would like more interaction with students;
- advanced age of teachers;
- remote work methods (DSTU, 2024).

In particular, the majority (99%) of master students believe that the amount of classroom time is sufficient for mastering the disciplines (97%), and the amount of time allocated for independent work (preparing scientific works/projects, individual tasks, processing of theoretical material, etc.) is sufficient.

Among the master students, 95% are satisfied with the equipment of classrooms and laboratories (see Fig. 19.8).

The reasons for not being completely satisfied were:

- outdated equipment and material base;
- lack of technology, multimedia equipment;
- lack of interactive whiteboards, projectors and computers in classrooms (DSTU, 2024).

The process of communication with teachers deserves special attention, even though 99% of master's level applicants are to some extent satisfied with it (see Fig. 19.9).

The reasons why they are not satisfied:

- some teachers do not have accounts in social networks or messengers;
- not all teachers are ready to communicate outside classrooms, there is no individual communication;
- there is no clear structure for communication, the university website is not an optimal instrument, it would be much better to allocate Google accounts for the educational process.
- we need a Google classroom environment to have the tasks available (DSTU, 2024).

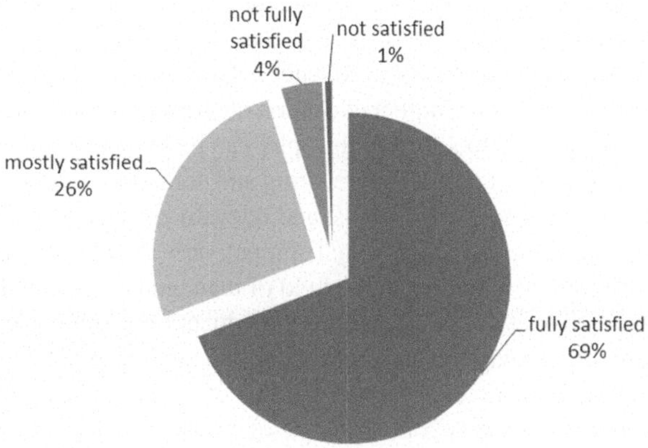

Fig. 19.8 Master students' satisfaction with the equipment, staffing of the auditoriums and laboratories. (Source: Own work based on (Rezultaty …, 2024))

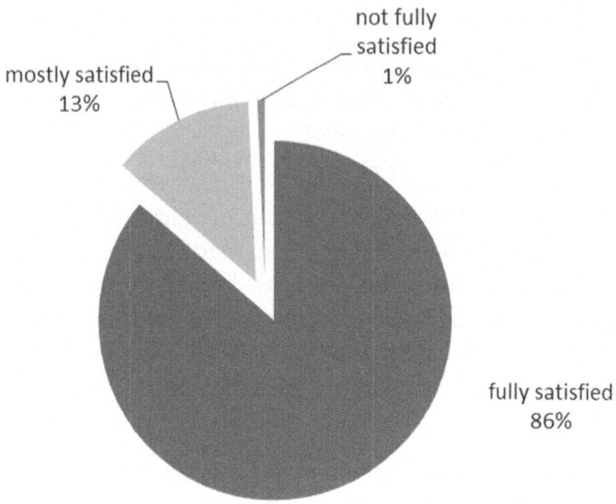

Fig. 19.9 Master students' satisfaction with the process of communication with professors. (Source: Own work based on (Rezultaty …, 2024))

According to the applicants, the learning process is complicated by technological challenges in the communication with teachers and the lack of some individual approach, which is also aggravated by overall difficult psychological and economic situation due to the war; anxiety and worry about the relatives. At the same time, students of almost all majors express the desire to improve their IT competences in order to achieve learning results (Karimov et al., 2023). A prior study has delineated strategies to augment students' learning outcomes through the intensified integration of IT components (Karimov et al., 2022).

19.4 General Regulations Behind Postgraduate Management Students' Teaching Practice

The section discusses the regulations behind postgraduate students' teaching practice, professional and integral competencies, as well as the expected learning outcomes. According to part six of Article 5 of the Law of Ukraine "On Higher Education", a doctor of philosophy is an educational and at the same time a scientific degree obtained at the third level of higher education on the basis of a master's degree. As it can be seen, the degree of Doctor of Philosophy can be obtained with competence-based approach to learning. In addition to preparing and defending a dissertation, a person must acquire theoretical knowledge, skills, and relevant competencies (Ministry…, 2021).

Acquiring the professional competence (PC03) "Ability to carry out scientific and pedagogical activities in the field of management in institutions of higher education" (Koreniuk et al., 2024) within the educational-scientific program "Management" is possible mainly through the student's teaching practice.

Teaching practice in the system of higher education at the third (educational-scientific) level is a component of professional training for scientific-pedagogical activity, which aims to consolidate the skills of postgraduate students' practical activity in the implementation of the educational process in institutions of higher education, including the teaching of special disciplines, the organization of educational students' activities, scientific and methodical work.

The terms of practice are determined by the schedule of the educational process, in particular, according to the Scheme of training for doctors of philosophy at DSTU (Skhema …, 2024), teaching practice is provided during the fourth semester of the II year of study. It is preceded by a 3-credit comprehensive course "Methodology of pedagogical activity".

During the course of teaching practice, postgraduate students are to master general, professional and integral competencies, as well as achieve certain learning outcomes. In particular, the "Management" training program envisages the achievement/deepening of the following program competencies within the framework of teaching practice:

- PROFESSIONAL COMPETENCE 02. The ability to present and discuss the results of scientific research and/or innovative developments in Ukrainian and English in oral and written forms, to study scientific literature on management and administration and to effectively use new information from various sources.
- PROFESSIONAL COMPETENCE 03. The ability to carry out scientific and pedagogical activities in the field of management in institutions of higher education.
- PROFESSIONAL COMPETENCE 04. The ability to initiate, develop, implement and manage scientific projects in management and related interdisciplinary areas and/or make proposals for financing scientific research, registration of intellectual property rights (Koreniuk et al., 2024).

The identified program competencies pave the way for achieving the following program learning outcomes (PLO):

- PLO 01. Apply modern tools and technologies to search, process and analyze information, in particular, statistical methods to analyze data of a large volume and/or complex structure, specialized databases and information systems.
- PLO 03. Develop and research conceptual, mathematical and computer models of processes and systems, effectively use them to obtain new knowledge and/or create innovative products in the field of management and related interdisciplinary areas.
- PLO 05. Develop full understanding of the general principles and methods of management sciences, as well as the methodology of scientific research, and apply them in one's own research in the field of management and in teaching practice.
- PLO 08. Develop and teach special educational disciplines in the area of management in educational institutions (Koreniuk et al., 2024).

According to the presented components, the immediate supervisor ensures organization and planning; approves the general schedule of practice, its position in the system of individual work planning of the applicant; determines the disciplines, study groups for conducting teaching practice (in the individual training plan the dates of classes, academic group, topic, type of classes are indicated). Later, the number of credit points for teaching practice (on a 100-point scale, on a national scale and on the ECTS scale) is entered into the individual plan of postgraduate students. When assigning a differentiated assessment, the level of theoretical training, the quality of performance of practice tasks, the level of mastery of pedagogical abilities and skills, attitude towards students, accuracy, discipline, quality of documentation and its submission time are taken into account. The generalized scheme of regulatory and methodological support of teaching practice is shown in Fig. 19.10.

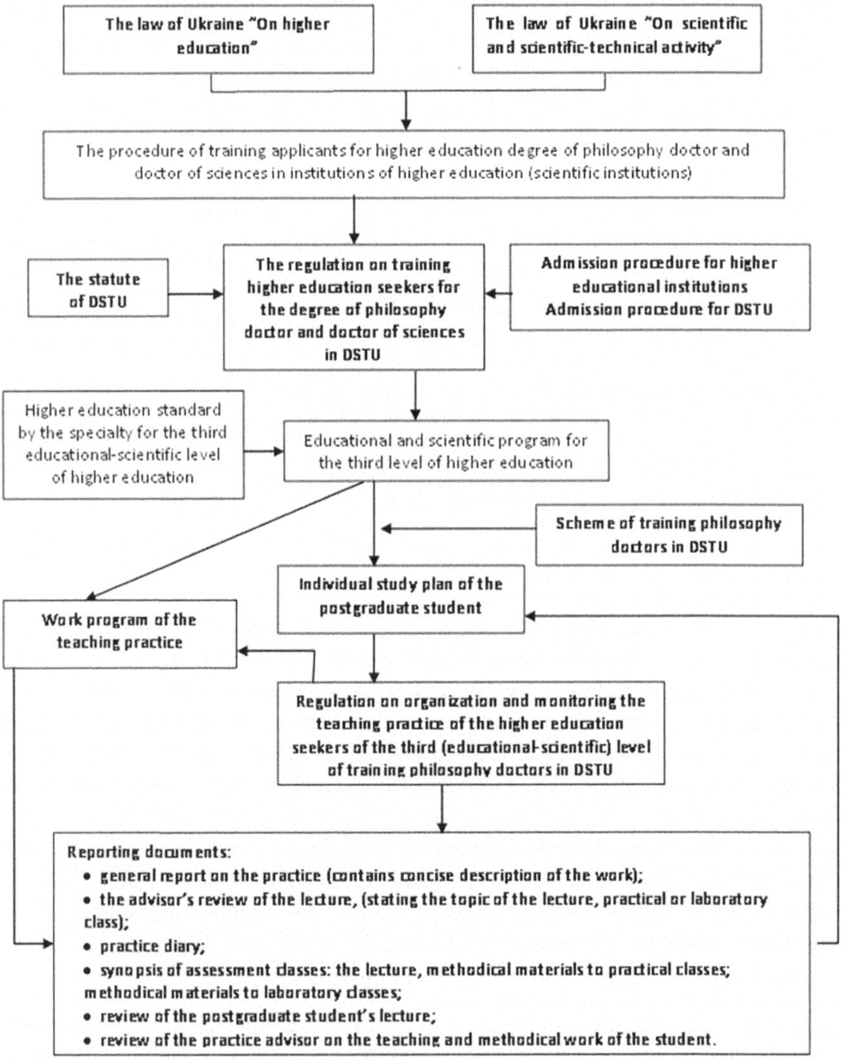

Fig. 19.10 Scheme of regulatory and methodological support of teaching practice. (Source: Own work)

The basis for the successful implementation of the tasks of teaching practice is the preliminary mastery of the academic disciplines of comprehensive scientific training, according to the structural and logical scheme of the educational and scientific program "Management":

- Philosophical dimensions of modern scientific research
- Methodology of pedagogical activity
- Foreign language for scientific activity
- Sociology of science
- Modeling and forecasting for economic processes

The organizational foundations of teaching practice are determined by the Regulations on the Training of Higher Education Candidates for the Doctor of Philosophy and Doctor of Science Degrees at DSTU (Hurzhyi & Medvedieva, 2024).

The resulting scheme should ensure the achievement of competencies that will allow future PhDs to meet the identified challenges of the time, among which, according to the study (Saltos-Rivas et al., 2023), a significant place is occupied by the teacher's digital competencies and mastery of European languages.

19.5 Some Practical Results About Developing Postgraduate Students' Competences During Teaching Practice

This section discusses empirical results from a questionnaire conducted among master's and bachelor's students, aimed at gathering their suggestions for improving the study process. The suggestions have been summarized and reduced to a few prominent trends that reflect students' primary learning needs.

Although many studies have been devoted to the role of pedagogical skill within the structure of the general competence of a higher school teacher, they still raise a number of questions, in particular, "regarding the current state of laying the foundations of pedagogical skill in the context of training teachers of a higher technical institution from among graduate students of technical universities who do not have basic higher pedagogical education, and are only beginning to acquire pedagogical knowledge during postgraduate studies and in further teaching activities at university departments in the chosen field" (Ignatyuk, 2019).

Since the expertise of a university teacher "affects the results of students' progress, their formation as individuals, competitive specialists" (Ignatyuk, 2019), it is advisable to consider the expectations of the potential audience more carefully.

Based on the results of an anonymous electronic survey of university students conducted by the sociological laboratory of DSTU (146 students of the second (master's) level and 386 students of the first (bachelor's) level, it is possible to single out the following suggestions for improving the level of satisfaction with the provision of educational services:

- to add more options for students—going abroad, competing in contests in Ukraine, various clubs;
- to conduct some of the classes remotely; to work in fully automated distance learning environment;
- to back-up the tasks on some online platform;
- to plan some feedback from the students, lecture is not enough, there must be a practical task afterwards;
- to provide more opportunities to students to go deeper into the topic during lectures and practical classes, so that they have the opportunity to test themselves as teachers, which in turn will increase respect and understanding towards the teacher, and help the teacher to look from the outside and find new, more effective tools and methods of delivering information and learn new tools;
- 20% of studying should be at the university and 80% must be self-study—let it be movies, videos, books in the field. It is not enough to have just the manuals for independent study;
- we need an option to take additional assignments to improve our points if it is not possible to attend classes regularly;
- more practical assignments and meetings with experts from enterprises;
- more presentations dealing with aspects of the chosen profession;
- more teachers' active cooperation with students;
- teacher-student interaction could be a bit more informal, lighter, perhaps with a little more humor. It probably applies to the entire structure of education;
- active involvement of young teachers;
- it would be really good if everything were newer, the visual aspect matters a lot;
- providing participants in the educational process with psychological support and help: the complex emotional state and psychological problems of those seeking education affect their willingness to learn;
- implement a blended learning system, where some classes are conducted online;
- much more communication;
- rework the information portal and communication with the teacher or switch to ready-made educational platforms;
- possibility to attend lectures online;
- full digitization to further facilitate the learning process (Rezultaty …, 2024).

From the above list, three main areas of improvement of educational services can be outlined: improvements to digital learning environment (including self-study components), enhancements in teachers' digital competence and pedagogical skills, and addressing the unique challenges posed by wartime are essential directions for development. However, the latter—overcoming wartime challenges—cannot reasonably be addressed by future teachers alone and therefore should not be considered their responsibility.

As one can see, pedagogical skills and innovative digital learning technologies complement each other, forming the competence and personality of a modern teacher, which must satisfy both the expected learning outcomes of future PhDs and the expectations of future consumers of educational services. Figure 19.11 summarizes the correlations between the needs and solutions applied in order to

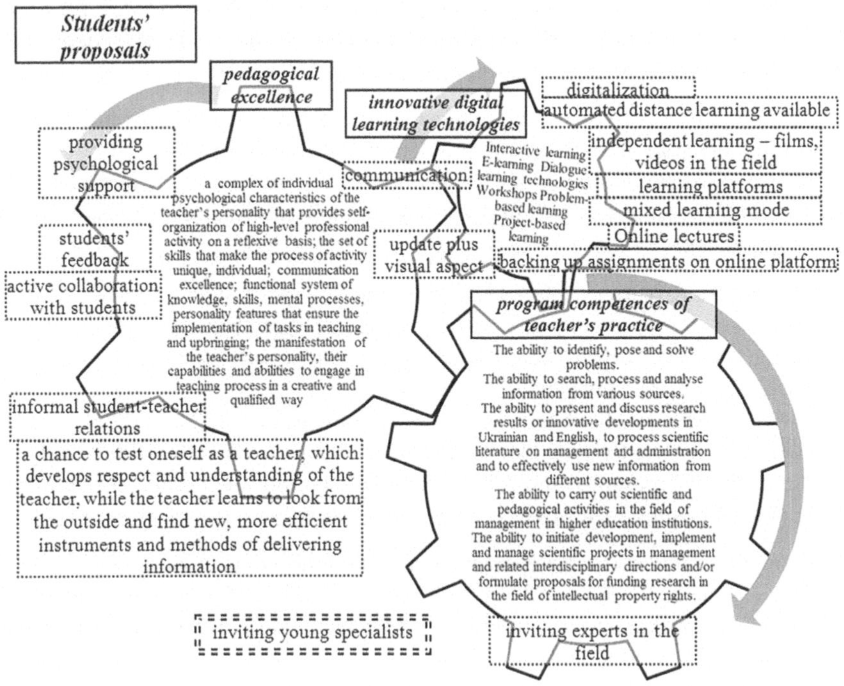

Fig. 19.11 Correlation between students' learning needs, teaching skills and program competences developed by teaching practice. (Source: Own work)

meet the expectations of potential postgraduate students, partially coming from the master student community.

The almost complete correlation of the program results of the teaching practice of future doctors of philosophy with the methods of innovative digital technologies and the expectations of consumers of educational services demonstrates the understanding of today's challenges to the system of higher education, both to state institutions (the Ministry of Education and Science) and to developers of educational and scientific programs (scientific and pedagogical university staff and the academic community).

– is solved automatically, taking into account the current age of future teachers—doctors of philosophy

19.6 Discussion

Postgraduate students have shown strong interest in integrating innovative digital technologies into the educational process, specifically "the skill and ability to implement and improve them in accordance with the needs of higher education and society" (Alexander et al., 2019). Notably, during the 2023–2024 academic year's teaching practice, second-year postgraduate students from the Department of Management contributed to developing methodological materials for independent work in various entrepreneurial disciplines, such as:

- "*Modeling and forecasting in management using MS Excel*"—innovative digital learning technologies (dialogical, problem-based learning);
- "*Marketing (presentation and visualization of the marketing project using MS Office tools*")—innovative digital learning technologies (dialogic project learning technologies).

In order to improve the postgraduate learning and teaching experience in areas highlighted in the previous section (improving the digital learning environment, advancing teachers' digital competence and pedagogical skills)—the following approaches should be considered:

Firstly, students' inclination towards e-learning can be fulfilled through various Massive Open Online Courses (MOOCs) and academic resources outside traditional academia. Below presented are a few examples, possibly enriching the informal component of postgraduate studies:

- IT company Genesis is an ecosystem of all-Ukrainian educational courses (internships via an LMS platform), in particular: "Creation and development of IT products", "Marketing of IT products", "Management in product IT", "Analytics in product IT" (Cooperation, 2024);
- the international IT Academy from CISCO company—an online course for mastering English in the context of information technologies "English for IT. Part 1" and others (Cisco, 2024);
- PJSC "ArcelorMittal Kryvyi Rih"—online training "Economic specialties" using classroom tools (ArcelorMittal, 2024).

Special attention should be paid to the assimilation and implementation of gamified interactive courses from a Ukrainian IT company Genesis, namely:

- an interactive gamified course "*Creation and development of IT products*" offering innovative digital learning technologies (interactive learning, e-learning; dialogue learning, workshop and problem-based learning, project-based learning); it covers a portion of student demands, namely: involving experts from enterprises, active cooperation with students, updating plus visual aspect, backup of assignments on an online platform, independent learning—films, thematic videos, learning platforms, mixed learning system, digitalization;
- an interactive gamified course "*Marketing of IT products*" offering innovative digital learning technologies as well (interactive learning, e-learning; dialogue learning, workshop and problem-based learning, project-based learning); it covers a good proportion of proposals from students: involving experts from enterprises, feedback from students, active cooperation with students, informal teacher-student relations, communication, updating plus visual aspect, backup of assignments on an online platform, automated distance learning, independent learning presented by films, thematic videos, learning platforms, mixed learning system, digitalization.

The cycle of courses is focused on the acquisition of non-technical competences in the IT sphere and is based on an interactive online platform (online tasks, videos, cases, testing) where learners get access to:

- LMS-platforms with an internship simulator in a product IT company.
- Pre-recorded video lectures with expert practitioners.
- Abstracts, infographics and a glossary of terms.
- Video lectures in the format of a dialogue with industry leaders.
- Interactive tasks and tests on the platform (Cooperation, 2024).

The given list almost fully takes into account the proposals of applicants to improve the level of satisfaction with the provision of educational services and promotes the acquisition of additional digital competences.

Secondly, improving digital literacy among teachers is essential, as this mandatory professional competence not only represents a flexible skill but also forms a core component of pedagogical professional training (Skakun, 2021, p. 46). Enhanced digital literacy equips scientific advisors to integrate innovative learning technologies into higher education through their students—young researchers.

Prior to COVID-19, studies indicated that higher education lecturers in Ukraine and Poland exhibited relatively low digital competence, highlighting a need for further professional development in this area (Khoruzha et al., 2019). There is positive Ukrainian university experience in designing the digital educational environment, launching a professional development program for both teachers and students. It dramatically improved the teachers digital expertise (Henseruk et al., 2021). Another collaborative research group (Kuzminska et al., 2022) confirmed that in order to develop postgraduate students' digital capacity it is crucial to provide them with instructors who possess a high level of digital competence within a relevant e-learning environment.

Thirdly, another area for potential growth is the curriculum for the postgraduate course "Methodology of pedagogical activity". The course should be dynamically reviewed and updated to meet the evolving needs of a diverse student community. Positive outcomes have been observed in initiatives to enrich course content with elements of e-learning, as evidenced by curriculum adjustments made specifically for graduate students (Klimova, 2021).

19.7 Conclusion

Training postgraduate business management students in Ukraine's current security and educational environment is rather challenging. Educators face pressure to continually update teaching tools and methods in response to developments in information channels and educational methodology (Voinova, 2023).

In order to improve the postgraduate students' digital and teaching skills, several growth points are suggested.

Incorporating informal learning elements can enrich current study programs. A cooperative model “State (Ministry of Education) – organizations (higher educational institutions and enterprises) – educational online platforms” (Karimov et al., 2023) has the potential to develop high-quality online courses that comply with educational standards and offer robust methodological support. As Saltos-Rivas et al. (2023) highlight, “the joint synergistic efforts of universities, scientific and pedagogical workers, higher education students and state institutions will contribute to the successful implementation of educational innovations and development in the modern digital world.” The COVID-19 pandemic accelerated this trend, prompting widespread adoption of online courses from corporations and governments globally (Umryk & Strutynska, 2022).

Strengthening the digital literacy of scientific advisors is essential, as they guide the development of young researchers. Advisors with high digital competence can better mentor students in applying innovative digital tools effectively.

Revisiting and enriching the course content of “Methodology of pedagogical activity” is critical, given the growing importance of digitalization in higher education and society.

Due to space constraints, this paper does not delve into the specific content of the postgraduate students’ course Methodology of pedagogical activity. However, the findings based on student feedback offer a foundation for expanding future research to include teacher perspectives. Involving postgraduate instructors would provide a more comprehensive view of the challenges and growth opportunities in digital competence, ultimately guiding further improvements in the “Methodology of pedagogical activity” course.

References

Alexander, B., Ashford-Rowe, K., Barajas-Murph, N., Dobbin, G., Knott, J., McCormack, M., Pomerantz, J., Seilhamer, R., & Weber, N. (2019). *Horizon Report 2019 Higher Education Edition*. EDU19. Retrieved from https://www.learntechlib.org/p/208644/. Accessed 13 July 2024.

ArcelorMittal Kryvyi Rih, Practice – Economic specialties. Retrieved from URL: https://classroom.google.com/w/Njk0MTkzODYxMjE0/t/all. Accessed 13 July 2024.

Cabinet of Ministers of Ukraine. (2012). Postanova pro zatverdzhennya Nacional’noï ramki kvalifikacij No. 1341 [*Resolution of the Cabinet of Ministers of Ukraine On approval of the National Qualifications Framework No. 1341*]. Official Gazette of Ukraine 101, 15.

Cisco Networking Academy, Teach with us. Retrieved from URL: https://www.netacad.com/educators. Accessed 13 July 2024.

Cooperation for shaping the future of Ukraine, Retrieved from URL: https://www.genesis-for-univ.com/. Accessed 13 July 2024.

Dniprovsky State Technical University (DSTU), Anketuvannia/Fakultet ekonomiky ta sotsialnykh komunikatsii [*Questionnaire/Faculty of Economics and Social Communications*]. Retrieved from URL: https://www.dstu.dp.ua/uni/index.html. Accessed 13 July 2024.

Hasgall, A., Saenen, B., Borrell-Damian, L., et al. (2019). *Survey. Doctoral education in Europe today: Approaches and institutional structures*. European University Association. Retrieved from https://www.researchgate.net/publication/334508578_Doctoral_education_in_Europe_today_approaches_and_institutional_structures

Henseruk, H., Buyak, B., Tereshchuk, H., Boyko, M., & Henseruk, Y. (2021). The teacher in the system of developing students' digital competence. In E. Smyrnova-Trybulska (Ed.), *E-learning in the time of COVID-19* (E-learning) (Vol. 13, pp. 238–248). Katowice–Cieszyn. https://doi.org/10.34916/el.2021.13.20

Hurzhyi, T., & Medvedieva, L. (Eds.). (2024). Polozhennia pro pidhotovku zdobuvachiv vyshchoi osvity stupenia doktora filosofii ta doktora nauk u Dniprovskomu derzhavnomu tekhnichnomu universyteti. [*Regulations on the preparation of higher education applicants for the degree of Doctor of Philosophy and Doctor of Sciences at the Dnipro State Technical University].* Dniprovsky State Technical University, 32 p.

Ignatyuk, O. (2019). Modeliuvannia vykladannia pedahohichnykh dystsyplin u protsesi pidhotovky maibutnikh doktoriv filosofii v umovakh tekhnichnoho universytetu. [*Modeling of teaching of pedagogical disciplines in the process of preparation the future doctors philosophy in the conditions of technical university*]. *Adaptive Management: Theory and Practice Pedagogy, 6*(11). https://doi.org/10.33296/2707-0255-6(11)-01

Joensuu-Salo, S., Peltonen, K., & Hämäläinen, M. (2023). The importance of HEI managerial practices in teachers' competence in implementing entrepreneurship education: Evidence from Finland. *The International Journal of Management Education., 21*(2), 100767.

Karimov, H., Romaniukha, M., Karimov, I., & Sorokina, L. (2022). Aspects of developing students' individual educational trajectory, *E-learning in the Transformation of Education in Digital Society. "E-learning" Series, 14*, 120–131. https://doi.org/10.34916/el.2022.14.09

Karimov, H., Romaniukha, M., Karimov, I., & Kuzmenko, N. (2023). Trends in developing digital skills of students in dual study mode. *E-learning & Artificial Intelligence (AI). "E-learning" Series, 15*, 280–294. https://doi.org/10.34916/el.2023.15.23

Khoruzha, L., Proshkin, V., & Kotenko, O. (2019). Digital competence: Abilities of a lecturer and expectations of students (Ukrainian-Polish context). In E. Smyrnova-Trybulska (Ed.), *E-learning and STEM Education* (E-learning) (Vol. 11, pp. 421–439). Katowice-Cieszyn. https://doi.org/10.34916/el.2019.11.27

Klimova, K. (2021). Vykorystannia onlain-platform u pidhotovtsi doktoriv filosofii z dystsypliny «Suchasni metodyky vykladannia u vyshchii shkoli ta pedahohichna maisternist». [*The use of online platforms in the training of doctors of philosophy in the discipline "Modern teaching methods in higher schools and pedagogical master's degree"*]. New Educational Technologies № 95. https://doi.org/10.52256/2710-3560.95.2021.10

Koreniuk, P., Plakhotnik, O., & Karavan, N., (eds.). (2024). Osvitno-naukova prohrama «Menedzhment» za spetsialnistiu 073 «Menedzhment» haluzi znan 07 «Upravlinnia ta administruvannia» dlia tretoho (osvitno-naukovoho) rivnia vyshchoi osvity. [*Educational-scientific program "Management" in the specialty 073 "Management" of the field of knowledge 07 "Management and administration" for the third (educational-scientific) level of higher education*]. Dniprovsky State Technical University, 16 p.

Kuzminska, O., Mazorchuk, M., Morze, N., Smyrnova-Trybulska, E., Stec, M., & Gutiérrez-Esteban, P. (2022). Graduate students' attitudes to the development of digital opportunities at the level of individual and educational organizations. In E. Smyrnova-Trybulska (Ed.), *E-learning in the transformation of education in digital society* (E-learning) (Vol. 14, pp. 191–204). Katowice–Cieszyn. https://doi.org/10.34916/el.2022.14.14

Law of Ukraine „On Higher Education" (1 July 2014 № 1556-VII, Revision on May 28, 2024. *The Official Bulletin of the Verkhovna Rada (BVR)*, 2014, No. 37–38, Article 2004, Retrieved from URL: https://zakon.rada.gov.ua/laws/show/1556-18?lang=en#Text. Accessed 10 July 2024.

Ministry of Education and Science of Ukraine. (2021). Informatsiino-analitychnyi zvit Ministerstva osvity i nauky Ukrainy pro provedennia eksperymentu z prysudzhennia stupenia doktora filosofii. [*Informational and analytical report of the Ministry of Education and Science of Ukraine on conducting an experiment on the awarding of the degree of Doctor of Philosophy*]. Ministry of Education and Science of Ukraine, 20 p.

Pidhotovka doktoriv filosofii (PhD) v umovakh reformuvannia vyshchoi osvity. [*Training of doctors of philosophy (PhD) in conditions of higher education reform*]. (2017). Materialy Vseukrainskoi naukovo-praktychnoi konferentsii (Zaporizhzhia, 5–6 zhovtnia 2017 r.) [*Materials of the All-Ukrainian scientific and practical conference (Zaporizhia, October 5–6, 2017)*]. Zaporizhia National University, 216 p.

Poriadok pidhotovky zdobuvachiv vyshchoi osvity stupenia doktora filosofii ta doktora nauk u vyshchykh navchalnykh zakladakh (naukovykh ustanovakh). [*The procedure for the preparation of higher education applicants for the degree of Doctor of Philosophy and Doctor of Sciences in higher educational institutions (scientific institutions)*]. (2016). Retrieved from URL: https://zakon.rada.gov.ua/laws/show/261-2016-%D0%BF?lang=en#Text. Accessed 10 July 2024.

Ratten, V. (2023). The post COVID-19 pandemic era: Changes in teaching and learning methods for management educators. *The International Journal of Management Education, 21*(2), 100777. https://doi.org/10.1016/j.ijme.2023.100777

Ratten, V., & Jones, P. (2021). Covid-19 and entrepreneurship education: Implications for advancing research and practice. *The International Journal of Management Education, 19*(1), 100432.

Rezultaty anketuvannia zdobuvachiv vyshchoi osvity shchodo yikh zadovolenosti nadanniam osvitnikh posluh v DDTU. [*Results of a survey of higher education applicants regarding their satisfaction with the provision of educational services at DDTU*]. (2024). Dniprovsky State Technical University, 27 p.

Saltos-Rivas, R., Novoa-Hernandez, P., & Serrano Rodriguez, R. (2023). Understanding university teachers' digital competencies: A systematic mapping study. *Educational and Information Technologies, 28*, 16771–16822. https://doi.org/10.1007/s10639-023-11669-w

Skakun, I. (2021). Digital competencies of the teacher of the future. *Futurity Education, 1*(2), 46–56. https://doi.org/10.57125/FED/2022.10.11.18

Skhema pidhotovky doktoriv filosofii v DDTU. [*Scheme of training of doctors of philosophy at DDTU*]. Retrieved from https://www.dstu.dp.ua/uni/downloads/shema_asp.pdf. Accessed 29 June 2024.

Truss, A., & Anderson, V. (2023). The navigational challenges of a blended learning approach to teaching in business and management. *The International Journal of Management Education, 21*(1), 100733. https://doi.org/10.1016/j.ijme.2022.100733

Umryk, M., & Strutynska, O. (2022). Experience in development of the university MOOCs environment to support pre-service teacher training. In E. Smyrnova-Trybulska (Ed.), *E-learning in the transformation of education in digital society scientific* (E-learning) (Vol. 14, pp. 75–88). Katowice–Cieszyn. https://doi.org/10.34916/el.2022.14.06

Vakaliuk, T., Spirin, O., Yevdokymov, V., Yatsyshyn, A. (2024). Dosvid vykorystannia khmarnykh servisiv dlia spilnoi roboty naukovoho kerivnyka ta aspirantiv. [*Experience of using cloud services for the joint work of the scientific supervisor and graduate students*]. Information Technologies and Teaching Aids, 100, №2, 122–140. https://doi.org/10.33407/itlt.v100i2.5321. ISSN: 2076-8184.

Voinova, S. O. (2023). Pro vplyv informatyzatsii na pidhotovku aspirantiv. [*On the influence of informatization on the training of graduate students*]. Herald of KhNTU № 4(87). https://doi.org/10.35546/kntu2078-4481.2023.4.4

Correction to: Examining M-Learning Trends Among University Students—Mobile Devices in Students' Study and Life

Eugenia Smyrnova-Trybulska, Małgorzata Przybyła-Kasperek, Marcin Gierczyk, and Dagmara Dobosz

Correction to:
Chapter 8 in: E. Smyrnova-Trybulska et al. (eds.),
***E-Learning and Enhancing Soft Skills*,**
https://doi.org/10.1007/978-3-031-82243-8_8

The original version of the chapter was inadvertently published with incorrect author name. The author name has now been corrected in the chapter to **Marcin** Gierczyk.

The updated version of this chapter can be found at
https://doi.org/10.1007/978-3-031-82243-8_8

E. Smyrnova-Trybulska et al. (eds.), *E-Learning and Enhancing Soft Skills*,
https://doi.org/10.1007/978-3-031-82243-8_20

Appendix A: A Student's Learning Log of Order Words with the Pedagogical AI Agent

Start

Great! Let's begin with Part 1 of our lesson.

Part 1: Explanation of the Concept of Order Words

Step 1: Let's talk about order words. Order words are special words we use in English to describe the sequence of events or steps. These include words like "first," "second," "third," "next," "then," and "finally." They help us organize our thoughts and communicate more clearly by showing the order in which things happen. For instance, when describing a morning routine, you might say, "First, I wake up. Second, I brush my teeth. Third, I eat breakfast." Here, "first, second, third" are order words that help us understand the sequence of actions.

Do you understand the meaning of order words? (Please answer "yes" or "no").

no

Alright, let's try another example to explain order words, focusing on a simple three-step process.

Imagine you're getting ready to watch a movie at home. Here's how you might use order words to describe the steps:

1. **First**, find a comfortable spot on the couch and get your favorite blanket.
2. **Second**, choose a movie you want to watch.
3. **Third**, turn on the TV and start the movie.

In this scenario, "first" is used to describe the initial step, which is getting cozy on the couch. "Second" is for choosing the movie, which is the next action. And "third" refers to the final step in this sequence, which is turning on the TV and starting the movie.

E. Smyrnova-Trybulska et al. (eds.), *E-Learning and Enhancing Soft Skills*,
https://doi.org/10.1007/978-3-031-82243-8

Does this example help clarify the concept of order words for you? (Please answer "yes" or "no").

yes

Great to hear that you understand now!

Part 2: Providing Examples About Using Order Words

Step 3: Let's use the life cycle of a butterfly as an example to further illustrate how to use order words effectively. The life cycle includes these stages: egg, caterpillar, pupa, and butterfly. Using order words, we could describe it like this:

- **First**, a butterfly lays her eggs on a leaf.
- **Next**, from these eggs, little caterpillars emerge. They are very hungry and start to eat leaves.
- **Then**, when the caterpillar is big enough, it forms a pupa around itself.
- **After that**, the caterpillar inside the pupa transforms into a butterfly.
- **Finally**, the butterfly emerges from the pupa.

In this example, "first" is used for the beginning step of the life cycle. "Next" and "then" indicate what happens after the previous steps. "After that" is typically used when several steps have already been mentioned, and you're describing later stages. "Finally" is for the last step, marking the completion of the cycle.

Do you understand this example of how to use order words? (Please answer "yes" or "no").

no

Okay, let's try a different example. This time, we'll look at how to bake chocolate chip cookies using order words in a five-step process:

1. **First**, gather all your ingredients like flour, sugar, chocolate chips, eggs, and butter.
2. **Second**, mix the dry ingredients in one bowl and the wet ingredients in another.
3. **Third**, combine the dry and wet ingredients together to make your cookie dough.
4. **Next**, scoop the dough onto a baking sheet in small portions.
5. **Finally**, bake the cookies in the oven until they are golden brown.

Here, we use "first" to begin the process with the initial step of gathering ingredients. "Second" and "third" describe the subsequent steps in making the dough. "Next" is used for the step of placing the dough on the baking sheet, and "finally" indicates the last step, which is baking the cookies.

Does this example help you understand how to use order words in describing a process? (Please answer "yes" or "no").

yes

Excellent! It's great that you understand now.

Part 3: Assessing Student's Understanding About the Concept of Order Words

Step 5: To confirm your understanding, let's do a quick multiple-choice question.

Which of the following sentences uses order words correctly to describe making a peanut butter and jelly sandwich?

(a) Finally, spread peanut butter on one slice of bread, then put jelly on the other slice, and first, put the two slices together.
(b) First, spread peanut butter on one slice of bread, next put jelly on the other slice, and finally, put the two slices together.
(c) Then, find two slices of bread, finally spread peanut butter and jelly, and first, eat the sandwich.

Please select the correct answer: (a), (b), or (c).

c

It looks like there's a bit of confusion in the choice. Let's review it:

In option (c)—"Then, find two slices of bread, finally spread peanut butter and jelly, and first, eat the sandwich"—the order words are not used correctly. The sentence starts with "Then," which is usually not used at the beginning. Also, it ends with "first," which should be at the start of a sequence, not at the end.

The correct answer is actually (b)—"First, spread peanut butter on one slice of bread, next put jelly on the other slice, and finally, put the two slices together." This sentence uses order words correctly to describe the sequence of making a sandwich: beginning with "first," following with "next," and ending with "finally."

Let's try another example to strengthen your understanding. We'll go back to Step 3 and provide a different scenario.

Step 3 (again): Let's consider the steps involved in planting a flower in a pot:

1. **First**, fill a pot with soil.
2. **Next**, make a small hole in the center of the soil.
3. **Then**, place the flower seed in the hole.
4. **After that**, cover the seed with a little bit of soil.
5. **Finally**, water the soil lightly.

In this example, "first" is used for starting the process with filling the pot with soil. "Next" and "then" indicate the subsequent steps of preparing and placing the seed. "After that" is used for the step following the placement of the seed, and "finally" indicates the last step, which is watering the soil.

Does this example help you understand the use of order words? (Please answer "yes" or "no").

yes

Great! I'm glad to hear that you understand the use of order words with this example.

Part 3: Assessing Student's Understanding About the Concept of Order Words (Continued)

Step 5 (Attempt 2): Let's try another multiple-choice question to confirm your understanding.

Which of the following sentences uses order words correctly to describe the process of getting ready for school?

(a) Next, put on your shoes, then eat breakfast, and finally, brush your teeth.
(b) First, wake up and get out of bed, next take a shower, and finally, put on your school uniform.
(c) Finally, wake up, then choose what to wear, and first, eat breakfast.

Please select the correct answer: (a), (b), or (c).

b

That's correct! Option (b)—"First, wake up and get out of bed, next take a shower, and finally, put on your school uniform"—uses order words correctly to describe the sequence of getting ready for school. It starts with "First" for the initial action of waking up, follows with "next" for taking a shower, and concludes with "finally" for putting on the uniform, indicating the last step in the process.

Part 4: Practice by Giving Some Concrete Story Writing or Planning Tasks and Provide Evaluation and Real-Time Feedback

Step 7: Now, let's practice using order words in a storytelling context. I'd like you to describe the story of Snow White in a 5-step story. Please use appropriate order words to tell your story. Remember to use both ordinal words (like first, second, third) and other order words (like next, then, after that).

Once you've completed your story, I will assess it based on the criteria of correct use of order words, logical sequence, and if requested, the exact number of steps.

Please share your 5-step story of Snow White when you're ready.

First: An evil witch envies the beauty of Snow White. Then: She tricked Snow White to eat a poisoned apple. After that: Snow White fell asleep for a long time, luckily seven dwarf men took good care of her. Next, a prince came and kissed Snow White, which brought Snow White back to life. Finally, the prince and Snow White live happily together.

Thank you for sharing your story. Here's the assessment:

Your story meets the criteria for an "Excellent" rating. You've correctly used order words, including both ordinal words and other order words. The story follows a logical sequence without repetition, and you've used the exact number of steps requested. Here's a breakdown:

1. **First**: Sets the stage with the witch's envy, a perfect beginning.
2. **Then**: Introduces a crucial plot point with the poisoned apple.
3. **After that**: Describes what happens following the key event, Snow White falling asleep and being cared for by the dwarfs.
4. **Next**: Leads towards the climax with the prince's arrival and action.
5. **Finally**: Concludes the story with a classic happy ending.

This is a well-structured narrative using order words effectively. Great job!

Now, let's continue practicing. Please describe the following scenarios in a 5-step story using appropriate order words:

1. A field trip.
2. Cleaning the classroom at the end of the semester.
3. Preparing for the final exam.

You can start with any of these scenarios. Remember to use order words to structure your story clearly.

To be well-prepared for the upcoming final exam, I will follow the following steps: First, I will review my own notes to refresh my memories and understanding about the learning contents. Next, I will write down points where I am not quite sure I fully understand them yet. After that, I will study the contents provided in this course and explore additional information which could help me better understand those uncertain points. Finally, I will use previous exam items for practice to self assess my level of understanding.

Thank you for sharing your plan for preparing for the final exam. Here's the assessment:

Your story also meets the criteria for an "Excellent" rating. You've used order words effectively to describe the steps of preparing for an exam. The narrative is clear and follows a logical sequence. Here's a summary of your steps:

1. **First**: Starting with a review of personal notes to refresh your memory sets a solid foundation.
2. **Next**: Identifying areas of uncertainty is a crucial second step.
3. **After that**: Studying course materials and seeking additional information to clarify doubts is an excellent third step.
4. **Finally**: Practicing with previous exam questions to assess understanding is a perfect conclusion to your preparation process.

This shows a well-thought-out plan using order words to structure the steps methodically and clearly.

Now, please choose either the field trip scenario or cleaning the classroom at the end of the semester, and describe it in a 5-step story using appropriate order words.

First, I will set the aim or learning goal of this field trip. Then, I will search for a suitable destination which can provide a rich environment for achieving the learning goal. Next, I will ask ChatGPT to provide a detailed program for this field trip. After that, I will discuss with ChatGPT to further improve this tentative program. Finally, I will send the program to all students to make them aware of this trip in advance.

Thank you for outlining your plan for organizing a field trip. Here's the assessment:

Your description once again meets the criteria for an "Excellent" rating. You've effectively used order words to lay out a clear and logical sequence for planning a field trip. Let's review your steps:

1. **First**: Setting the aim or learning goal is an ideal starting point.
2. **Then**: Choosing a suitable destination aligned with the learning goal is a logical next step.
3. **Next**: Asking for assistance to draft a detailed program shows planning and resourcefulness.
4. **After that**: Collaborating for program improvement demonstrates thoroughness and commitment to quality.
5. **Finally**: Distributing the program to students in advance ensures preparedness and communication.

This plan is well-structured and thoughtfully organized, demonstrating a good understanding of using order words to describe a process.

Now, let's move to the final scenario: describing the process of cleaning the classroom at the end of the semester in a 5-step story using appropriate order words. Please go ahead when you're ready.

I want to skip this one and move to Part 5.

Absolutely, we can move on to Part 5.

Part 5: Apply with Learners' Own Chosen Topics and Provide Evaluation and Real-Time Feedback

Step 8: Self-Directed Learning

Now, you have the opportunity to choose your own topic and describe a process or story using order words. This is a chance for you to apply what you've learned in a context that interests you.

Please describe any process or story in a structured way using order words. Once you've shared it, I will evaluate it based on the criteria we've been using: correct use of order words, logical sequence, and adherence to the requested number of steps.

Feel free to select any topic you like for this exercise.

I want to make a plan for preparing my panel presentation slides. First, I will write down some key ideas or main concepts I would like to share. Next, I will use MS Copilot in PPT to generate a draft based on the contents from step 1. Then, I will tell Copilot what additional information I want to add or texts which need to be refined. After that, I will go through the whole slides to make sure they are presented in a logical sequence and consistent in terms of terms and formats.

Thank you for sharing your plan for preparing panel presentation slides. Here's the assessment:

Your plan is rated as "Excellent." You've successfully used order words to describe a clear and logical process for creating presentation slides. Let's review your steps:

1. **First**: Starting with writing down key ideas or main concepts is an excellent initial step for content creation.
2. **Next**: Utilizing MS Copilot in PowerPoint to generate a draft is a resourceful and efficient second step.
3. **Then**: Directing Copilot to refine the content ensures that your specific needs and ideas are incorporated.
4. **After that**: Reviewing the entire slide deck for logical flow and consistency in terminology and format is a crucial final check.

Your plan demonstrates a structured and thoughtful approach to preparing presentation slides, showing good use of order words to organize the steps effectively.

If you have another topic you'd like to explore using order words, please feel free to share it. Otherwise, we can proceed to the next part of our lesson.

Here is a revised version: First, I will write down some key ideas or main concepts I would like to share. Next, I will use MS Copilot in PPT to generate a draft based on the contents from step 1. Then, I will tell Copilot what additional information I want to add or texts which need to be refined. After that, I will go through the whole slides to make sure they are presented in a logical sequence and consistent in terms of terms and formats. Finally, I will do a rehearsal using my prepared slide to make sure it can be delivered with the given time frame.

Thank you for sharing the revised version of your plan for preparing panel presentation slides. Here's the updated assessment:

This revised plan maintains an "Excellent" rating. The addition of a final step enhances the thoroughness of your preparation. Let's review your steps:

1. **First**: Writing down key ideas or main concepts is a strong foundation for your presentation.
2. **Next**: Generating a draft with MS Copilot in PowerPoint based on your initial concepts is an effective way to start shaping your presentation.
3. **Then**: Instructing Copilot to add additional information or refine texts ensures that your presentation is tailored to your specific needs.
4. **After that**: Reviewing the slides for logical sequence and consistency in terms and format is crucial for clarity and professionalism.
5. **Finally**: Rehearsing with the prepared slides to ensure they fit within the given time frame is an excellent last step, vital for a polished and effective presentation.

The addition of rehearsal as the final step shows a comprehensive approach to preparing for a presentation, considering not only the content and design of the slides but also the delivery aspect.

If you have any more topics or scenarios you'd like to explore using order words, feel free to share. Otherwise, we can move on to the next part of our session.

No

Alright, since you don't have more topics or scenarios to explore, we can proceed to the next part.

Part 6: Report Student's Overall Performance and Suggest for Further Study

Based on our session today, you've demonstrated a strong understanding of how to use order words effectively. You've successfully applied these words in different contexts, from explaining natural processes to planning personal tasks. Your ability to structure information logically and clearly is commendable.

For further study, I suggest:

1. **Practice Regularly**: Continue applying order words in your daily activities or while writing. This can be as simple as describing your day or planning an event.
2. **Explore More Complex Scenarios**: As you get more comfortable, try using order words in more complex or multi-step processes. This will help you enhance your ability to organize and articulate intricate sequences.
3. **Engage in Storytelling**: Storytelling is a great way to practice using order words. Try narrating stories or events to friends or family, focusing on the sequence of events.
4. **Read and Listen**: Pay attention to how order words are used in books, articles, and conversations. This can give you a better understanding of how these words fit into different contexts and narratives.

Step 10: Congratulations on mastering the usage of order words! You've done a great job in understanding and applying these important linguistic tools. Keep practicing, and you'll continue to improve your communication skills. Goodbye and best of luck in your future language endeavors!

The manufacturer's authorised representative in the EU is Springer Nature Customer Service Centre GmbH, Europaplatz 3, 69115 Heidelberg, Germany. If you have any concerns regarding our products, please contact ProductSafety@springernature.com

Printed and bound by CPI Group (UK) Ltd, Croydon, CR0 4YY
01/10/2026
02225956-0011